Elliptic
Polynomials

Elliptic Polynomials

J.S. Lomont
John Brillhart

CRC Press
Taylor & Francis Group
Boca Raton London New York

CRC Press is an imprint of the
Taylor & Francis Group, an **informa** business

CRC Press
Taylor & Francis Group
6000 Broken Sound Parkway NW, Suite 300
Boca Raton, FL 33487-2742

First issued in paperback 2019

© 2001 by Taylor & Francis Group, LLC
CRC Press is an imprint of Taylor & Francis Group, an Informa business

No claim to original U.S. Government works

ISBN-13: 978-1-58488-210-7 (hbk)
ISBN-13: 978-0-367-39820-0 (pbk)

Library of Congress Cataloging-in-Publication Data

Lomont, John S., 1924-
    Elliptic polynomials / J.S. Lomont, John Brillhart.
        p. cm.
    Includes bibliographical references and indexes.
    ISBN 1-58488-210-7 (alk. paper)
    1. Elliptic functions. 2. Polynomials. I. Brillhart, John, 1930- II. Title.

QA343 .L68 2000
515'.983--dc21   00-055483

Library of Congress Card Number 00-055483

Visit the Taylor & Francis Web site at
http://www.taylorandfrancis.com

and the CRC Press Web site at
http://www.crcpress.com

This book is dedicated to the memory of Leonard Carlitz,

a remarkable mathematician and an outstanding

editor of the *Duke Mathematical Journal*

Contents

Tables

Preface

This work is a combination of two fields of mathematics: elliptic functions and orthogonal polynomials. The remarkable interplay here between the two fields provides an interesting development of some basic aspects of each; there is new material about various classes of polynomials on the one hand and the odd Jacobi elliptic functions and their inverses on the other hand.

This work is centered on and is developed around two main subjects. The first is the problem of proving a conjecture about the Jacobi elliptic functions $sn (t, k), 0 < k < 1$: In the Maclaurin expansion

$$sn (t, k) = t - (k^2 + 1) \frac{t^3}{3!} + (k^4 + 14k^2 + 1) \frac{t^5}{5!}$$

$$- (k^2 + 1)(k^4 + 134k^2 + 1) \frac{t^7}{7!} + (k^8 + 1228k^6 + 5478k^4 + 1228k^2 + 1) \frac{t^9}{9!}$$

$$- (k^2 + 1)(k^4 + 14k^2 + 1)(k^4 + 11054k^2 + 1) \frac{t^{11}}{11!} + \cdots,$$

the polynomials $k^2 + 1$ and $k^4 + 14k^2 + 1$, respectively, divide the coefficients of the t^{4n+3} and t^{6n+5} terms in this expansion, as the initial terms already indicate.

An examination of the Maclaurin expansions of two other odd elliptic functions [WhW, p. 494]

$$sc (t, k) = \frac{sn (t, k)}{cn (t, k)} \text{ and } sd (t, k) = \frac{sn (t, k)}{dn (t, k)}$$

reveals that the same divisibility pattern holds, i.e., the coefficients of the t^{4n+3} and t^{6n+5} terms are, respectively, divisible by $k^2 - 2$ and $k^4 - 16k^2 + 16$ and by $2k^2 - 1$ and $16k^4 - 16k^2 + 1$. (There is also a divisibility pattern for the squares of these functions.)

These examples suggest that there exists a class of functions containing these three functions where the Maclaurin expansion of each function in the class has this divisibility pattern, and that there is a formula for the coefficients in such expansions. Each of these suggestions is true, as we show in this work. In fact, the class of functions is the set of odd elliptic functions $f^{-1}(t)$, where $f(t)$ is given by the elliptic integral

$$(\star) \quad f(t) = a_1 \int_0^t \frac{du}{\sqrt{1 - c_1 u^2 + c_2 u^4}}, \quad \text{where } a_1, c_1, c_2 \in \mathbb{R}, \ a_1 \neq 0,$$

and the formula for the coefficients is given in terms of a certain two-variable polynomial $P_n(x,y)$, which has the properties: $x \parallel P_{2n+1}(x,y)$ and $y \parallel P_{3n+2}(x,y)$. The expansion formula is

$$(\star\star) \qquad f^{-1}(t) = \frac{1}{a_1} \sum_{n=0}^{\infty} P_n\left(-\frac{c_1}{a_1^2}, \frac{c_1^2 + 12c_2}{a_1^4}\right) \frac{t^{2n+1}}{(2n+1)!}.$$

The second problem is the investigation of some unusual orthogonal sequences of polynomials. Such sequences are obtained as follows from any function f defined in $(\star)$, but with the additional positivity condition $c_2 > 0$, holding: Generate the polynomial sequence $\{f_n(x)\}_{n=0}^{\infty}$ from the expansion

$$e^{xf(t)} = \sum_{n=0}^{\infty} f_n(x) \frac{t^n}{n!}.$$

Then define two secondary sequences $\{G_m(z)\}_{m=0}^{\infty}$ and $\{H_m(z)\}_{m=0}^{\infty}$ by the equations

$$G_m(x^2) = \frac{f_{2m+1}(x)}{x} \quad \text{and} \quad H_m(x^2) = \frac{f_{2m+2}(x)}{x^2}, \quad (z = x^2).$$

These secondary sequences are orthogonal and their moments are found to be certain values of the same polynomial $P_n(x,y)$ and its companion polynomial $Q_n(x,y)$, respectively. The $P_n(x,y)$ and $Q_n(x,y)$ polynomials are themselves an important associated area of research, since they are the coefficients of the Maclaurin expansions of odd Jacobi elliptic functions and their squares. In fact, besides $(\star\star)$, we have

$$(\star\star\star) \qquad \left(f^{-1}(t)\right)^2 = \frac{2}{a_1^2} \sum_{n=0}^{\infty} Q_n\left(-\frac{c_1}{a_1^2}, \frac{c_1^2 + 12c_2}{a_1^4}\right) \frac{t^{2n+2}}{(2n+2)!}.$$

Example. For values $a_1 = 1$, $c_1 = 0$, and $c_2 = -1$, the formula $(\star)$ gives $f(t) = \mathrm{sl}^{-1}(t)$, the inverse sinelemniscate function. Also, the formulas $(\star\star)$ and $(\star\star\star)$, respectively, give the series

$$\mathrm{sl}(t) = \sum_{n=0}^{\infty} P_n(0,-12) \frac{t^{2n+1}}{(2n+1)!} = \sum_{n=0}^{\infty} (-12)^n a_n \frac{t^{4n+1}}{(4n+1)!}$$

and

$$\mathrm{sl}^2(t) = 2\sum_{n=0}^{\infty} Q_n(0,-12)\frac{t^{2n+2}}{(2n+2)!} = 2\sum_{n=0}^{\infty}(-12)^n \beta_n \frac{t^{4n+2}}{(4n+2)!},$$

where α_n and β_n are obtained recursively from the formulas

$$\beta_{n+1} = \sum_{j=0}^{n}\binom{4n+4}{4j+2}\beta_j\beta_{n-j}, \quad \beta_0 = 1,$$

and

$$\alpha_{n+1} = \frac{1}{3}\sum_{j=0}^{n}\binom{4n+3}{4j+1}\alpha_j\beta_{n-j}, \quad \alpha_0 = 1.$$

The title of this research monograph, *Elliptic Polynomials*, refers to the polynomials that occur in this investigation, which come from elliptic integrals and functions. In studying these, we consider such things as orthogonality and the construction of weight functions and measures, the finding of structure constants and interesting inequalities, and deriving numerous useful formulas and evaluations. Although much of the material is familiar, this is a new mathematical area whose development intersects with classical subjects at many points.

Introduction

At the beginning of the 1960s, L. Carlitz published two papers [Ca2], [Ca4] dealing with the "first level" or "primary" polynomial sequence $\{\delta_n(x, k)\}_{n=0}^{\infty}$ generated by

$$e^{x\,\mathrm{sn}^{-1}(t,k)} = \sum_{n=0}^{\infty} \delta_n(x, k) \frac{t^n}{n!}, \; 0 < k < 1,$$

and also the two "second level" or "secondary" sequences $\{A_m(z, k)\}_{m=0}^{\infty}$ and $\{B_m(z, k)\}_{m=0}^{\infty}$ derived from $\{\delta_n(x, k)\}_{n=0}^{\infty}$ by the equations for $m \geq 0$,

$$A_m(x^2, k) = \frac{\delta_{2m+1}(x, k)}{x} \text{ and } B_m(x^2, k) = \frac{\delta_{2m+2}(x, k)}{x^2}, \; (z = x^2).$$

(Here $\mathrm{sn}^{-1}(t, k)$ is the inverse Jacobi elliptic function.) Although the sequence $\{\delta_n(x, k)\}_{n=0}^{\infty}$ is not orthogonal, Carlitz showed the two secondary sequences derived from it are orthogonal and constructed a discrete measure over the interval $(-\infty, 0]$ for each of them, in terms of which the inner products for the A and B sequences could be represented.

In 1965, Al-Salam [Al], a student of Carlitz, investigated a generalization of Carlitz's work. From his analysis, he concluded that the Jacobi functions $f(t, k) = \mathrm{sn}^{-1}(t, k), 0 < k < 1$, are the only solutions to the following problem: For what functions f are the two secondary sequences derived from the generating function $e^{xf(t)}$ orthogonal? Unfortunately, there was an error in his paper which led him to omit a large number of solutions from his list of solutions to this problem. We will say more about Al-Salam's investigation later in this introduction. In the study presented below, we give the complete solution to this problem and describe the broad theory of the "elliptic" polynomials that we have developed around it.

We begin the discussion of our work by observing that it deals with certain classes of functions $f(t)$ and the sequences of polynomials they generate. To understand how this work is organized, it is first necessary to examine what these classes are and how they are related to each other.

Initially, there are eight classes, but this number will shrink to six when we show that two of the classes are the same as two others. The first four of the eight classes are denoted by $\mathcal{F}$, $\mathcal{F}_0$, $\mathcal{F}_1$, and $\mathcal{F}_2$. They form a nested sequence $\mathcal{F} \supset \mathcal{F}_0 \supset \mathcal{F}_1 \supset \mathcal{F}_2$.

The first class, $\mathcal{F}$, consists of all real functions f that are analytic in a neighborhood of 0 with the property that $f(0) = 0$ and $f'(0) \neq 0$, i.e., the Maclaurin expansion of f has the form

(†) $$f(t) = a_1 t + a_2 t^2 + a_3 t^3 + \cdots,$$

where $a_k \in \mathbb{R}$ and $a_1 \neq 0$. A function $f \in \mathcal{F}$ is used to generate a "first level" or "primary" sequence of polynomials $\{f_n(x)\}_{n=0}^{\infty}$ by the generating function

(††) $$e^{x\, f(t)} = \sum_{n=0}^{\infty} f_n(x) \frac{t^n}{n!}.$$

The sequences of polynomials $\{f_n(x)\}_{n=0}^{\infty}$ generated in this way are the familiar "binomial" sequences of polynomials.

The second class, $\mathcal{F}_0$, is the set of all odd functions in $\mathcal{F}$, i.e., $\mathcal{F}_0 = \{f \in \mathcal{F} : f \text{ odd}\}$. Here

(‡) $$f(t) = a_1 t + a_3 t^3 + a_5 t^5 + \cdots.$$

When $f \in \mathcal{F}_0$, the polynomials in the primary sequence satisfy the identity $f_n(-x) = (-1)^n f_n(x)$, which implies that alternate terms in $f_n(x)$ are zero. Following Carlitz [Ca2], [Ca4], it is then possible to define two "second level" or "secondary" sequences of polynomials $\{G_m(z)\}_{m=0}^{\infty}$ and $\{H_m(z)\}_{m=0}^{\infty}$ by the equations

$$G_m(x^2) = \frac{f_{2m+1}(x)}{x} \quad \text{and} \quad H_m(x^2) = \frac{f_{2m+2}(x)}{x^2}, \quad m \geq 0, \quad \text{with } z = x^2.$$

The classes $\mathcal{F}$ and $\mathcal{F}_0$ are discussed in Chapter 1, parts (a) and (b), respectively. Part (a) contains an effective algorithm (Theorem 1.4) for computing the primary sequence, given the coefficients a_k of f, and the elegant and much-used "reversing" formula (Theorem 1.5): For $f \in \mathcal{F}$, if $f_n(x) = \sum_{j=0}^{n} \varphi_j^{(n)} x^j$, $n \geq 0$, and $[f(t)]^k = \sum_{n=k}^{\infty} a_n^{(k)} t^n$, $k \geq 0$, then

$$k!\, \varphi_k^{(n)} = n!\, a_n^{(k)}, \quad 0 \leq k \leq n.$$

Part (a), like many other parts of this work, contains useful evaluations, as in Corollaries 1.6, 1.13, and 1.11. In part (b) there are two determinant formulas for $f_n(x)$ (Theorem 1.18), generating functions for $G_m(z)$ and $H_m(z)$ (Theorem 1.24), and other results such as formulas for $G_m(4z)$ and $H_m(4z)$.

The third class, $\mathcal{F}_1$, is a much smaller set of functions and is central to our investigation. It is the set of functions

$$\mathcal{F}_1 = \left\{ f \in \mathcal{F}_0 : f(t) = a_1 \int_0^t \frac{du}{\sqrt{1 - c_1 u^2 + c_2 u^4}} \right\},$$

where $c_1 = \dfrac{6a_3}{a_1}$ and $c_2 = \dfrac{1}{a_1^2}(27a_3^2 - 10a_1 a_5)$. Thus, $\mathcal{F}_1$ is the three-parameter family of odd functions defined by this elliptic integral, where the parameters $a_1 \neq 0$, c_1, and c_2 range over all real numbers (because a_3 and a_5 do). Since the sequences of polynomials $\{f_n(z)\}_{n=0}^\infty$, $\{G_m(z)\}_{m=0}^\infty$, and $\{H_m(z)\}_{m=0}^\infty$ are generated directly and secondarily from elliptic integrals, we call them "elliptic polynomials of the first kind."

This class is discussed in Chapter 3, where it is shown that the primary and secondary sequences derived from an $f \in \mathcal{F}_1$ satisfy the recursions (3.11), (3.14), and (3.15), respectively, and it is shown in Theorem 3.5 that f being in $\mathcal{F}_1$ is a necessary and sufficient condition for these recursions to hold. It is also shown in this chapter (Corollary 3.6) that for $f \in \mathcal{F}_1$, the first three coefficients a_1, a_3, and a_5 in (‡) determine all the coefficients of f, thus demonstrating in another way that the set of functions in $\mathcal{F}_1$ is a three-parameter family of functions.

At the end of Chapter 3, the two parameters c_2 and (the discriminant) $\Delta = 9a_3^2 - 5a_1 a_5$ are used to partition $\mathcal{F}_1$ into six sub-classes. That none of these sub-classes is empty is shown by exhibiting at least one function in each, viz.,

$$f(t) = \mathrm{sn}^{-1}(t, k), \quad \mathrm{slh}^{-1}(t) = \int_0^t \frac{du}{\sqrt{1 + u^4}}, \quad \tanh^{-1} t,$$

$$t, \quad \sin^{-1} t, \quad \text{and} \quad \mathrm{sl}^{-1}(t) = \int_0^t \frac{du}{\sqrt{1 - u^4}}.$$

The fourth class, $\mathcal{F}_2$, is the set of functions in $\mathcal{F}_1$ for which $c_2 > 0$, i.e., $\mathcal{F}_2 = \{f \in \mathcal{F}_1 : c_2 > 0\}$. This is the preeminent class in our work, since these are the functions (and the only functions) that generate orthogonal secondary sequences (Theorem 7.3). We should also mention that the primary sequence generated by an $f \in \mathcal{F}$ is never orthogonal. We will prove this fact in Theorem 6.10.

As it happens, the condition $c_2 > 0$ that defines the class $\mathcal{F}_2$ within $\mathcal{F}_1$, holds in only three of the six sub-classes in $\mathcal{F}_1$. In Chapter 7, which is devoted to $\mathcal{F}_2$, these three classes are labeled by "Class $(\epsilon_{c_2}, \epsilon_\Delta)$" as follows: I $(+,-)$, II $(+,+)$, and III $(+,0)$. Of the six functions listed above, the three that lie in these classes are $f(t) = \mathrm{sn}^{-1}(t, k)$, $\mathrm{slh}^{-1}(t)$, and $\tanh^{-1} t$, which is to say, the other three functions

(‡‡) $f(t) = t, \quad \sin^{-1} t, \quad \text{and} \quad \mathrm{sl}^{-1}(t),$

are not in $\mathcal{F}_2$ and so do not generate orthogonal secondary sequences. The final formulas (7.35)–(7.44) in Chapter 7 give canonical forms for the functions in Classes I, II, and III.

There are also some formulas in Chapter 7 that relate to orthogonal secondary sequences, such as inner products of G's and H's (Theorem 7.4), and recursions for the G and H moments (Theorem 7.6). There is other important material in this chapter, but we will defer discussing it until after we have introduced some other ideas.

At this point we can say more about the paper of Al-Salam. The general problem he investigated is the following: Find all real functions $A(t)$ and $f(t)$ so that the two secondary sequences generated by $A(t)\,e^{zf(t)}$ are orthogonal (in the generalized sense). His necessary condition on $f(t)$ is the same as our (3.24) with $a_1 = 1$, viz.,

$$f(t) = \int_0^t \frac{du}{\sqrt{1 - c_1\,u^2 + c_2\,u^4}}.$$

After proving this result, he unaccountably and wrongly says [Al, p. 81] that we may assume without loss of generality that $c_1 = c_2 + 1$, where $c_1^2 + c_2^2 \neq 0$ and $c_2 \geq 0$. He concludes from this that the only solutions for $f(t)$ are the functions $f(t) = \mathrm{sn}^{-1}(t, k)$. This linear relation between c_1 and c_2 actually reduces the two-parameter family of solutions to a one-parameter family, thereby leaving out a large collection of solutions, such as all functions in Class II. Thus, although his initial analysis was correct, his listing of solutions with $A(t) = 1$ does not go beyond what was already in Carlitz's work.

The second four classes of functions are denoted by $\mathcal{F}^{-1}$, $\mathcal{F}_0^{-1}$, $\mathcal{F}_1^{-1}$, and $\mathcal{F}_2^{-1}$, respectively. These also form a nested sequence $\mathcal{F}^{-1} \supset \mathcal{F}_0^{-1} \supset \mathcal{F}_1^{-1} \supset \mathcal{F}_2^{-1}$. As the notation suggests, these four classes correspondingly consist of the inverses of the functions in the first four classes.

How do these four new classes relate to the previous four? First, if $f \in \mathcal{F}$, it follows from (†) and elementary analysis that $f^{-1} \in \mathcal{F}$. Thus, $\mathcal{F}^{-1} = \mathcal{F}$. The same is true for $\mathcal{F}_0$, since the inverse of an odd function in $\mathcal{F}$ is odd (Corollary 2.2). Thus, $\mathcal{F}_0^{-1} = \mathcal{F}_0$. Further, in Chapter 5, which deals with $\mathcal{F}_1^{-1}$, it is shown that $\mathcal{F}_1 \cap \mathcal{F}_1^{-1} = \{a_1\,t, a_1 \neq 0\}$, so $\mathcal{F}_1$ and $\mathcal{F}_1^{-1}$ are almost disjoint (Theorem 5.2). Finally, in Chapter 7 (Theorem 7.2), we see that $\mathcal{F}_2 \cap \mathcal{F}_2^{-1} = \emptyset$. Thus, of the original eight classes, there are just six that are distinct.

We next discuss the important idea of using f^{-1} to generate polynomial sequences. Since $f^{-1} \in \mathcal{F}$ when $f \in \mathcal{F}$, we can use f^{-1} in place of f in (††) to generate a primary polynomial sequence, denoted by $\{\bar{f}_n(x)\}_{n=0}^\infty$ (the bar is not conjugation). If $f \in \mathcal{F}_0$, then $f^{-1} \in \mathcal{F}_0$, so f^{-1} will also generate two secondary sequences, denoted by $\{\bar{G}_m(z)\}_{m=0}^\infty$ and $\{\bar{H}_m(z)\}_{m=0}^\infty$. If $f \in \mathcal{F}_1$, then the sequences $\{\bar{f}_n(x)\}_{n=0}^\infty$, $\{\bar{G}_m(z)\}_{m=0}^\infty$, $\{\bar{H}_m(z)\}_{m=0}^\infty$, generated by $e^{zf^{-1}(f)}$, where f^{-1} is now an elliptic function, are called "elliptic polynomials of the second kind." The natural question to ask then is whether there are any relationships between the functions and polynomials on the two "sides," the f-side and the f^{-1}-side. We call these sides, respectively, the "first" and "second" sides. As we will see, there exist at least three different types of such relationships.

The first type of relationship involves the coefficients of the polynomials $f_n(x)$ and $\bar{f}_n(x)$. There are two examples in Chapter 2. One consists of orthogonality relations between the coefficients expressed in Corollary 2.6. The other is the important result (Theorem 2.3) that the coefficients of $f_n(x)$ and $\bar{f}_n(x)$ are *inversion* coefficients, i.e., if

$$f_n(x) = \sum_{j=0}^{n} \varphi_j^{(n)} x^j \text{ and } \bar{f}_n(x) = \sum_{j=0}^{n} \bar{\varphi}_j^{(n)} x^j,$$

then for $f \in \mathcal{F}$, we have that

$$x^n = \sum_{k=0}^{n} \varphi_k^{(n)} \bar{f}_k(x) = \sum_{k=0}^{n} \bar{\varphi}_k^{(n)} f_k(x), \ n \geq 0.$$

This inversion formula is quite useful in later developments.

The second type of relationship is a correspondence between recursions on the two sides, such as (3.11) and (5.9), as well as (3.37) and (5.20) and the specializations of these, (10.8) and (12.4). In the first example, (3.11) gives a fourth-order recursion for $f_n(x)$, while (5.9) contains a second-order recursion for $\bar{f}_n(x)$ which involves a fourth-order derivative operator. The specializations (10.8) and (12.4) give a simple example: If we denote the primary sequences for $f(t) = \tanh^{-1} t$ by $\{\delta_n(x)\}_{n=0}^{\infty}$ and for $f^{-1}(t) = \tanh t$ by $\{\lambda_n(x)\}_{n=0}^{\infty}$, then for $n \geq 0$, (10.8) and (12.4) are, respectively,

$$\delta_{n+2}(x) = x\,\delta_{n+1}(x) + n(n+1)\,\delta_n(x) \rightsquigarrow \lambda_{n+1}(x) = x[\lambda_n(x) - \lambda_n''(x)],$$

where $\delta_0(x) = 1$, $\delta_1(x) = x$, and $\lambda_0(x) = 1$. The exact nature of this correspondence has not been worked out.

Another example comes from the fact that $\mathcal{F}_2 \cap \mathcal{F}_2^{-1} = \emptyset$. Since a function $f \in \mathcal{F}_2$ (and only in $\mathcal{F}_2$) generates orthogonal secondary sequences, then f^{-1}, which is not in $\mathcal{F}_2$, does not. Thus, for example, the recursion formula (3.14) for $\{G_m(z)\}_{m=0}^{\infty}$ would be different in kind from a recursion formula for $\{\check{G}_m(z)\}_{m=0}^{\infty}$, since the latter sequence is not orthogonal.

The third relationship between the two sides is one that is unexpected, intriguing, and valuable. It occurs in $\mathcal{F}_1$ (Theorem 5.7) and in $\mathcal{F}_2$ (Theorem 7.8). Theorem 5.7 gives a formula for the Maclaurin expansions for f^{-1} and $(f^{-1})^2$. It states that if $f \in \mathcal{F}_0$, then for a certain polynomial $P_n(x, y)$,

$$f \in \mathcal{F}_1 \iff f^{-1}(t) = \frac{1}{a_1} \sum_{n=0}^{\infty} P_n\left(-\frac{c_1}{a_1^2}, \frac{c_1^2 + 12c_2}{a_1^4}\right) \frac{t^{2n+1}}{(2n+1)!}.$$

There is a similar result for $\left(f^{-1}(t)\right)^2$ involving $Q_n(x, y)$. We will further discuss these polynomials below. Note that the arguments of $P_n(x, y)$ depend only on a_1, a_3, and a_5 through c_1 and c_2.

In Theorem 5.12, we give a collection of Maclaurin expansions obtained from using this result which includes the three expansions mentioned in the Preface. There are also the expansions of the sinelemniscate sl (t) and the hyperbolic sinelemniscate slh (t) and their squares, whose coefficients Hurwitz [Hu] showed have such remarkable properties.

The moment polynomials $P_n(x,y)$ and $Q_n(x,y)$ that occur in these formulas are introduced and studied in Chapter 4. They are defined by complicated recursion formulas whose motivation will be discussed when we consider Theorem 7.8. Nonetheless, they are a great improvement over the general reversion coefficients given by the determinant formula in (2.7).

Some of their basic properties are proved in Theorem 4.2. Their respective generating functions g and h are introduced there and are shown in Theorem 4.5 and Theorem 4.6 to satisfy various simple differential equations. In particular, g satisfies the differential equation (4.22), viz.,

$$(g')^2 = 1 + x\,g^2 + \frac{y - x^2}{12}\,g^4$$

and h satisfies (4.27), viz.,

$$(h')^2 = 2\,h + 4x\,h^2 + \frac{2(y - x^2)}{3}\,h^3.$$

Thus, g and h are elliptic functions that depend on x and y. Theorem 4.7 translates these derivative relationships into a collection of identities involving the P's and Q's.

Further, Theorems 4.8 and 4.9 deal with arithmetic properties of the coefficients of $P_n(x,y)$ and $Q_n(x,y)$. Theorem 4.8 is used in Theorem 4.10 to establish when x and y will divide these polynomials. It is these results, when applied to the Maclaurin expansions of the three odd elliptic functions and their squares mentioned in the Preface (cf. (5.33), (5.41), and (5.42)), which prove that certain polynomials periodically divide coefficients in these expansions. The chapter ends with the evaluation of two rather elegant determinant formulas that involve the P's and Q's.

Chapter 6 is a presentation of basic material relating to inner products and moments, which is included here to make this work more self-contained. The key theorem here is the familiar theorem of Favard (Theorem 6.9), which says that a given sequence of polynomials is orthogonal if and only if it satisfies a specific type of second-order recursion formula. This theorem is the theoretical basis for the orthogonality proofs in this work. In particular, the recursions (3.14) and (3.15) are of this type and are equivalent by Theorems 3.2 and 3.4 to the function f being in $\mathcal{F}_1$. In Chapter 6, when $f \in \mathcal{F}$, there is a simple proof that the primary sequence generated from f is never orthogonal (Theorem 6.10). This fact, of course, already follows from the well-known result of Meixner [Me], who lists all generating functions of the form $A(t)\,e^{z\,f(t)}$ that produce orthogonal primary sequences. The generating function in (††) is not on this list.

Associated with any "nice" sequence of real orthogonal polynomials is a unique Borel probability measure μ, and associated with this measure is a sequence $\{\mu_n\}_{n=0}^{\infty}$ of moments given by $\mu_n = \int_{\mathbf{R}} x^n d\mu(x)$. To facilitate the treatment of moments, we introduced three sequences of "moment" polynomials in Definition 4.1. These polynomials have the following property: If $\{G_n\}_{n=0}^{\infty}$ is some orthogonal sequence with the associated moments $\{\mu_G(n)\}_{n=0}^{\infty}$ (where $\mu_G(0) = 1$), then $\mu_G(n) = T_n(\mu_G(1), \mu_G(2))$, $n \geq 0$, where $\{T_n\}_{n=0}^{\infty}$ is one of the sequences of moment polynomials (Theorem 7.7). Since this theorem identifies the moments with $P_n(x, y)$ and $Q_n(x, y)$, we see the recursions that the moments satisfy, as in (7.23), are satisfied by $P_n(x, y)$ and $Q_n(x, y)$ as well. This is the origin of the recursive definitions for $P_n(x, y)$ and $Q_n(x, y)$ in Definition 4.1.

Theorem 7.8 is comparable to Theorem 5.7, but deals with functions $f \in \mathcal{F}_2$. This theorem shows the following surprising relationship between the two sides: For an $f \in \mathcal{F}_2$, generate the secondary sequences $\{G_m(z)\}_{m=0}^{\infty}$ and $\{H_m(z)\}_{m=0}^{\infty}$ and consider the moments $\mu_G(n)$ and $\mu_H(n)$ associated with these two sequences. Then we have the two expansions

$$f^{-1}(t) = \frac{1}{a_1} \sum_{n=0}^{\infty} \mu_G(n) \frac{t^{2n+1}}{(2n+1)!}$$

and

$$\left(f^{-1}(t)\right)^2 = \frac{2}{a_1^2} \sum_{n=0}^{\infty} \mu_H(n) \frac{t^{2n+2}}{(2n+2)!}.$$

These are the formulas from which the polynomials $P_n(x, y)$ and $Q_n(x, y)$ were initially derived, for, as Theorem 7.7 shows, these are the polynomials that relate the moments $\mu_G(n)$ and $\mu_H(n)$ to the moments $\mu_G(1)$ and $\mu_G(2)$. Once the polynomials had been introduced, it was proved without using any idea of a more general orthogonality (cf. [ShT],[Vi]) that they also appear as coefficients in Theorem 5.7 with f in $\mathcal{F}_1 - \mathcal{F}_2$, i.e., where the secondary sequences need not be orthogonal (at least in the usual sense).

We have now completed the overview of that part of our work that deals with classes of functions and the ideas and means that are used throughout. In the remainder of this introduction we will discuss results relating to particular functions in Classes I, II, and III.

In part (a) of Chapter 8, we investigate the Class I family of Carlitz,

$$f(t, k) = \mathrm{sn}^{-1}(t, k) = \int_0^t \frac{du}{\sqrt{(1 - u^2)(1 - k^2 u^2)}}, \quad 0 < k < 1.$$

Of special note in part (a) are the Maclaurin expansions of $\mathrm{sn}\,(t, k)$ and $\mathrm{sn}^2(t, k)$ in (8.25a) and (8.25b), respectively, where Theorem 8.11 specifies which of the coefficients in these expansions the factors $k^2 + 1$ and $k^4 + 14k^2 + 1$ will divide. These results and similar ones for the functions $\mathrm{sc}\,(t, k)$ and $\mathrm{sd}\,(t, k)$, mentioned in the Preface, are given in Theorem 5.12.

At the end of part (a), we have included Carlitz's two discrete measures for the secondary sequences (with typos corrected). In Chapter 16 (Theorem 16.2) we prove that these measures are unique (up to normalization), while in Corollary 16.3 we prove that each of the secondary sequences is an orthonormal basis for the corresponding L_2 space.

In part (b) of Chapter 8, we investigate a Class II family of functions

$$f^\star(t, k) = \int_0^t \frac{du}{\sqrt{1 - 2(1 - 2k^2)u^2 + u^4}}$$

$$= \sum_{n=0}^\infty P_m(1 - 2k^2)\frac{t^{2m+1}}{2m + 1}, \ 0 < k < 1,$$

where the final series involves the Legendre polynomial $P_m(x)$ (see (3.29)). One of the main results of this part of the work is a construction, modeled on Carlitz's construction mentioned in part (a), of measures on $\mathbb{R}$ for the two orthogonal secondary sequences $\{A_m^\star(z, k)\}_{m=0}^\infty$ and $\{B_m^\star(z, k)\}_{m=0}^\infty$ in this part.

The portion of our work that relates to Class III deals mostly with the function $f(t) = \tanh^{-1} t$ and a little with the related function $f(t) = \tan^{-1} t$. These two functions are the "endpoint" functions ($k = 0$ or 1) for the two families in Classes I and II that we examined in (8.1) and (8.49). Here $f(t, 1) = \text{sn}^{-1}(t, 1) = \tanh^{-1} t$, $f^\star(t, 0) = \tanh^{-1} t$ and $f^\star(t, 1) = \tan^{-1} t$. (The function $f(t, 0) = \text{sn}^{-1}(t, 0) = \sin^{-1} t$ is not in $\mathcal{F}_2$, as we saw in (‡‡), so it is not included here.)

These endpoint cases are mentioned in [Ca4, pp. 123–124], where Carlitz does not explicitly discuss orthogonality of the secondary sequences, nor does he discuss the matter of discrete measures or weight functions for the secondary sequences in these cases. He does remark, however, that the primary sequence derived from $f(t) = \tan^{-1} t$ is orthogonal, which is incorrect, since no primary sequence is orthogonal.

Chapter 10 deals with the primary sequence (denoted by $\delta_n(x)$) generated by $f(t) = \tanh^{-1} t$. As it happens, the polynomials in the primary sequence, which are monic with positive integer coefficients, are a modified form of the Mittag–Leffler polynomials, which we will refer to here as just the "Mittag–Leffler polynomials." In working with these polynomials, which occur throughout the literature (see references after Table 10.1), we discovered they have many intriguing forms. Besides basic recursions (10.8) and (10.14) for these polynomials, there are operator formulas (10.20)–(10.30) and representations of the polynomials in various ways. These involve rising and falling factorial polynomials; Stirling, Lah, and Delannoy numbers; binomial coefficient polynomials; hypergeometric functions; Jacobi polynomials; and integral representations. There are double and half-argument formulas (Theorems 10.22 and 10.23) and the curious formula in (10.84). The structure constants for this sequence of polynomials are in Theorem 10.49. A formula for $\delta_{m+n}(x)$ is proved in Theorem

10.52 and a Rodrigues–Toscano type formula for $\delta_n(x)$ is given in Theorem 10.59.

Chapter 11 discusses numerical properties of the coefficients of $\delta_n(x)$. There is a formula for the rth derivative of $\delta_n(x)$ proved here as well.

The results in Chapter 12 are on the "second" side relative to Chapter 11. The results in Chapter 13 are important (containing the $A(z)$ structure constants) and are written out in detail, because they are used in the developments relating to the weight functions in Chapter 14. Chapter 12 also contains the proof that the zeros of the polynomial $\lambda_n(x)$ are all real and distinct.

Chapter 17 is devoted to inequalities involving $\delta_n(x)$, $A_m(z)$, and $B_m(z)$. Some of these inequalities are similar to Turán's inequality for the Legendre polynomials.

In our work, the orthogonal secondary sequences for $f(t) = \tanh^{-1} t$, denoted by $\{A_m(z)\}_{m=0}^{\infty}$ and $\{B_m(z)\}_{m=0}^{\infty}$, are the focus of Chapter 13. Here the structure constants for $\{A_m(z)\}_{m=0}^{\infty}$ are obtained. Chapter 14 presents one of the basic in this work, namely the weight functions $w_A(z)$ and $w_B(z)$ for these two orthogonal sequences (Definition 14.1) and associated formulas (Theorem 14.6 and Theorem 14.7) dealing with integral representations of inner products. In this case, we can evaluate the moments explicitly (Theorem 14.2), viz., $\mu_A(n) = -C_{2n+1}$ and $\mu_B(n) = \frac{1}{2}C_{2n+3}$, where the C_n's are the tangent numbers. The structure constants for $\{B_m(z)\}_{m=0}^{\infty}$ are obtained as well (Theorem 14.13).

In conclusion, we remark that this work deals with many formulas, some from the literature and others devised for use in this work, such as (1.27). Since this is an investigation into special functions and polynomials, there are many new formulas that result from this investigation. Finally, there are formulas that have interesting, even elegant simplicity, such as (1.8), (1.31), (10.35), (10.115), (10.116), (11.11), and (12.4).

Acknowledgments. We would like to express our gratitude to Robert Condon of the Mathematics Department at the University of Arizona and to Richard Blecksmith and Eric Behr of the Mathematics Department at Northern Illinois University for their assistance in helping us solve many of our AMS-TEX problems. Their knowledge, advice, and support have been invaluable.

Chapter 1

Binomial Sequences of Polynomials

This is the first of two introductory chapters that establish the general setting for most of the later developments.

In this chapter we introduce two types of polynomial sequences which we call "first level" or "primary" and "second level" or "secondary." The primary sequence $\{f_n(x)\}_{n=0}^{\infty}$ is obtained from a certain type of real function f by the generating function

$$e^{xf(t)} = \sum_{n=0}^{\infty} f_n(x)\frac{t^n}{n!}.$$

In part (a), we give some of the basic properties of this sequence, such as the familiar addition formula and a practical recursion formula for computing the f's, as well as a collection of evaluations, a very useful identity, and some formulas.

In part (b), the function f is assumed to be odd, which implies the alternate terms of the polynomial $f_n(x)$ are zero. We can then derive a pair of secondary sequences $\{G_m(z)\}_{m=0}^{\infty}$ and $\{H_m(z)\}_{m=0}^{\infty}$ from the primary sequence by the equations

$$G_m(x^2) = \frac{f_{2m+1}(x)}{x} \quad \text{and} \quad H_m(x^2) = \frac{f_{2m+2}(x)}{x^2},$$

so these polynomials have as their zeros the squares of the non-zero zeros of $f_{2n+1}(x)$ and $f_{2n+2}(x)$, respectively. We also obtain some evaluations and formulas, such as the generating functions for the G and H sequences.

The evaluations and formulas obtained in this initial chapter will be used throughout the work.

1 (a) The functions $\mathcal{F}$

In what follows, we will use $\mathcal{F}$ to stand for the set of real functions f where

$$\mathcal{F} = \{f : f \text{ is analytic at } 0, \ f(0) = 0, \text{ and } f'(0) \neq 0\}.$$

We can then write

$$(1.1) \qquad\qquad f(t) \overset{\text{def}}{=} \sum_{n=0}^{\infty} a_n t^n,$$

where $a_0 = 0$, $a_1 \neq 0$, and $a_n = \dfrac{f^{(n)}(0)}{n!}$, and the series converges in some non-zero neighborhood of 0 (cf. [KrP]).

For $f \in \mathcal{F}$, we also have the expansion

$$(1.2) \qquad\qquad [f(t)]^k \overset{\text{def}}{=} \sum_{n=k}^{\infty} a_n^{(k)} t^n, \ k \geq 0,$$

and the generating function

$$(1.3) \qquad\qquad e^{x f(t)} \overset{\text{def}}{=} \sum_{n=0}^{\infty} f_n(x) \frac{t^n}{n!}.$$

Since $a_1 \neq 0$, it is easy to see by expanding $e^{x f(t)}$ that $f_n(x)$ is a polynomial of degree n. We can therefore write

$$(1.4) \qquad\qquad f_n(x) \overset{\text{def}}{=} \sum_{j=0}^{n} \varphi_j^{(n)} x^j, \ n \geq 0,$$

where $\varphi_j^{(n)} = \dfrac{f_n^{(j)}(0)}{j!}$.

Note the special values, which are readily computed from (1.2)–(1.4) for $n \geq 0$:

$$a_n^{(0)} = \delta_{n,0}, \ a_n^{(1)} = a_n, \text{ and } a_n^{(n)} = a_1^n;$$
$$f_0(x) = 1, \ f_1(x) = a_1 x, \ f_2(x) = a_1^2 x^2 + 2 a_2 x;$$
$$\varphi_0^{(n)} = \delta_{n,0} = f_n(0).$$

Example. The function $f(t) = e^t - 1 = \sum_{n=1}^{\infty} \dfrac{1}{n!} t^n$ provides a simple example of the preceding. Then $a_n = \dfrac{1}{n!}$ in (1.1), while in (1.4), we have

$f_0(x) = 1$ and $f_n(x) = \sum\limits_{j=1}^{n} S(n,j) x^j$, where $S(n,j)$ is the Stirling number of the second kind (see [Ri1]). To show this, note that equation (1.3) and [AbS, p. 824] give

$$\sum_{n=0}^{\infty} f_n(x)\frac{t^n}{n!} = e^{xf(t)} = 1 + \sum_{j=1}^{\infty}\frac{x^j}{j!}(e^t-1)^j = 1 + \sum_{j=1}^{\infty}x^j\Big\{\sum_{n=j}^{\infty}S(n,j)\frac{t^n}{n!}\Big\}$$

$$= 1 + \sum_{n=1}^{\infty}\Big\{\sum_{j=1}^{n}S(n,j)x^j\Big\}\frac{t^n}{n!},$$

from which the result follows by equating corresponding coefficients. Here $\varphi_j^{(n)} = S(n,j)$.

We begin our discussion by establishing the familiar addition formula for $f_n(x)$.

Theorem 1.1. For $f \in \mathcal{F}$ and $n \geq 0$,

(1.5)
$$f_n(x+y) = \sum_{k=0}^{n}\binom{n}{k}f_k(x)f_{n-k}(y).$$

Proof. From (1.3) we have that

$$\sum_{n=0}^{\infty} f_n(x+y)\frac{t^n}{n!} = e^{(x+y)f(t)} = e^{xf(t)}e^{yf(t)} = \sum_{n=0}^{\infty} f_n(x)\frac{t^n}{n!}\sum_{n=0}^{\infty} f_n(y)\frac{t^n}{n!}$$

$$= \sum_{n=0}^{\infty}\Big(\sum_{k=0}^{n}\binom{n}{k}f_k(x)f_{n-k}(y)\Big)\frac{t^n}{n!}.$$

The result follows by equating the coefficients of $\dfrac{t^n}{n!}$. □

Definition 1.2 [Ai, p. 99] [Ro, p. 9] A sequence $\{P_n(x)\}_{n=0}^{\infty}$ of polynomials is called "binomial," or " of binomial type," if and only if $P_0(x) = 1$, $\deg P_n = n$, and

$$P_n(x+y) = \sum_{k=0}^{n}\binom{n}{k}P_k(x)P_{n-k}(y).$$

Theorem 1.1 shows that the sequence of polynomials generated in (1.3) is binomial. What is somewhat surprising is that the converse is true.

Theorem 1.3. A sequence of polynomials $\{f_n(x)\}_{n=0}^{\infty}$ is binomial if and only if it is generated by $e^{xf(t)}$, where $f \in \mathcal{F}$.

Proof. See [Ai, p. 115, Theorem 3.59]. □

We next consider the question of how to compute the $f_n(x)$'s efficiently. The following theorem, which rests on a result of Pourahmadi [Po, p. 304], provides us with recursive algorithms for computing $f_n(x)$ and $\varphi_k^{(n)}$.

Theorem 1.4. For $f \in \mathcal{F}$ and $n \geq 0$, we have

(1.6)
$$f_{n+1}(x) = n!\, x \sum_{j=0}^{n} \frac{n+1-j}{j!}\, a_{n+1-j}\, f_j(x),$$

where $f_0(x) = 1$. Also, for $0 \leq r \leq n$,

(1.7)
$$\varphi_{r+1}^{(n+1)} = n! \sum_{j=r}^{n} \frac{n+1-j}{j!}\, a_{n+1-j}\, \varphi_r^{(j)},$$

where, as before, $\varphi_0^{(n)} = \delta_{n,0}$.

Proof. From (1.1) we have $x f(t) = \displaystyle\sum_{n=1}^{\infty} a_n x t^n$. Then using the recursion formula in [Po, p. 304], with $c_j = \dfrac{f_j(x)}{j!}$ from (1.3), we have

$$\frac{f_{n+1}(x)}{(n+1)!} = \sum_{j=0}^{n} \left(1 - \frac{j}{n+1}\right) a_{n+1-j} x\, \frac{f_j(x)}{j!}$$

or

$$f_{n+1}(x) = n!\, x \sum_{j=0}^{n} \frac{n+1-j}{j!}\, a_{n+1-j}\, f_j(x).$$

which is (1.6).

Using (1.6) and the fact that $\varphi_0^{(n+1)} = 0$ for $n \geq 0$, we obtain

$$\sum_{r=0}^{n} \varphi_{r+1}^{(n+1)} x^r = \sum_{r=1}^{n+1} \varphi_r^{(n+1)} x^{r-1} = \frac{f_{n+1}(x)}{x}$$

$$= n! \sum_{j=0}^{n} \frac{n+1-j}{j!}\, a_{n+1-j} \sum_{r=0}^{j} \varphi_r^{(j)} x^r$$

$$= \sum_{r=0}^{n} \left\{ n! \sum_{j=r}^{n} \frac{n+1-j}{j!}\, a_{n+1-j}\, \varphi_r^{(j)} \right\} x^r.$$

Equating the corresponding coefficients of x^r gives (1.7). $\square$

The next theorem gives an interesting and useful relationship between the coefficients of $f_n(x)$ and $[f(t)]^k$.

Theorem 1.5. Reversing Formula. If $f \in \mathcal{F}$ and $0 \leq k \leq n$, then

(1.8)
$$k!\, \varphi_k^{(n)} = n!\, a_n^{(k)}.$$

Proof. Using (1.4), (1.3), and (1.2), we have

$$\sum_{n=0}^{\infty}\left(\sum_{k=0}^{n}\frac{\varphi_k^{(n)}}{n!}x^k\right)t^n = \sum_{n=0}^{\infty}f_n(x)\frac{t^n}{n!} = e^{xf(t)} = \sum_{k=0}^{\infty}\frac{(xf(t))^k}{k!}$$

$$= \sum_{k=0}^{\infty}\left(\sum_{n=k}^{\infty}a_n^{(k)}t^n\right)\frac{x^k}{k!} = \sum_{n=0}^{\infty}\left(\sum_{k=0}^{n}\frac{a_n^{(k)}}{k!}x^k\right)t^n.$$

Equating corresponding coefficients of t^n and then of x^n gives the result. □

Equation (1.8) is readily verified in the case of example following (1.4), where $\varphi_k^{(n)} = S(n,k)$, $1 \le k \le n$. The equations (1.2) and [AbS, p. 824] give $\sum_{n=k}^{\infty}a_n^{(k)}t^n = (e^t-1)^k = k!\sum_{n=k}^{\infty}S(n,k)\frac{t^n}{n!}$, $k \ge 1$. Thus, equating the corresponding coefficients, we find that $a_n^{(k)} = \dfrac{k!}{n!}S(n,k)$, $1 \le k \le n$, which implies that $k!\varphi_k^{(n)} = k!S(n,k) = n!a_n^{(k)}$.

The following corollary is a collection of useful results about $\varphi_j^{(n)}$.

Corollary 1.6. If $f \in \mathcal{F}$, we have for $n \ge 0$ that

$$(1.9) \qquad\qquad \varphi_n^{(n)} = a_1^n,$$

$$(1.10) \qquad\qquad \varphi_n^{(n+1)} = n(n+1)\,a_1^{n-1}a_2,$$

$$(1.11) \qquad \varphi_n^{(n+2)} = \tfrac{1}{2}n(n+1)(n+2)\,a_1^{n-2}(2a_1a_3 + (n-1)\,a_2^2),$$

$$(1.12) \qquad\qquad \varphi_0^{(n)} = \delta_{n,0}.$$

Also,

$$(1.13) \qquad\qquad \varphi_1^{(n)} = n!\,a_n, \ n \ge 1,$$

$$(1.14) \qquad\qquad \varphi_2^{(n)} = \frac{n!}{2}\sum_{k=1}^{n-1}a_k a_{n-k}, \ n \ge 2,$$

$$(1.15) \qquad \varphi_3^{(n)} = \frac{(n-1)!}{2}\sum_{j=2}^{n-1}(n-j)\,a_{n-j}\sum_{k=1}^{j-1}a_k\,a_{j-k}, \ n \ge 3,$$

and for $n \ge 4$,

$$(1.16) \qquad \varphi_4^{(n)} = \frac{(n-1)!}{2}\sum_{j=3}^{n-1}\frac{n-j}{j}\,a_{n-j}\sum_{k=2}^{j-1}(j-k)\,a_{j-k}\sum_{r=1}^{k-1}a_r\,a_{k-r}.$$

Proof. (1.9)–(1.11) From (1.8) we obtain

(1.17)
$$\varphi_n^{(n)} = a_n^{(n)}, \quad \varphi_n^{(n+1)} = (n+1)\, a_{n+1}^{(n)},$$
$$\varphi_n^{(n+2)} = (n+1)(n+2)\, a_{n+2}^{(n)}.$$

On the other hand, from (1.2) we find that

$$a_n^{(n)} t^n + a_{n+1}^{(n)} t^{n+1} + a_{n+2}^{(n)} t^{n+2} + \cdots = (a_1 t + a_2 t^2 + a_3 t^3 + \cdots)^n$$
$$\equiv a_1^n t^n + n a_1^{n-1} a_2 t^{n+1} + \tfrac{1}{2} n a_1^{n-2} \left(2 a_1 a_3 + (n-1) a_2^2\right) t^{n+2} \pmod{t^{n+3}}.$$

Equating corresponding coefficients and using (1.17) give the results.

(1.12) See special value calculations after (1.4).

(1.13) Put $k = 1$ into (1.8).

(1.14) Putting $k = 2$ into (1.8) gives $\varphi_2^{(n)} = \tfrac{1}{2} n! \, a_n^{(2)}$. The final factor is computed from (1.2) using $k = 2$.

(1.15) In (1.7) put $k = 2$ and replace n by $n - 1$. Then substitute (1.14) for $\varphi_2^{(j)}$.

(1.16) Use (1.7) with $n - 1$ for n and $k = 3$ and then (1.15).   □

Note. There is a formula for $\varphi_k^{(n)}$, where $\varphi_k^{(n)} = B_{n,k}$ as in [Co, p. 134, [3d]].

Theorem 1.7. For $0 \le r \le n$,

(1.18)
$$\varphi_{r+1}^{(n+1)} = \sum_{j=r}^{n} \binom{n}{j} \varphi_r^{(j)} \varphi_1^{(n+1-j)}.$$

Proof. In (1.13), solve for a_n and substitute the result into (1.7).   □

Theorem 1.8. For $n \ge 1$,

(1.19)
$$f_n'(x) = n! \sum_{k=0}^{n-1} \frac{a_{n-k}}{k!} f_k(x).$$

Proof. From (1.3) we have

$$\sum_{n=1}^{\infty} \frac{f_n'(x)}{n!} t^n = f(t)\, e^{x f(t)} = \left(\sum_{n=0}^{\infty} a_n t^n\right)\left(\sum_{n=0}^{\infty} \frac{f_n(x)}{n!} t^n\right)$$
$$= \sum_{n=1}^{\infty} \left(\sum_{k=0}^{n-1} a_{n-k} \frac{f_k(x)}{k!}\right) t^n.$$

The result follows by equating corresponding coefficients of t^n.   □

The next result is a curious variant of (1.7).

Corollary 1.9. [Co, p. 136 [3k]] For $0 \leq r \leq n$,

(1.20) $$(r+1)\,\varphi_{r+1}^{(n+1)} = (n+1)! \sum_{j=r}^{n} \frac{a_{n+1-j}}{j!}\,\varphi_r^{(j)}.$$

Proof. For $n = r = 0$: True by (1.9) and (1.13). For $r = 0$, $n \geq 1$: True by (1.12) and (1.13). For $1 \leq r \leq n$: Substituting (1.4) into (1.19), we obtain for $n \geq 1$,

$$\sum_{r=1}^{n+1} r\,\varphi_r^{(n+1)} x^{r-1} = (n+1)! \left\{ a_{n+1} + \sum_{j=1}^{n} \frac{a_{n+1-j}}{j!} \sum_{r=1}^{j} \varphi_r^{(j)} x^r \right\}$$

or

$$\sum_{r=0}^{n} (r+1)\,\varphi_{r+1}^{(n+1)} x^r = (n+1)! \left\{ a_{n+1} + \sum_{r=1}^{n} \left\{ \sum_{j=r}^{n} \frac{a_{n+1-j}}{j!} \varphi_r^{(j)} \right\} x^r \right\}.$$

Using equation (1.13) and equating corresponding coefficients of x^r gives the result. $\square$

In Theorem 1.11, which follows the next lemma, we derive a formula for the rth derivative of $f_n(x)$.

Lemma 1.10. For $n \geq 0$, $0 \leq r \leq n$, and $0 \leq s \leq n - r$, we have

(1.21) $$\binom{r+s}{r} \varphi_{r+s}^{(n)} = \sum_{j=s}^{n-r} \binom{n}{j} \varphi_r^{(n-j)} \varphi_s^{(j)}.$$

Proof. Substituting (1.4) into the two sides of (1.5) gives

$$f_n(x+y) = \sum_{j=0}^{n} \varphi_j^{(n)} (x+y)^j = \sum_{j=0}^{n} \varphi_j^{(n)} \sum_{r=0}^{j} \binom{j}{r} x^r y^{n-r}$$

$$= \sum_{r=0}^{n} x^r \sum_{j=r}^{n} \binom{j}{r} \varphi_j^{(n)} y^{j-r} = \sum_{r=0}^{n} \sum_{s=0}^{n-r} \left\{ \binom{r+s}{r} \varphi_{r+s}^{(n)} \right\} x^r y^s$$

and

$$\sum_{j=0}^{n} \binom{n}{j} f_{n-j}(y) f_j(x) = \sum_{j=0}^{n} \binom{n}{j} f_{n-j}(y) \sum_{r=0}^{j} \varphi_r^{(j)} x^r$$

$$= \sum_{r=0}^{n} x^r \sum_{j=r}^{n} \binom{n}{j} \varphi_r^{(j)} f_{n-j}(y) = \sum_{r=0}^{n} x^r \sum_{j=0}^{n-r} \binom{n}{j+r} \varphi_r^{(j+r)} f_{n-j-r}(y)$$

$$= \sum_{r=0}^{n} x^r \sum_{j=0}^{n-r} \binom{n}{j+r} \varphi_r^{(j+r)} \sum_{s=0}^{n-j-r} \varphi_s^{(n-j-r)} y^s$$

$$= \sum_{r=0}^{n} x^r \sum_{s=0}^{n-r} y^s \sum_{j=0}^{n-r-s} \binom{n}{j+r} \varphi_r^{(j+r)} \varphi_s^{(n-j-r)}$$

$$= \sum_{r=0}^{n} \sum_{s=0}^{n-r} \left\{ \sum_{j=s}^{n-r} \binom{n}{j} \varphi_r^{(n-j)} \varphi_s^{(j)} \right\} x^r y^s.$$

Equating the coefficients of $x^r y^s$ gives the result. $\square$

Theorem 1.11. For $n \geq 0$ and $0 \leq r \leq n$, we have that

$$(1.22) \qquad f_n^{(r)}(x) = r! \sum_{j=0}^{n-r} \binom{n}{j} \varphi_r^{(n-j)} f_j(x) = r! \sum_{j=r}^{n} \binom{n}{j} \varphi_r^{(j)} f_{n-j}(x).$$

Proof. Differentiating (1.4) r times and using (1.21) gives

$$\frac{1}{r!} f_n^{(r)}(x) = \sum_{s=r}^{n} \binom{s}{r} \varphi_s^{(n)} x^{s-r} = \sum_{s=0}^{n-r} \binom{r+s}{r} \varphi_{r+s}^{(n)} x^s$$

$$= \sum_{s=0}^{n-r} x^s \sum_{j=s}^{n-r} \binom{n}{j} \varphi_r^{(n-j)} \varphi_s^{(j)} = \sum_{j=0}^{n-r} \binom{n}{j} \varphi_r^{(n-j)} \sum_{s=0}^{j} \varphi_s^{(j)} x^s$$

$$= \sum_{j=0}^{n-r} \binom{n}{j} \varphi_r^{(n-j)} f_j(x).$$

The second form in (1.22) follows from the substitution $j \to n - j$. $\square$

1 (b) The functions $\mathcal{F}_0$ – The sequences $\{G_m(z)\}$ and $\{H_m(z)\}$

We next single out the subset $\mathcal{F}_0$ in $\mathcal{F}$ defined by

$$\mathcal{F}_0 = \{f \in \mathcal{F} : f \ odd\}.$$

The binomial sequences $\{f_n(x)\}_{n=0}^\infty$ we will primarily be concerned with in this monograph are those that are obtained from functions f in $\mathcal{F}_0$.

The next theorem shows how the oddness of an $f \in \mathcal{F}_0$ translates into a property of $f_n(x)$.

Theorem 1.12. Let $f \in \mathcal{F}$. Then $f \in \mathcal{F}_0$ if and only if

$$(1.23) \qquad\qquad f_n(-x) = (-1)^n f_n(x), \ n \geq 0.$$

Proof. ($\Longrightarrow$) We have in (1.3) that

$$\sum_{n=0}^{\infty} f_n(-x) \frac{t^n}{n!} = e^{-xf(t)} = e^{xf(-t)} = \sum_{n=0}^{\infty} f_n(x) \frac{(-t)^n}{n!} = \sum_{n=0}^{\infty} (-1)^n f_n(x) \frac{t^n}{n!}.$$

The result follows from equating corresponding coefficients of t^n.

($\Longleftarrow$) We have that

$$e^{xf(-t)} = \sum_{n=0}^{\infty} f_n(x) \frac{(-t)^n}{n!} = \sum_{n=0}^{\infty} (-1)^n f_n(x) \frac{t^n}{n!}$$

$$= \sum_{n=0}^{\infty} f_n(-x) \frac{t^n}{n!} = e^{-xf(t)}.$$

Differentiating this identity with respect to x and then setting $x = 0$ implies $f \in \mathcal{F}_0$. $\square$

Table 1.1 $f_n(x)$, $0 \le n \le 11$, $f \in \mathcal{F}_0$

n	$f_n(x)$
0	1
1	$a_1 x$
2	$a_1^2 x^2$
3	$x(a_1^3 x^2 + 6a_3)$
4	$x^2(a_1^4 x^2 + 24a_1 a_3)$
5	$x(a_1^5 x^4 + 60a_1^2 a_3 x^2 + 120 a_5)$
6	$x^2[a_1^6 x^4 + 120a_1^3 a_3 x^2 + 360(a_3^2 + 2a_1 a_5)]$
7	$x[a_1^7 x^6 + 210a_1^4 a_3 x^4 + 2520 a_1(a_3^2 + a_1 a_5)x^2 + 5040 a_7]$
8	$x^2[a_1^8 x^6 + 336a_1^5 a_3 x^4 + 3360a_1^2(3a_3^2 + 2a_1 a_5)x^2 + 40320(a_1 a_7$ $+ a_3 a_5)]$
9	$x[a_1^9 x^8 + 504a_1^6 a_3 x^6 + 15120a_1^3(a_1 a_5 + 2a_3^2)x^4 + 60480(3a_1^2 a_7$ $+ 6a_1 a_3 a_5 + a_3^3)x^2 + 362880 a_9]$
10	$x^2[a_1^{10} x^8 + 720a_1^7 a_3 x^6 + 15120a_1^4(2a_1 a_5 + 5a_3^2)x^4 + 604800 a_1$ $\cdot(a_1^2 a_7 + 3a_1 a_3 a_5 + a_3^3)x^2 + 1814400(2a_1 a_9 + 2a_3 a_7 + a_5^2)]$
11	$x[a_1^{11} x^{10} + 990a_1^8 a_3 x^8 + 55440a_1^5(a_1 a_5 + 3a_3^2)x^6 + 1663200a_1^2$ $\cdot(a_1^2 a_7 + 4a_1 a_3 a_5 + 2a_3^3)x^4 + 19958400(a_1^2 a_9 + 2a_1 a_3 a_7 + a_1 a_5^2$ $+ a_3^2 a_5)x^2 + 39916800 a_{11}] + 39916800 a_{11}]$

The following corollary gives some properties of various coefficients when $f \in \mathcal{F}_0$.

Corollary 1.13. If $f \in \mathcal{F}_0$, $0 \leq j \leq n$ and $n - j$ odd, then

$$(1.24) \qquad\qquad a_{2j} = 0,$$

$$(1.25) \qquad\qquad \varphi_j^{(n)} = 0,$$

and

$$(1.26) \qquad\qquad a_n^{(j)} = 0.$$

Proof. (1.24) This follows from f being an odd function.
(1.25) Using (1.4), we have

$$\sum_{j=0}^{n} (-1)^n \varphi_j^{(n)} x^j = (-1)^n f_n(x) = f_n(-x) = \sum_{j=0}^{n} (-1)^j \varphi_j^{(n)} x^j,$$

so $((-1)^n - (-1)^j)\varphi_j^{(n)} = 0$, from which the result follows.
(1.26) Use (1.25) in (1.8). □

Corollary 1.14. If $f \in \mathcal{F}_0$, then for $n \geq 0$,

$$(1.27) \qquad \varphi_2^{(2n+2)} = \sum_{j=0}^{n} \binom{2n+1}{2j+1} \varphi_1^{(2j+1)} \varphi_1^{(2n+1-2j)}$$

$$= \frac{1}{2} \sum_{j=0}^{n} \binom{2n+2}{2j+1} \varphi_1^{(2j+1)} \varphi_1^{(2n+1-2j)}.$$

Proof. In (1.18) put $r = 1$ and replace n by $2n + 1$, which gives

$$\varphi_2^{(2n+2)} = \sum_{j=1}^{2n+1} \binom{2n+1}{j} \varphi_1^{(j)} \varphi_1^{(2n+2-j)} = \sum_{j=0}^{n} \binom{2n+1}{2j+1} \varphi_1^{(2j+1)} \varphi_1^{(2n+1-2j)},$$

using (1.25). To obtain the second expression, solve (1.13) for a_n and substitute the result into (1.14), using (1.26). This gives

$$\varphi_2^{(2n+2)} = \frac{(2n+2)!}{2} \sum_{j=1}^{2n+1} a_j a_{2n+2-j} = \frac{(2n+2)!}{2} \sum_{j=0}^{n} a_{2j+1} a_{2n+1-2j}$$

$$= \frac{(2n+2)!}{2} \sum_{j=0}^{n} \frac{\varphi_1^{(2j+1)}}{(2j+1)!} \frac{\varphi_1^{(2n+1-2j)}}{(2j+1)!}$$

$$= \frac{1}{2} \sum_{j=0}^{n} \binom{2n+2}{2j+1} \varphi_1^{(2j+1)} \varphi_1^{(2n+1-2j)},$$

where we have used (1.24). □

We will use the notation for the falling factorial polynomial [Jo, p. 45]

$$(1.28) \qquad (x)_n \stackrel{\text{def}}{=} \prod_{j=1}^{n}(x-j+1), \ n \geq 1,$$

and the inverse difference operator, defined on $\{(x)_n\}$ [Jo, pp. 100, 104], by

$$(1.29) \qquad \Delta^{-1}(x)_n = \frac{1}{n+1}(x)_{n+1} + C.$$

Corollary 1.15. If $f \in \mathcal{F}_0$, then for $n \geq 1$, we have

$$(1.30) \qquad \varphi_n^{(n+4)} = 60\binom{n+4}{5}a_1^{n-2}\left(2a_1 a_5 + (n-1)a_3^2\right).$$

Proof. Substituting $n+3$ for n and $n-1$ for r in (1.18), we find that

$$\varphi_n^{(n+4)} = \binom{n+3}{4}\varphi_{n-1}^{(n-1)}\varphi_1^{(5)} + \binom{n+3}{2}\varphi_{n-1}^{(n+1)}\varphi_1^{(3)} + \varphi_{n-1}^{(n+3)}\varphi_1^{(1)}$$
$$= 5(n+3)_4 a_1^{n-1}a_5 + 3(n+3)_5 a_1^{n-2}a_3^2 + a_1\varphi_{n-1}^{(n+3)},$$

using (1.9), (1.11), (1.13), and (1.28). Setting $\varphi_n^{(n+4)} = a_1^n b_{n+1}$, we obtain the difference equation (see [Jo, pp. 2, 3] with $h = 1$),

$$a_1^n(b_{n+1} - b_n) = 5(n+3)_4 a_1^{n-1}a_5 + 3(n+3)_5 a_1^{n-2}a_3^2,$$

which is that

$$\Delta b_n = 5(n+3)_4\left(\frac{a_5}{a_1}\right) + 3(n+3)_5\left(\frac{a_3}{a_1}\right)^2.$$

From (1.29) we find that

$$b_n = (n+3)_5\left(\frac{a_5}{a_1}\right) + \frac{1}{2}(n+3)_6\left(\frac{a_3}{a_1}\right)^2 + C$$

or

$$b_{n+1} = (n+4)_5\left(\frac{a_5}{a_1}\right) + \frac{1}{2}(n+4)_6\left(\frac{a_3}{a_1}\right)^2 + C$$

so

$$\varphi_n^{(n+4)} = (n+4)_5 a_1^{n-1}a_5 + \frac{1}{2}(n+4)_6 a_1^{n-2}a_3^2 + a_1^n C.$$

Setting $n = 2$ and using (1.14) and (1.24), we find that $C = 0$, which gives the result. □

Lemma 1.16. Let $\{u_j\}_{j=0}^{\infty} \subset \mathbb{C}$ be an arbitrary sequence. Then, for $n, r, s \geq 0$,

$$(1.31) \qquad \sum_{j=0}^{n} \binom{rn + 2s + 2}{rj + s + 1} u_j\, u_{n-j} = 2 \sum_{j=0}^{n} \binom{rn + 2s + 1}{rj + s + 1} u_j\, u_{n-j}.$$

Proof. We have for $0 \leq j \leq n$ that

$$f_{r,s}(n,j) \overset{\text{def}}{=} \binom{rn + 2s + 2}{rj + s + 1} - 2\binom{rn + 2s + 1}{rj + s + 1}$$
$$= -\frac{r(n - 2j)}{rn + 2s + 2}\binom{rn + 2s + 2}{rj + s + 1}.$$

Then

$$f_{r,s}(n, n - j) = \frac{r(n - 2j)}{rn + 2s + 2}\binom{rn + 2s + 2}{rn - rj + s + 1}$$
$$= \frac{r(n - 2j)}{rn + 2s + 2}\binom{rn + 2s + 2}{rj + s + 1} = -f_{r,s}(n,j).$$

Thus, re-indexing by $j \to n - j$, we have

$$S = \sum_{j=0}^{n} f_{r,s}(n,j) u_j\, u_{n-j} = \sum_{j=0}^{n} f_{r,s}(n, n - j) u_j\, u_{n-j} = -S,$$

so $S = 0$, which implies the lemma. □

For $f \in \mathcal{F}_0$, we can easily compute a table of $f_n(x)$'s from (1.6). (The calculation in (1.6) is reduced by a half because of (1.24).)

Scholium 1.17. Using (1.31), we can reconcile the two forms of $\varphi_2^{(2n+2)}$ in Corollary 1.14. If we put $u_j = \varphi_1^{(2j+1)}$, $r = 1$, and $s = 0$ into Lemma 1.16, we find that

$$(1.32) \qquad \frac{1}{2}\sum_{j=0}^{n}\binom{2n + 2}{2j + 1}\varphi_1^{(2j+1)}\varphi_1^{(2n+1-2j)}$$

$$= \sum_{j=0}^{n}\binom{2n + 1}{2j + 1}\varphi_1^{(2j+1)}\varphi_1^{(2n+1-2j)}. \qquad \square$$

Formulas (1.33) and (1.34) that follow express the polynomial $f_n(x)$ as an $n \times n$ determinant.

Theorem 1.18. If $f \in \mathcal{F}_0$, then

$$(1.33) \quad f_n(x) = \det \mathcal{A}_n(x) =
\begin{vmatrix}
a_1 x & 0 & 6a_3 x & 0 & 120a_5 x & 0 & \cdots \\
-1 & a_1 x & 0 & 18a_3 x & 0 & 600a_5 x & \cdots \\
0 & -1 & a_1 x & 0 & 36a_3 x & 0 & \cdots \\
0 & 0 & -1 & a_1 x & 0 & 60a_3 x & \cdots \\
0 & 0 & 0 & -1 & a_1 x & 0 & \cdots \\
0 & 0 & 0 & 0 & -1 & a_1 x & \cdots \\
\vdots & \vdots & \vdots & \vdots & \vdots & \vdots & \vdots
\end{vmatrix},$$

where $\mathcal{A}_n(x) = [\alpha_{ij}]_{n \times n}$, $n \geq 1$, and

$$\alpha_{ij} = \begin{cases}
(j - i + 1)\dfrac{(j-1)!}{(i-1)!}a_{j-i+1}x, & j - i \geq 0,\ j - i \text{ even}, \\
0, & j - i \geq 0,\ j - i \text{ odd}, \\
-1, & i - j = 1, \\
0, & \text{otherwise.}
\end{cases}$$

Also,

$$(1.34) \quad f_n(x) = \det \mathcal{B}_n(x) =
\begin{vmatrix}
a_1 x & 1 & 0 & 0 & 0 & \cdots \\
0 & a_1 x & 2 & 0 & 0 & \cdots \\
3a_3 x & 0 & a_1 x & 3 & 0 & \cdots \\
0 & 3a_3 x & 0 & a_1 x & 4 & \cdots \\
5a_5 x & 0 & 3a_3 x & 0 & a_1 x & \cdots \\
\vdots & \vdots & \vdots & \vdots & \vdots & \vdots
\end{vmatrix},$$

where $\mathcal{B}_n(x) = [\beta_{ij}]_{n \times n}$, $n \geq 1$, and

$$\beta_{ij} = \begin{cases}
(i - j + 1)a_{i-j+1}x, & i - j \geq 0,\ i - j \text{ even}, \\
0, & i - j \geq 0,\ i - j \text{ odd}, \\
j - 1, & j - i = 1, \\
0, & j - i > 1.
\end{cases}$$

Proof. (1.33) Let $g(y) = e^{xy}$ and write $e^{xf(t)} = (g \circ f)(t)$.
Now Ivanoff's formula [Iv, p. 212] in operator form is

$$(1.35) \quad (g \circ f)^{(n)}(0) =$$

$$\left[
\begin{vmatrix}
f'(0)D & f''(0)D & f'''(0)D & \cdots & f^{(n)}(0)D \\
-1 & f'(0)D & 2f''(0)D & \cdots & \binom{n-1}{1}f^{(n-1)}(0)D \\
0 & -1 & f'(0)D & \cdots & \binom{n-1}{2}f^{(n-2)}(0)D \\
\vdots & \vdots & \vdots & \ddots & \vdots \\
0 & 0 & 0 & \cdots & f'(0)D
\end{vmatrix} g(y)
\right]_{y=0},$$

where $D = \dfrac{d}{dy}$ and the (i,j)-th entry is

$$(i,j) = \begin{cases} \binom{j-1}{i-1} f^{(j-i+1)}(0)D, & 1 \le i \le j \le n, \\ -1, & i-j = 1, \\ 0, & \text{otherwise.} \end{cases}$$

To obtain (1.33) from (1.35), note that using (1.3) we have

$$(g \circ f)^{(n)}(0) = f_n(x),$$

and using (1.1) and $a_{2k} = 0$ that

$$f^{(r)}(0) = \begin{cases} r!\,a_r, & r \text{ odd}, \\ 0, & r \text{ even}. \end{cases}$$

Also, $Dg(y)\Big]_{y=0} = x$ and $g(0) = 1$.

(1.34) Consider the diagonal matrix

$$\mathcal{D}_n = \text{diag}\left[0!,\ 1!,\ 2!, \cdots ,\ (n-1)!\right]$$

and the matrix $A_n^T(x)$, where the (i,j)-th element is

$$\alpha_{ij}^T = \alpha_{ji} = \begin{cases} (i-j+1)\dfrac{(i-1)!}{(j-1)!}\,a_{i-j+1}\,x, & i-j \ge 0, i-j \text{ even}, \\ 0, & i-j \ge 0, i-j \text{ odd}, \\ -1, & j-i = 1, \\ 0, & \text{otherwise.} \end{cases}$$

Then

$$C_n(x) \overset{\text{def}}{=} \mathcal{D}_n^{-1} A^T(x)\mathcal{D}_n = \begin{bmatrix} a_1 x & -1 & 0 & 0 & 0 & 0 & \cdots \\ 0 & a_1 x & -2 & 0 & 0 & 0 & \cdots \\ 3a_3 x & 0 & a_1 x & -3 & 0 & 0 & \cdots \\ 0 & 3a_3 x & 0 & a_1 x & -4 & 0 & \cdots \\ 5a_5 x & 0 & 3a_3 x & 0 & a_1 x & -5 & \cdots \\ \vdots & \vdots & \vdots & \vdots & \vdots & \vdots & \vdots \end{bmatrix},$$

where the (i,j)-th element is

$$\dfrac{(j-1)!}{(i-1)!}\,\alpha_{ij}^T = \begin{cases} (i-j+1)\,a_{i-j+1}\,x, & i-j \ge 0,\ i-j \text{ even}, \\ 0, & i-j \ge 0,\ i-j \text{ odd}, \\ -i, & j-i = 1, \\ 0, & \text{otherwise.} \end{cases}$$

But then

$$f_n(x) = \det A_n(x) = \det \left(D_n^{-1} A_n^T(x) D_n \right) = \det C_n(x),$$

so

$$f_n(x) = (-1)^n f_n(-x) = \det \left(-C_n(-x) \right) = \det B_n(x). \quad \square$$

We next consider the two sets of polynomials derived from $\{f_n(x)\}_{n=0}^{\infty}$ by dropping out the zero terms. From Table 1.1, we note that $x^{-1} f_{2m+1}(x)$ and $x^{-2} f_{2m+2}(x)$ are polynomials in x^2. That this is true in general is clear from (1.12) and (1.25). Thus, except for $x = 0$, the zeros of $G_m(z)$ and $H_m(z)$ are the squares of the zeros of $f_{2m+1}(x)$ and $f_{2m+2}(x)$, respectively.

Definition 1.19. For $f \in \mathcal{F}_0$ and $m \geq 0$, define

$$(1.36) \qquad G_m(z) \stackrel{\text{def}}{=} \sum_{j=0}^{m} \varphi_{2j+1}^{(2m+1)} z^j$$

and

$$(1.37) \qquad H_m(z) \stackrel{\text{def}}{=} \sum_{j=0}^{m} \varphi_{2j+2}^{(2m+2)} z^j.$$

It follows from these definitions, (1.4), and (1.25), putting $z = x^2$, that

$$(1.38) \qquad G_m(x^2) = \frac{f_{2m+1}(x)}{x}$$

and

$$(1.39) \qquad H_m(x^2) = \frac{f_{2m+2}(x)}{x^2}.$$

Lemma 1.20. For $m \geq 0$,

$$(1.40) \qquad G_m(0) = (2m+1)!\, a_{2m+1}$$

and

$$(1.41) \qquad H_m(0) = \frac{1}{2}(2m+2)! \sum_{k=0}^{m} a_{2k+1}\, a_{2m+1-2k}.$$

Proof. From (1.36) and (1.13) we get

$$G_m(0) = \varphi_1^{(2m+1)} = (2m+1)!\, a_{2m+1},$$

while from (1.37) and (1.14),

$$H_m(0) = \varphi_2^{(2m+2)} = \frac{1}{2}(2m+2)! \sum_{k=0}^{m} a_{2k+1} a_{2m+1-2k}. \quad \square$$

In the next two theorems we give recursion formulas that connect the G's and the H's.

Theorem 1.21. For $m \geq 1$,

$$(1.42) \qquad G_m(z) = (2m)! z \sum_{k=0}^{m-1} \frac{2m-1-2k}{(2k+2)!} a_{2m-1-2k} H_k(z) + G_m(0)$$

and for $m \geq 0$,

$$(1.43) \qquad H_m(z) = (2m+1)! \sum_{k=0}^{m} \frac{2m+1-2k}{(2k+1)!} a_{2m+1-2k} G_k(z).$$

Proof. Putting $n = 2m$ in (1.6) and using (1.24) gives

$$f_{2m+1}(x) = (2m)! x \sum_{k=0}^{2m} \frac{2m+1-k}{k!} a_{2m+1-k} f_k(x)$$

$$= (2m)! x \sum_{k=0}^{m} \frac{2m+1-2k}{(2k)!} a_{2m+1-2k} f_{2k}(x).$$

Then

$$\frac{f_{2m+1}(x)}{x} = (2m+1)! a_{2m+1}$$

$$+ (2m)! x^2 \sum_{k=0}^{m-1} \frac{2m-1-2k}{(2k+2)!} a_{2m-1-2k} \frac{f_{2k+2}(x)}{x^2},$$

so, using (1.38)–(1.40) with $z = x^2$, we have

$$G_m(z) = (2m)! \left(z \sum_{k=0}^{m-1} \frac{2m-1-2k}{(2k+2)!} a_{2m-1-2k} H_k(z) + (2m+1) a_{2m+1} \right).$$

A similar argument establishes (1.43). $\square$

Corollary 1.22. For $m \geq 1$,

$$(1.44) \quad G_m(z) = G_m(0) + (2m)! \frac{z}{2}$$

$$\times \sum_{k=0}^{m-1} \left(\sum_{j=k}^{m-1} \frac{(2m-2j-1)(2j-2k+1)}{j+1} a_{2m-2j-1} a_{2j-2k+1} \right) \frac{G_k(z)}{(2k+1)!}$$

and

$$(1.45) \quad H_m(z) = (2m+1)! \left\{ z \sum_{k=0}^{m-1} \left(\sum_{j=k}^{m-1} \frac{(2m-2j-1)(2j-2k+1)}{2k+3} \right. \right.$$

$$\left. \times\, a_{2m-2j-1} a_{2j-2k+1} \right) \frac{H_k(z)}{(2k+2)!} + \sum_{j=0}^{m} (2m-2j+1)\, a_{2j+1}\, a_{2m-2j+1} \right\}.$$

Proof. These results follow from combining (1.42) and (1.43). □

The next theorem deals with the derivatives of $G_m(z)$ and $H_m(z)$.

Theorem 1.23. We have for $m \geq 1$ that

$$(1.46) \quad (2z\, D_z + 1) G_m(z)$$

$$= (2m)! \left\{ \frac{G_m(0)}{(2m)!} + (2m+1) z \sum_{k=0}^{m-1} \frac{a_{2m-2k-1}}{(2k+2)!} H_k(z) \right\}$$

and for $m \geq 0$,

$$(1.47) \quad (z D_z + 1) H_m(z) = \frac{(2m+2)!}{2} \sum_{k=0}^{m} \frac{a_{2m-2k+1}}{(2k+1)!} G_k(z).$$

Proof. Differentiating (1.38), we obtain

$$2x^2 G'_m(x^2) = -f_{2m+1}(x) + x\, f'_{2m+1}(x)$$

$$= -G_m(x^2) + (2m+1)! \sum_{k=0}^{m} \frac{a_{2m-2k+1}}{(2k)!} f_{2k}(x),$$

using (1.38), (1.19), (1.24), and canceling x. Thus,

$$2x^2 G'_m(x^2) + G_m(x^2) = (2m+1)!\, a_{2m+1}$$

$$+ x^2 (2m+1)! \sum_{k=1}^{m} \frac{a_{2m-2k+1}}{(2k)!} H_{k-1}(x^2).$$

Using (1.40) and replacing x^2 by z gives the result. Formula (1.47) is proved in a similar way. □

It is also simple to find generating functions for $G_m(z)$ and $H_m(z)$.

Theorem 1.24. If $f \in \mathcal{F}_0$ and $z < 0$, then

$$(1.48) \quad \frac{1}{\sqrt{|z|}} \sin\left(\sqrt{|z|}\, f(t)\right) = \sum_{m=0}^{\infty} G_m(z) \frac{t^{2m+1}}{(2m+1)!}$$

and

$$(1.49) \quad \frac{1}{z} \left\{ \cos\left(\sqrt{|z|}\, f(t)\right) - 1 \right\} = \sum_{m=0}^{\infty} H_m(z) \frac{t^{2m+2}}{(2m+2)!}.$$

Proof. (1.48) Using (1.3), (1.23), and (1.38), we have that

$$\sin[\sqrt{|z|}\,f(t)] = \frac{1}{2i}\left\{e^{i\sqrt{|z|}f(t)} - e^{-i\sqrt{|z|}f(t)}\right\}$$

$$= \frac{1}{2i}\sum_{n=0}^{\infty}\left\{f_n(i\sqrt{|z|}) - (-1)^n f_n(i\sqrt{|z|})\right\}\frac{t^n}{n!}$$

$$= \frac{1}{i}\sum_{n=0}^{\infty} f_{2n+1}(i\sqrt{|z|})\frac{t^{2n+1}}{(2n+1)!} = \sqrt{|z|}\sum_{n=0}^{\infty} G_n(z)\frac{t^{2n+1}}{(2n+1)!}.$$

The proof of (1.49) is similar. $\square$

The next two tables are readily obtained from Table 1.1 or (1.42) and (1.43).

Table 1.2 $G_m(z)$, $0 \le m \le 4$, $f \in \mathcal{F}_0$

n	$G_m(z)$
0	a_1
1	$a_1^3 z + 6a_3$
2	$a_1^5 z^2 + 60a_1^2 a_3 z + 120a_5$
3	$a_1^7 z^3 + 210a_1^4 a_3 z^2 + 2520a_1(a_3^2 + a_1 a_5)z + 5040a_7$
4	$a_1^9 z^4 + 504a_1^6 a_3 z^3 + 15120a_1^3(a_1 a_5 + 2a_3^2)z^2$
	$\quad + 60480(3a_1^2 a_7 + 6a_1 a_3 a_5 + a_3^3)z + 362880a_9$

Table 1.3 $H_m(z)$, $0 \le m \le 4$, $f \in \mathcal{F}_0$

n	$H_m(z)$
0	a_1^2
1	$a_1^4 z + 24a_1 a_3$
2	$a_1^6 z^2 + 120a_1^3 a_3 z + 360(a_3^2 + 2a_1 a_5)$
3	$a_1^8 z^3 + 336a_1^5 a_3 z^2 + 3360a_1^2(3a_3^2 + 2a_1 a_5)z + 40320(a_1 a_7 + a_3 a_5)$
4	$a_1^{10} z^4 + 720a_1^7 a_3 z^3 + 15120a_1^4(2a_1 a_5 + 5a_3^2)z^2$
	$+604800a_1(a_1^2 a_7 + 3a_1 a_3 a_5 + a_3^3)z + 1814400(2a_1 a_9 + 2a_3 a_7 + a_5^2)$

The next three theorems deal with properties of $G_m(z)$ and $H_m(z)$.

Theorem 1.25. Let $x, y < 0$, $u = x + y - 2\sqrt{xy}$, and $v = x + y + 2\sqrt{xy}$. Then we have

$$(1.50) \quad \frac{vH_m(v) - uH_m(u)}{v - u} = \frac{1}{2}\sum_{j=0}^{m}\binom{2m+2}{2j+1}G_j(x)G_{m-j}(y), \ m \geq 0,$$

$$(1.51) \quad G_m(x) + y\sum_{j=0}^{m-1}\binom{2m+1}{2j+1}G_j(x)H_{m-j-1}(y)$$

$$= \begin{cases} \dfrac{\sqrt{|u|}G_m(u) + \sqrt{|v|}G_m(v)}{\sqrt{|u|} + \sqrt{|v|}}, & m \geq 1, \ x \leq y, \\[4mm] \dfrac{\sqrt{|u|}G_m(u) - \sqrt{|v|}G_m(v)}{\sqrt{|u|} - \sqrt{|v|}}, & m \geq 1, \ x > y, \end{cases}$$

and for $m \geq 1$,

$$(1.52) \quad \frac{uH_m(u) + vH_m(v)}{(u-v)^2}$$

$$= \frac{1}{8}\left\{\frac{H_m(x)}{y} + \frac{H_m(y)}{x} + \sum_{j=0}^{m-1}\binom{2m+2}{2j+2}H_j(x)H_{m-j-1}(y)\right\}.$$

Proof. (1.50) To begin, note that $u < 0$ and $\sqrt{|u|} = \sqrt{|x|} + \sqrt{|y|}$. Also, since $(\sqrt{|x|} - \sqrt{|y|})^2 > 0$, then $v < 0$ and $\sqrt{p|v|} = |\sqrt{|x|} - \sqrt{|y|}|$. Finally, $v - u = 4\sqrt{xy} > 0$.

Now,

$$\sum_{m=0}^{\infty}\left\{\sum_{j=0}^{m}\binom{2m+2}{2j+1}G_j(x)G_{m-j}(y)\right\}\frac{t^{2m+2}}{(2m+2)!}$$

$$= \sum_{m=0}^{\infty}\left\{\sum_{j=0}^{m}\frac{G_j(x)}{(2j+1)!}\frac{G_{m-j}(y)}{(2m-2j+1)!}\right\}t^{2m+2}$$

$$= \left(\sum_{m=0}^{\infty}G_m(x)\frac{t^{2m+1}}{(2m+1)!}\right)\left(\sum_{m=0}^{\infty}G_m(y)\frac{t^{2m+1}}{(2m+1)!}\right)$$

$$= \frac{1}{\sqrt{xy}}\sin\left(\sqrt{|x|}\,f(t)\right)\sin\left(\sqrt{|y|}\,f(t)\right)$$

$$= \frac{1}{2\sqrt{xy}}\left\{\cos\left(\sqrt{|v|}\,f(t)\right) - \cos\left(\sqrt{|u|}\,f(t)\right)\right\}$$

$$= \frac{1}{2\sqrt{xy}}\left\{v\sum_{m=0}^{\infty}H_m(v)\frac{t^{2m+2}}{(2m+2)!} - u\sum_{m=0}^{\infty}H_m(u)\frac{t^{2m+2}}{(2m+2)!}\right\}$$

$$= 2\sum_{m=0}^{\infty}\left\{\frac{vH_m(v) - uH_m(u)}{v - u}\right\}\frac{t^{2m+2}}{(2m+2)!},$$

where we have used (1.48) and (1.49). Equating the corresponding coefficients gives the result.

(1.51) Using (1.48) and (1.49), we have

$$\sum_{m=1}^{\infty}\Big\{\sum_{j=0}^{m-1}\binom{2m+1}{2j+1}G_j(x)H_{m-j-1}(y)\Big\}\frac{t^{2m+1}}{(2m+1)!}$$

$$=\sum_{m=0}^{\infty}\Big\{\sum_{j=0}^{m}\binom{2m+3}{2j+1}G_j(x)H_{m-j}(y)\Big\}\frac{t^{2m+3}}{(2m+3)!}$$

$$=\Big(\sum_{m=0}^{\infty}G_m(x)\frac{t^{2m+1}}{(2m+1)!}\Big)\Big(\sum_{m=0}^{\infty}H_m(y)\frac{t^{2m+1}}{(2m+1)!}\Big)$$

$$=\frac{1}{y\sqrt{x}}\sin\Big(\sqrt{|x|}\,f(t)\Big)\Big[\cos\Big(\sqrt{|y|}\,f(t)\Big)-1\Big]$$

$$=\frac{1}{y\sqrt{|x|}}\Big\{\sin\Big(\sqrt{|x|}\,f(t)\Big)\cos\Big(\sqrt{|y|}\,f(t)\Big)-\sin\Big(\sqrt{|x|}\,f(t)\Big)\Big\}.$$

Assume now that $x\le y$. Then the RHS equals

$$\frac{1}{y\sqrt{|x|}}\Big\{\frac{1}{2}\Big[\sin\Big(\sqrt{|u|}\,f(t)\Big)+\sin\Big(\sqrt{|v|}\,f(t)\Big)\Big]-\sin\Big(\sqrt{|x|}\,f(t)\Big)\Big\}$$

$$=\frac{\sqrt{|u|}}{2y\sqrt{|x|}}\sum_{m=0}^{\infty}G_m(u)\frac{t^{2m+1}}{(2m+1)!}+\frac{\sqrt{|v|}}{2y\sqrt{|x|}}\sum_{m=0}^{\infty}G_m(v)\frac{t^{2m+1}}{(2m+1)!}$$

$$-\frac{1}{y}\sum_{m=0}^{\infty}G_m(x)\frac{t^{2m+1}}{(2m+1)!}$$

$$=\frac{1}{2y}\sum_{m=0}^{\infty}\Big\{\frac{\sqrt{|u|}G_m(u)}{\sqrt{|x|}}+\frac{\sqrt{|v|}G_m(v)}{\sqrt{|x|}}-2G_m(x)\Big\}\frac{t^{2m+1}}{(2m+1)!}.$$

But the $m=0$ term in the last sum is 0, so equating corresponding coefficients gives the result. The change in sign in the second form when $x>y$ occurs in $\sin\Big(\sqrt{|v|}\,f(t)\Big)$ above, since the sine is an odd function.

(1.52) We have

$$\sum_{m=1}^{\infty}\Big\{\sum_{j=0}^{m-1}\binom{2m+2}{2j+2}H_j(x)H_{m-j-1}(y)\Big\}\frac{t^{2m+2}}{(2m+2)!}$$

$$=\sum_{m=0}^{\infty}\Big\{\sum_{j=0}^{m}\binom{2m+4}{2j+2}H_j(x)H_{m-j}(y)\Big\}\frac{t^{2m+4}}{(2m+4)!}$$

$$=\Big(\sum_{m=0}^{\infty}\frac{H_m(x)}{(2m+2)!}t^{2m+2}\Big)\Big(\sum_{m=0}^{\infty}H_m(y)\frac{t^{2m+2}}{(2m+2)!}\Big)$$

$$=\frac{1}{xy}\Big\{\cos\Big(\sqrt{|x|}\,f(t)\Big)-1\Big\}\Big\{\cos\Big(\sqrt{|y|}\,f(t)\Big)-1\Big\}$$

$$= \frac{1}{xy}\Big\{ \cos\left(\sqrt{|x|}\,f(t)\right)\cos\left(\sqrt{|y|}\,f(t)\right)$$

$$- \cos\left(\sqrt{|x|}\,f(t)\right) - \cos\left(\sqrt{|y|}\,f(t)\right) + 1\Big\}$$

$$= \frac{1}{xy}\Big\{\frac{1}{2}\Big[\cos\left(\sqrt{|u|}\,f(t)\right) + \cos\left(\sqrt{|v|}\,f(t)\right)\Big]$$

$$- \cos\left(\sqrt{|x|}\,f(t)\right) - \cos\left(\sqrt{|y|}\,f(t)\right) + 1\Big\}$$

$$= \frac{1}{2xy}\Big(1 + u\sum_{m=0}^{\infty} H_m(u)\frac{t^{2m+2}}{(2m+2)!}\Big) + \frac{1}{2xy}\Big(1 + v\sum_{m=0}^{\infty} H_m(v)\frac{t^{2m+2}}{(2m+2)!}\Big)$$

$$- \frac{1}{y}\Big(\frac{1}{x} + \sum_{m=0}^{\infty} H_m(x)\frac{t^{2m+2}}{(2m+2)!}\Big) - \frac{1}{x}\Big(\frac{1}{y} + \sum_{m=0}^{\infty} H_m(y)\frac{t^{2m+2}}{(2m+2)!}\Big) + \frac{1}{xy}$$

$$= \sum_{m=0}^{\infty}\Big\{\frac{uH_m(u) + vH_m(v)}{2xy} - \frac{H_m(x)}{y} - \frac{H_m(y)}{x}\Big\}\frac{t^{2m+2}}{(2m+2)!}$$

$$= \frac{8}{(u-v)^2}\sum_{m=0}^{\infty}\Big\{uH_m(u) + vH_m(v) - 2xH_m(x) - 2yH_m(y)\Big\}\frac{t^{2m+2}}{(2m+2)!}.$$

But the $m = 0$ term in the last sum is zero, so equating the corresponding coefficients gives the result. $\square$

Corollary 1.26. We have

(1.53) $\qquad H_m(4z) = \frac{1}{2}\sum_{j=0}^{m}\binom{2m+2}{2j+1}G_j(z)G_{m-j}(z),\ m \geq 0,$

(1.54) $\qquad G_m(4z) = G_m(z) + z\sum_{j=0}^{m-1}\binom{2m+1}{2j+1}G_j(z)H_{m-j-1}(z),\ m \geq 1,$

and

(1.55) $\qquad H_m(4z) = H_m(z) + \frac{z}{2}\sum_{j=0}^{m-1}\binom{2m+2}{2j+2}H_j(z)H_{m-j-1}(z),\ m \geq 1.$

Proof. If we put $x = y = z$ (so that $u = 4z$ and $v = 0$) in Theorem 1.25, the results will follow for $z < 0$. Since these polynomial equations hold for infinitely many values of z, they will also be true for all $z \in \mathbb{R}$ and so will be polynomial identities in $\mathbb{R}[z]$. $\square$

The next theorem shows that the polynomial sequences $\{G_m(z)\}_{m=0}^{\infty}$ and $\{H_m(z)\}_{m=0}^{\infty}$ are binomial only when $f(t) = t$ in Theorem 1.27.

Theorem 1.27. For $m \geq 0$, we have that

(1.56) $\qquad$ the sequence $\{G_m(z)\}_{m=0}^{\infty}$ is binomial $\iff G_m(z) = z^m$

and

(1.57)   the sequence $\{H_m(z)\}_{m=0}^{\infty}$ is binomial $\Longleftrightarrow H_m(z) = z^m$.

Proof. (1.56), (1.57) ($\Longleftarrow$) This is the binomial expansion.
(1.56) ($\Longrightarrow$) Putting $m = 0$ in

(1.58) $$G_m(x+y) = \sum_{k=0}^{m} \binom{m}{k} G_k(x) G_{m-k}(y)$$

gives $G_0(x+y) = G_0(x)G_0(y) \Longrightarrow a_1 = a_1^2, \Longrightarrow a_1 = 1$, since $a_1 \neq 0$ (cf. (1.5)).

For $m = 1$ and $x = y = 0$, equation (1.58) becomes $G_1(0) = 2G_1(0) \Longrightarrow G_1(0) = 0$.

Now assume that $G_r(0) = 0$ for all r such that $1 \leq r \leq m$, for some $m \geq 1$. Then for $x = y = 0$, equation (1.58) becomes

$$G_{m+1}(0) = \sum_{r=0}^{m+1} \binom{m+1}{r} G_r(0) G_{m+1-r}(0) = 2G_{m+1}(0),$$

so again $G_{m+1}(0) = 0$. Thus, by induction, $G_m(0) = 0$ for all $m \geq 1$. But then $a_{2m+1} = 0$, $m \geq 1$, since by (1.40)

$$(2m+1)! \, a_{2m+1} = G_m(0).$$

Also, since $a_{2k} = 0$, this implies by (1.1) that $f(t) = t$, so (1.4) gives $f_n(x) = x^n$, $n \geq 0$, or finally by (1.38) it follows that $G_m(z) = z^m$, $m \geq 0$.
(1.57) ($\Longrightarrow$) Putting $m = 0$ in

(1.59) $$H_m(x+y) = \sum_{k=0}^{m} \binom{m}{k} H_k(x) H_{m-k}(y)$$

gives that $a_1^4 = a_1^2$ or $a_1 = \pm 1$.

In the same way as in part (a), we find for $m \geq 1$ that

(1.60) $H_m(0) = 0,$

so using (1.41), we find for $m \geq 1$ that

(1.61) $$\sum_{k=0}^{m} a_{2k+1} a_{2m+1-2k} = 0.$$

When $m = 1$, equation (1.61) gives $24a_1 a_3 = 0 \Longrightarrow a_3 = 0$.

Now assume that $a_{2r+1} = 0$ for all r, $1 \le r \le m$, for some $m \ge 1$. Then

$$\sum_{r=0}^{m+1} a_{2r+1}a_{2m+3-2r} = 0 \implies a_{2m+3}a_1 = -\sum_{r=0}^{m} a_{2r+1}a_{2m+3-2r} = 0.$$

Hence, by induction, $a_{2m+1} = 0$ for $m \ge 1$. Thus, by (1.24) we have that $f(t) = \epsilon t$, where $\epsilon = \pm 1$, so (1.4) implies that $f_n(x) = \epsilon^n x^n$, $n \ge 0$. This implies by (1.39) that $H_m(z) = z^m$, $m \ge 0$. $\square$

Note. When $G_m(z) = H_m(z) = z^m$, then $x^{2m} = G_m(x^2) = \dfrac{f_{2n+1}(x)}{x}$, so $f_{2n+1}(x) = x^{2n+1}$. Similarly, $f_{2m+2}(x) = x^{2m+2}$, so for $n \ge 0$, $f_n(x) = x^n$. Thus, $e^{xf(t)} = \sum_{n=0}^{\infty} \dfrac{(xt)^n}{n!}$, which implies $f(t) = t$.

Chapter 2

The Binomial Sequences Generated from f^{-1}, $f \in \mathcal{F}_0$

In Chapter 1 we generated two kinds of polynomial sequences, primary and secondary, from a particular kind of real function f. In this chapter we repeat this development using f^{-1} in place of f, thereby creating a parallel but "distorted" world, much like the looking-glass world in *Alice in Wonderland*. As we will see later, there are important relationships between the f-side and the f^{-1}-side, which we call the "first" and "second" sides.

One of these is the interesting fact that the coefficients of the polynomials on one side are the coefficients in inverting the sums for primary and secondary polynomials on the other side (Theorem 2.3 and Corollary 2.4), a fact that will be useful in later developments. The coefficients of corresponding polynomials on the two sides also satisfy certain orthogonality relations (Corollary 2.6). There are other important relationships between the two sides that we will discuss later.

Theorem 2.1. If $f \in \mathcal{F}$, then f^{-1} exists and $f^{-1} \in \mathcal{F}$.

Proof. This follows from standard analysis. $\square$

Since $f^{-1} \in \mathcal{F}$ when $f \in \mathcal{F}$, the formulas that were derived for f will also hold for f^{-1}. For simplicity, we will use the same letters for both, but we will put a bar over those that are derived from f^{-1}. For example,

$$(2.1) \qquad f^{-1}(t) \overset{\text{def}}{=} \sum_{n=0}^{\infty} \bar{a}_n t^n, \ (\bar{a}_0 = 0)$$

$$(2.2) \qquad [f^{-1}(t)]^k \overset{\text{def}}{=} \sum_{n=k}^{\infty} \bar{a}_n^{(k)} t^n, \ k \geq 1$$

$$(2.3) \qquad e^{x f^{-1}(t)} \overset{\text{def}}{=} \sum_{n=0}^{\infty} \bar{f}_n(x) \frac{t^n}{n!},$$

$$(2.4) \qquad \bar{f}_n(x) \overset{\text{def}}{=} \sum_{k=0}^{n} \bar{\varphi}_k^{(n)} x^k, \ n \geq 0.$$

Also, note that placing a bar over the appropriate letters in one formula produces another true formula (omitting a double bar) as, for example, in Corollaries 1.6 and 1.8. We will refer to this operation as "inverse barring." The secondary sequences derived from $\bar{f}_n(x)$ will be denoted by $\bar{G}_m(z)$ and $\bar{H}_m(z)$.

When $f \in \mathcal{F}$, the coefficients $\bar{a}_n$ of f^{-1} can be expressed in terms of the a_n's (see [AbS, p. 16, 3.6.25] or [Ka]) using the recursion formula

$$(2.5) \qquad \bar{a}_n = -\frac{1}{a_1^n} \sum_{k=1}^{n-1} a_n^{(k)} \bar{a}_k, \ n \geq 2,$$

which follows directly from the identity $f^{-1}(f(t)) = t$, when the series for $f(t)$ is substituted into the series for $f^{-1}(t)$.

Table 2.1 $\bar{a}_n$, $1 \leq n \leq 5$, $f \in \mathcal{F}$

n	$\bar{a}_n$
1	$\dfrac{1}{a_1}$
2	$-\dfrac{a_2}{a_1^3}$
3	$\dfrac{1}{a_1^5}(2a_2^2 - a_1 a_3)$
4	$\dfrac{1}{a_1^7}(5a_1 a_2 a_3 - a_1^2 a_4 - 5a_2^3)$
5	$\dfrac{1}{a_1^9}(-a_1^3 a_5 + 6a_1^2 a_2 a_4 + 3a_1^2 a_3^2 - 2(a_1 a_2^2 a_3 + 14a_2^4))$

Recall, when $f \in \mathcal{F}_0$, then in these formulas $a_{2k} = 0$ by (1.24). (Here we would normally compute the $a_n^{(k)}$'s using a recursion formula obtained from (2.2). However, in the present case, we can alternatively use the relationship in (1.8) and the recursion (1.7) for the numbers $\varphi_k^{(n)}$ for the same purpose.)

There is also the $(n-1) \times (n-1)$ determinant formula for $\bar{a}_n$ ([Ka, p. 204] or [Wa, p. 4]):

For $n \geq 2$ and $1 \leq i, j \leq n-1$, we have

$$(2.6) \quad \bar{a}_n = \frac{(-1)^{n-1}}{n! \, a_1^{2n-1}} \times$$

$$\begin{vmatrix} na_2 & 2na_3 & 3na_4 & \cdots & n(n-2)a_{n-1} & n(n-1)a_n \\ a_1 & (n+1)a_2 & (2n+1)a_3 & \cdots & (n^2-3n+1)a_{n-2} & (n-1)^2 a_{n-1} \\ 0 & 2a_1 & (n+2)a_2 & \cdots & (n^2-4n+2)a_{n-3} & (n-1)(n-2)a_{n-2} \\ \vdots & \vdots & \vdots & \vdots & \vdots & \vdots \\ 0 & 0 & 0 & \cdots & (n-2)a_1 & 2(n-1)a_2 \end{vmatrix},$$

where the $(i,j)^{th}$ entry is $((n-1)(j-i+1)+j)\,a_{j-i+2}$ and $a_k = 0$ when $k \leq 0$.

When $f \in \mathcal{F}_0$, then $\bar{a}_{2m} = 0$ from (1.24) and (2.6) simplifies to

$$(2.7) \quad \bar{a}_{2n+1} = \frac{(-1)^n}{(2n+1)} n! \, a_1^{3n+1} \times$$

$$\begin{vmatrix} (2n+1)a_3 & 2(2n+1)a_5 & \cdots & (n-1)(2n+1))a_{2n-1} & n(2n+1)a_{2n+1} \\ a_1 & 2(n+1)a_3 & \cdots & (2n^2-3n-1)a_{2n-3} & n(2n-1)a_{2n-1} \\ 0 & 2a_1 & \cdots & (2n^2-5n-1)a_{2n-5} & n(2n-3)a_{2n-3} \\ \vdots & \vdots & \vdots & \vdots & \vdots \\ 0 & 0 & \cdots & (n-1)a_1 & 3n\,a_3 \end{vmatrix},$$

where the (i,j)-th entry is $(2n(j-i+1)+j)\,a_{2j-2i+3}$, where $n \geq 1$ and $a_k = 0$, when $k \geq 0$. In this we have used the formula from [Pr, 458]:

$$\begin{vmatrix} a_{11} & 0 & a_{12} & 0 & \cdots & a_{1n} & 0 \\ 0 & b_{11} & 0 & b_{12} & \cdots & 0 & b_{1n} \\ a_{21} & 0 & a_{22} & 0 & \cdots & a_{2n} & 0 \\ \vdots & \vdots & \vdots & \vdots & \cdots & \vdots & \vdots \\ a_{n1} & 0 & a_{n2} & 0 & \cdots & a_{nn} & 0 \\ 0 & b_{n1} & 0 & b_{n2} & \cdots & 0 & b_{nn} \end{vmatrix}$$

$$= \begin{vmatrix} a_{11} & a_{12} & \cdots & a_{1n} \\ \vdots & \vdots & \cdots & \vdots \\ a_{n1} & a_{n2} & \cdots & a_{nn} \end{vmatrix} \begin{vmatrix} b_{11} & b_{12} & \cdots & b_{1n} \\ \vdots & \vdots & \cdots & \vdots \\ b_{n1} & b_{n2} & \cdots & b_{nn} \end{vmatrix}.$$

Corollary 2.2. If $f \in \mathcal{F}_0$, then $f^{-1} \in \mathcal{F}_0$.

Proof. By Theorem 2.1, we know that $f^{-1} \in \mathcal{F}$. But then $f^{-1}(-t) = f^{-1}\left(-f(f^{-1}(t))\right) = f^{-1}\left(f(-f^{-1}(t))\right) = -f^{-1}(t)$, so $f^{-1} \in \mathcal{F}_0$. $\square$

Note that this theorem implies we can use inverse barring when $f \in \mathcal{F}_0$.

Table 2.2 $\bar{a}_{2n+1}$, $0 \le n \le 4$, $f \in \mathcal{F}_0$

n	$\bar{a}_{2n+1}$
1	$\dfrac{1}{a_1}$
3	$-\dfrac{a_3}{a_1^4}$
5	$\dfrac{1}{a_1^7}(3a_1^2a_3^2 - a_1^3a_5)$
7	$\dfrac{1}{a_1^{10}}(8a_1a_3a_5 - a_1^2a_7 - 12a_3^3)$
9	$\dfrac{1}{a_1^{13}}(55a_3^4 + 10a_1^2a_3a_7 + 5a_1^2a_5^2 - 55a_1a_3^2a_5 - a_1^3a_9)$

Since the sequences $\{f_n(x)\}_{n=0}^{\infty}$ and $\{\bar{f}_n(x)\}_{n=0}^{\infty}$ contain a single polynomial of each non-negative degree (such a sequence is called "simple"), we can invert equations (1.4) and (2.4). The next theorem shows the role the φ's and $\bar{\varphi}$'s play in the inversion of $\bar{f}$ and f, respectively.

Theorem 2.3. If $f \in \mathcal{F}$, then for $n \ge 0$,

$$(2.8) \qquad x^n = \sum_{k=0}^{n} \varphi_k^{(n)} \bar{f}_k(x) = \sum_{k=0}^{n} \bar{\varphi}_k^{(n)} f_k(x).$$

Proof. Using Corollary 2.2 and (1.8), we have that

$$\sum_{n=0}^{\infty} x^n \frac{t^n}{n!} = \sum_{n=0}^{\infty} \frac{(xt)^n}{n!} = e^{xt} = e^{xf^{-1}(f(t))} = \sum_{k=0}^{\infty} \frac{\bar{f}_k(x)}{k!} [f(t)]^k$$

$$= \sum_{k=0}^{\infty} \frac{\bar{f}_k(x)}{k!} \sum_{n=k}^{\infty} a_n^{(k)} t^n = \sum_{n=0}^{\infty} \left\{ \sum_{k=0}^{n} \frac{a_n^{(k)}}{k!} \bar{f}_k(x) \right\} t^n$$

$$= \sum_{n=0}^{\infty} \left\{ \sum_{k=0}^{n} \frac{\varphi_k^{(n)}}{n!} \bar{f}_k(x) \right\} t^n = \sum_{n=0}^{\infty} \left\{ \sum_{k=0}^{n} \varphi_k^{(n)} \bar{f}_k(x) \right\} \frac{t^n}{n!}.$$

The first result follows by equating corresponding coefficients. A similar proof, with $e^{xf(f^{-1}(t))}$ replacing the fourth term in the above argument, establishes the second result. $\square$

The next corollary gives the inversion formulas for the G's and H's.

Table 2.3 $\bar{f}_n(x)$, $0 \leq n \leq 5$, $f \in \mathcal{F}_0$

n	$\bar{f}_n(x)$
0	1
1	$\dfrac{1}{a_1} x$
2	$\dfrac{1}{a_1^2} x^2$
3	$\dfrac{1}{a_1^4} x(a_1 x^2 - 6a_3)$
4	$\dfrac{1}{a_1^5} x^2 (a_1 x^2 - 24a_3)$
5	$\dfrac{1}{a_1^7} x[a_1^2 x^4 - 60 a_1 a_3 x^2 + 120(3a_3^2 - a_1 a_5)]$

Corollary 2.4. If $f \in \mathcal{F}_0$, then for $n \geq 0$,

$$(2.9) \qquad z^n = \sum_{k=0}^{n} \bar{\varphi}_{2k+1}^{(2n+1)} G_k(z) = \sum_{k=0}^{n} \bar{\varphi}_{2k+2}^{(2n+2)} H_k(z)$$

and

$$z^n = \sum_{k=0}^{n} \varphi_{2k+1}^{(2n+1)} \bar{G}_k(z) = \sum_{k=0}^{n} \varphi_{2k+2}^{(2n+2)} \bar{H}_k(z).$$

Proof. Using Corollary 2.2, replacing n by $2n+1$ in the second equation of (2.8), and using (1.25) with a bar, we obtain

$$x^{2n+1} = \sum_{k=0}^{2n+1} \bar{\varphi}_k^{(2n+1)} f_k(x) = \sum_{k=0}^{n} \bar{\varphi}_{2k+1}^{(2n+1)} f_{2k+1}(x).$$

Then (1.38) implies (2.9), setting $x^2 = z$. The second equation in (2.9) is proved in a similar way, while the third and fourth equations are the barred versions of the first and second. $\square$

The next two results show that orthogonality conditions hold between the φ's and the $\bar{\varphi}$'s.

Corollary 2.5. If $f \in \mathcal{F}$, then for $0 \leq j \leq n$,

$$(2.10) \qquad \sum_{k=j}^{n} \varphi_k^{(n)} \bar{\varphi}_j^{(k)} = \sum_{k=j}^{n} \bar{\varphi}_k^{(n)} \varphi_j^{(k)} = \delta_{nj}.$$

Proof. If we substitute the sum in (1.4) into the first equation of (2.8), we obtain

$$x^n = \sum_{k=0}^{n} \varphi_k^{(n)} \sum_{j=0}^{k} \bar{\varphi}_j^{(k)} x^j = \sum_{j=0}^{n} \left(\sum_{k=j}^{n} \varphi_k^{(n)} \bar{\varphi}_j^{(k)} \right) x^j.$$

Equating the coefficients on the two sides gives the result. The second equation can be derived from the first by putting a bar on the suitable letters in the first equation. □

Corollary 2.6. Let $f \in \mathcal{F}_0$, then for $0 \leq j \leq n$,

$$(2.11) \qquad \sum_{k=j}^{n} \varphi_{2k+1}^{(2n+1)} \bar{\varphi}_{2j+1}^{(2k+1)} = \delta_{nj}$$

and

$$(2.12) \qquad \sum_{k=j}^{n} \varphi_{2k+2}^{(2n+2)} \bar{\varphi}_{2j+2}^{(2k+2)} = \delta_{nj}.$$

Proof. The equations in (2.11) and (2.12) follow from (2.10) by replacing n by $2n+1$ and $2n+2$, respectively, and then using (1.25). □

Theorem 2.7. Suppose that $f, g \in \mathcal{F}$ and that

$$e^{xf(t)} = \sum_{n=0}^{\infty} f_n(x) \frac{t^n}{n!} \quad \text{and} \quad e^{xg(t)} = \sum_{n=0}^{\infty} g_n(x) \frac{t^n}{n!},$$

where

$$f_n(x) = \sum_{k=0}^{n} \varphi_k^{(n)} x^k \quad \text{and} \quad g_n(x) = \sum_{k=0}^{n} \psi_k^{(n)} x^k.$$

Also, let

$$\Phi_n = \begin{bmatrix} \varphi_1^{(1)} & 0 & 0 & \cdots & 0 \\ \varphi_1^{(2)} & \varphi_2^{(2)} & 0 & \cdots & 0 \\ \vdots & \vdots & \vdots & \vdots & \vdots \\ \varphi_1^{(n)} & \varphi_2^{(n)} & \varphi_3^{(n)} & \cdots & \varphi_n^{(n)} \end{bmatrix}$$

and

$$\Psi_n = \begin{bmatrix} \psi_1^{(1)} & 0 & 0 & \cdots & 0 \\ \psi_1^{(2)} & \psi_2^{(2)} & 0 & \cdots & 0 \\ \vdots & \vdots & \vdots & \vdots & \vdots \\ \psi_1^{(n)} & \psi_2^{(n)} & \psi_3^{(n)} & \cdots & \psi_n^{(n)} \end{bmatrix}.$$

Then for $n \geq 1$,

$$(2.13) \qquad \Psi_n = \Phi_n^{-1} \iff g = f^{-1}.$$

THE BINOMIAL SEQUENCES GENERATED FROM f^{-1}

Proof. ($\Longrightarrow$) It is clear from (2.10) that the entries in Φ_n^{-1} are $\bar{\varphi}_k^{(n)}$, so $\psi_k^{(n)} = \bar{\varphi}_k^{(n)}$. Thus, $g_n(x) = \sum_{k=0}^{n} \bar{\varphi}_k^{(n)} x^k$. Using (2.10), we then have that

$$e^{xg[f(t)]} = \sum_{k=0}^{\infty} \frac{g_k(x)}{k!}[f(t)]^k = \sum_{k=0}^{\infty} \frac{g_k(x)}{k!} \sum_{n=k}^{\infty} a_n^{(k)} t^n$$

$$= \sum_{n=0}^{\infty} t^n \left(\sum_{k=0}^{n} \frac{a_n^{(k)}}{k!} g_k(x) \right) = \sum_{n=0}^{\infty} \frac{t^n}{n!} \left(\sum_{k=0}^{n} \varphi_k^{(n)} g_k(x) \right)$$

$$= \sum_{n=0}^{\infty} \frac{t^n}{n!} \left(\sum_{k=0}^{n} \varphi_k^{(n)} \sum_{j=0}^{k} \bar{\varphi}_j^{(k)} x^j \right) = \sum_{n=0}^{\infty} \frac{t^n}{n!} \sum_{j=0}^{n} \left(\sum_{k=j}^{n} \varphi_k^{(n)} \bar{\varphi}_j^{(k)} \right) x^j$$

$$= \sum_{n=0}^{\infty} \frac{t^n}{n!} x^n = e^{xt}.$$

Differentiating with respect to x and setting $x = 0$ gives $g(f(t)) = t$. ($\Longleftarrow$) Since $g = f^{-1}$, (2.4) implies that $\Phi_n = \bar{\Phi}_n$. But (2.10) implies that $\bar{\Phi}_n = \Phi_n^{-1}$. $\square$

Chapter 3

The Functions $\mathcal{F}_1$ –
Elliptic Polynomials of the First Kind

In this chapter we determine an important subset $\mathcal{F}_1$ of odd functions which will be central to our investigation of orthogonal secondary sequences. This set of functions is defined in Definition 3.3 by a certain elliptic integral of the first kind containing three parameters. It is these functions that produce the two types of elliptic polynomials of the first kind. Corollary 3.6 shows that if $f \in \mathcal{F}_1$, then its Maclaurin expansion is completely determined once its first three coefficients are specified. Following this, we partition $\mathcal{F}_1$ into six classes, three of which, as we will see later, contain functions that generate orthogonal secondary sequences. The other three classes contain functions that do not generate sequences of orthogonal polynomials, although these latter polynomials are somewhat interesting because they possess some of the properties of orthogonal polynomials. However, we will not discuss these polynomials here.

Lemma 3.1. Let $\lambda \neq 0$ be real. Then the two sets of functions

(3.1) $\{x^4 \sinh(\lambda x),\ x^3 \cosh(\lambda x),\ x^2 \sinh(\lambda x),\ x \cosh(\lambda x)\}$

and

(3.2) $\{x^4 \cosh(\lambda x),\ x^3 \sinh(\lambda x),\ x^2 \cosh(\lambda x),\ x \sinh(\lambda x)\}$

are linearly independent.

Proof. (3.1) Suppose for certain numbers α_1, α_2, α_3, and α_4 that the equation

$$\alpha_4 x^4 \sinh(\lambda x) + \alpha_3 x^3 \cosh(\lambda x) + \alpha_2 x^2 \sinh(\lambda x) + \alpha_1 x \cosh(\lambda x) = 0$$

holds for all real x. Then it follows that

$$(\alpha_1 + \alpha_2 x + \alpha_3 x^2 + \alpha_4 x^3)e^{\lambda x} = (\alpha_1 - \alpha_2 x + \alpha_3 x^2 - \alpha_4 x^3)e^{-\lambda x}.$$

As $x \to \infty$, then depending on the sign of λ, one side of this equation approaches $\pm\infty$ while the other side approaches 0, a contradiction unless $\alpha_1 = \alpha_2 = \alpha_3 = \alpha_4 = 0$.

(3.2) The proof is similar. □

To prepare the way for what follows, we set

$$(3.3) \quad c_1 = \frac{6a_3}{a_1}, \quad c_2 = \frac{1}{a_1^2}(27a_3^2 - 10a_1a_5) = \frac{c}{a_1^2}, \quad \text{and} \quad \Delta = 9a_3^2 - 5a_1a_5,$$

so that

$$(3.4) \quad\quad\quad c = 9a_3^2 + 2\Delta \quad \text{and} \quad \Delta = \frac{a_1^2}{8}(4c_2 - c_1^2).$$

Note that the values of c_1 and c_2 do not change when f is multiplied by a non-zero number. Also, define the partial derivative operator

$$(3.5) \quad \mathcal{D} \stackrel{\text{def}}{=} D_t^4 - [a_1^2 x^2 + c_1(tD_t + 2)^2]D_t^2 + c_2(tD_t)(tD_t + 1)^2(tD_t + 2),$$

which, after some expanding, becomes

$$(3.6) \quad\quad \mathcal{D} = (c_2 t^4 - c_1 t^2 + 1)D_t^4 + (10c_2 t^3 - 5c_1 t)D_t^3$$
$$+ (24c_2 t^2 - 4c_1 - a_1^2 x^2)D_t^2 + 12c_2 t D_t.$$

For simplicity write (3.6) as

$$(3.7) \quad\quad \mathcal{D} = AD_t^4 + BD_t^3 + (C - a_1^2 x^2)D_t^2 + DD_t,$$

where

$$(3.8) \quad\quad \begin{aligned} A &= A(t) = c_2 t^4 - c_1 t^2 + 1, \quad B = 10c_2 t^3 - 5c_1 t, \\ C &= 24c_2 t^2 - 4c_1, \quad D = 12c_2 t. \end{aligned}$$

Also, combining (1.3), (1.38), and (1.39), we obtain the useful equation

$$(3.9) \quad e^{xf(t)} = \sum_{n=0}^{\infty} f_n(x)\frac{t^n}{n!} = 1 + x\sum_{m=0}^{\infty} G_m(x^2)\frac{t^{2m+1}}{(2m+1)!}$$
$$+ x^2 \sum_{m=0}^{\infty} H_m(x^2)\frac{t^{2m+2}}{(2m+2)!}.$$

We now prove one of the basic theorems of this work.

Theorem 3.2. Six-Part Theorem Suppose that $f \in \mathcal{F}_0$. Then the following six statements are equivalent:

$$(3.10) \qquad \qquad \mathcal{D}e^{zf(t)} = 0,$$

$$(3.11) \qquad f_{n+4}(x) = [a_1^2 x^2 + c_1(n+2)^2]f_{n+2}(x)$$
$$- n(n+1)^2(n+2) c_2 f_n(x), \ n \geq 0,$$

$$(3.12) \qquad \qquad A(t)[f'(t)]^2 = a_1^2,$$

$$(3.13) \qquad \qquad [(f^{-1})'(t)]^2 = \frac{1}{a_1^2}(A \circ f^{-1})(t),$$

$$(3.14) \quad G_{m+2}(z) = [a_1^2 z + c_1(2m+3)^2]G_{m+1}(z)$$
$$- 4(m+1)^2(2m+1)(2m+3) c_2 G_m(z), \ m \geq 0,$$

and

$$(3.15) \quad H_{m+2}(z) = [a_1^2 z + 4c_1(m+2)^2]H_{m+1}(z)$$
$$- 4(m+1)(m+2)(2m+3)^2 c_2 H_m(z), \ m \geq 0.$$

Proof. Note that parts (3.12), (3.14), and (3.15) relate an elliptic integral and its secondary sequences of polynomials.

(3.10) $\iff$ (3.11) Applying $\mathcal{D}$ to the first two terms in (3.9) produces, after some calculation, the equation

$$(3.16) \quad \mathcal{D}e^{zf(t)} = \mathcal{D}\sum_{n=0}^{\infty} f_n(x)\frac{t^n}{n!} = \sum_{n=0}^{\infty}\{f_{n+4}(x)$$
$$- [a_1^2 x^2 + c_1(n+2)^2]f_{n+2}(x) + n(n+1)^2(n+2) c_2 f_n(x)\}\frac{t^n}{n!},$$

which clearly shows the two sides vanish together. (Note how the form of $\mathcal{D}$ in (3.5) is related to the form of the recursion in (3.11).)

(3.10) $\iff$ (3.12) Applying $\mathcal{D}$ to $e^{zf(t)}$ gives, after some calculation, the equation

$$(3.17) \qquad \mathcal{D}e^{zf(t)} = \{Rx^4 + Sx^3 + Tx^2 + Ux\}e^{zf(t)},$$

where

(3.18a) $R = A(f')^4 - a_1^2(f')^2,$

(3.18b) $S = 6A(f')^2 f'' + B(f')^2 - a_1^2 f'',$

(3.18c) $T = Af'f''' + 3A(f'')^2 + 3Bf'f'' + C(f')^2,$

(3.18d) $U = Af^{(iv)} + Bf''' + Cf'' + Df'.$

Then clearly from (3.17), $De^{xf(t)} = 0 \iff$

(3.19) $R = S = T = U = 0.$

Since $f' \neq 0$, we can cancel $(f')^2$ in the equation $R = 0$ from (3.18a), which gives $A(f')^2 = a_1^2$. Thus, (3.10) $\implies$ (3.12).

(3.12) $\implies$ (3.10) Assume (3.12). From (3.8) we readily compute that

(3.20) $A' = \dfrac{2}{5}B, \ B' = \dfrac{5}{4}C, \ \text{and} \ C' = 4D.$

Also, differentiating the equation $(f')^2 = \dfrac{a_1^2}{A}$ and using (3.20), we obtain

$$f'' = -\frac{B}{5A}f', \ f''' = \frac{12B^2 - 25AC}{100A^2}f'$$

$$f^{(iv)} = \frac{45ABC - 12B^3 - 100A^2D}{100A^3}f'.$$

Substituting these results into (3.18b)–(3.18d), we find that the value of each is zero, which completes the proof in this case, using the equivalence in (3.19).

(3.12) $\iff$ (3.13) From the standard formula for the derivative of f^{-1}, we obtain

(3.21) $f'(f^{-1}(t)) = \dfrac{1}{(f^{-1})'(t)}.$

Replacing f by f^{-1} in (3.12) and using the above formula, we get (3.13). The reverse argument is immediate.

(3.11) $\iff$ (3.14) That (3.11) $\implies$ (3.14) is immediate by substituting $2m + 1$ for n in (3.11) and using (1.38). To prove the converse, apply $\mathcal{D}$ to the first and third terms in (3.9), so, after considerable calculation, we find

(3.22)
$$De^{xf(t)} = x \sum_{m=0}^{\infty} \left\{ G_{m+2}(x^2) - [a_1^2 x^2 + c_1(2m+3)^2]G_{m+1} \right.$$
$$\left. + 4(m+1)^2(2m+1)(2m+3)\, c_2\, G_m(x^2) \right\} \frac{t^{2m+1}}{(2m+1)!}$$
$$+ x^2 \sum_{m=0}^{\infty} \left\{ H_{m+2}(x^2) - [a_1^2 x^2 + 4c_1(m+2)^2]H_{m+1}(x^2) \right.$$
$$\left. + 4(m+1)(m+2)(2m+3)^2 c_2 H_m(x^2) \right\} \frac{t^{2m+2}}{(2m+2)!}.$$

Now, if we assume (3.14), the first series on the right side of (3.22) drops out and the resulting equation shows that $\mathcal{D}e^{xf(t)}$ is an even function of x. On the other hand, using (3.17), we can express this same function as the sum of an even and an odd function of x, respectively, viz.,

$$(3.23) \quad \mathcal{D}e^{xf(t)} = (Rx^4 + Sx^3 + Tx^2 + Ux)(\sinh{(xf)} + \cosh{(xf)})$$
$$= (Rx^4 \cosh(xf) + Sx^3 \sinh{(xf)} + Tx^2 \cosh(xf) + Ux \sinh{(xf)})$$
$$+ (Rx^4 \sinh{(xf)} + Sx^3 \cosh(xf) + Tx^2 \sinh{(xf)} + Ux \cosh(xf)).$$

Since, as we have concluded, $\mathcal{D}e^{xf(t)}$ is an even function of x, it follows that the odd part in (3.23) vanishes, i.e.,

$$Rx^4 \sinh{(xf)} + Sx^3 \cosh(xf) + Tx^2 \sinh{(xf)} + Ux \cosh(xf) = 0$$

for all x. But then Lemma 3.1 implies that $R = S = T = U = 0$, which is (3.19), which in turn is equivalent to (3.10) which is equivalent to (3.11).

(3.10) $\Longleftrightarrow$ **(3.15)** This is proved in a similar way. $\square$

Remark. The initial terms from Tables 1.1–1.3 are, respectively, $f_0(x) = 1$, $f_1(x) = a_1 x$; $f_2(x) = a_1^2 x^2$; $f_3(x) = x(a_1^3 x^2 + 6a_3)$; $G_0(z) = a_1$, $G_1(z) = a_1^3 z + 6a_3$; $H_0(z) = a_1^2$, $H_1(z) = a_1^4 z + 24a_1 a_3$.

The subset of functions in $\mathcal{F}_0$ for which the relations in Theorem 3.2 are true is readily determined by condition (3.12), which suggests the following definition of the family $\mathcal{F}_1$ of functions, which is parameterized by the set $\{(a_1, a_3, a_5) \in \mathbb{R}^3 : a_1 \neq 0\}$.

Definition 3.3. [ByF] Let

$$(3.24) \quad \mathcal{F}_1 =$$
$$\left\{ f \in \mathcal{F}_0 : f(t) = f(t; a_1, a_3, a_5) \stackrel{\text{def}}{=} a_1 \int_0^t \frac{du}{\sqrt{1 - c_1 u^2 + c_2 u^4}} \right\}.$$

Here $a_1 \neq 0$, $a_3, a_5 \in \mathbb{R}$, and t is in a small non-zero neighborhood of 0.

Definition 3.4. If $f \in \mathcal{F}_1$, we call $f_n(x)$ and $G_m(z)$, $H_m(z)$ "primary and secondary elliptic polynomials of the first kind," respectively.

Note. We do not need to use $\pm a_1$ in front of the integral in (3.24) because a change in sign in f can be effected by replacing a_1, a_3, and a_5 by their negatives.

The next theorem formalizes the previous development.

Theorem 3.5. The six results in (3.10)–(3.15) are true $\Longleftrightarrow f \in \mathcal{F}_1$.

Proof. Since the condition $f \in \mathcal{F}_1$ is the same as (3.12) and the six results in Theorem 3.2 are logically equivalent, the theorem holds. $\square$

The next corollary is important because it shows how the coefficients a_n, $n \geq 7$, in the Maclaurin expansion of an $f \in \mathcal{F}_1$ are recursively determined from the initial coefficients a_1, a_3, and a_5.

Corollary 3.6. Let $f \in \mathcal{F}_0$. Then for $m \geq 0$,

$$(3.25) \quad f \in \mathcal{F}_1 \Longleftrightarrow a_{2n+5} =$$

$$\frac{1}{(m+2)(2m+5)}\left\{\frac{c_1}{2}(2m+3)^2 a_{2m+3} - c_2(m+1)(2m+1)a_{2m+1}\right\},$$

where $a_1 \neq 0$ and a_3 are arbitrary. If $c_2 \neq 0$, then a_5 is also arbitrary, while if $c_2 = 0$, then $a_5 = \dfrac{27a_3^2}{10a_1}$.

Proof. ($\Longrightarrow$) Since $f \in \mathcal{F}_1$, then by Theorem 3.5, we can set $y = 0$ in (3.14), and using (1.40) obtain the recursion in (3.25), a recursion which starts at $m = 0$. When $m = 0$, (3.25) reduces to the statement $a_5 = a_5$ when $c_2 \neq 0$ and to $a_5 = \dfrac{27a_3^2}{10a_1}$ when $c_2 = 0$. Thus, when $c_2 \neq 0$, the coefficients a_1, a_3, and a_5 are independent initial values.

($\Longleftarrow$) Since $f \in \mathcal{F}_0$, then $f(t) = \displaystyle\sum_{m=0}^{\infty} a_{2m+1}t^{2m+1}$, where $\{a_{2m+1}\}_{m=0}^{\infty}$ satisfies (3.25). Then $f'(t) = \displaystyle\sum_{m=0}^{\infty}(2m+1)a_{2m+1}t^{2m}$. If we set $b_m = (2m+1)a_{2m+1}$, then for $m \geq 0$ it follows from (3.25) that

$$(3.26) \qquad (m+2)b_{m+2} = \frac{c_1}{2}(2m+3)b_{m+1} - c_2(m+1)b_m,$$

where $b_0 = a_1$ and $b_1 = \dfrac{a_1 c_1}{2}$.

Now, if we show that

$$(3.27) \qquad\qquad\qquad \frac{a_1}{\sqrt{A(t)}} = f'(t),$$

then squaring this equation will give (3.12) and the converse will be established using Theorem 3.5. But (3.27) is

$$(3.28) \qquad\qquad\qquad \frac{a_1}{\sqrt{1 - c_1 t^2 + c_2 t^4}} = \sum_{m=0}^{\infty} b_{2m+1}t^{2m}.$$

Next, recall that the Legendre polynomials $\{P_m(x)\}_{m=0}^{\infty}$ [Ke, p. 480], [GrR, p. 1026, 8.914, 1], which are generated by

$$(3.29) \qquad\qquad\qquad \frac{1}{\sqrt{1 - 2xt + t^2}} = \sum_{m=0}^{\infty} P_m(x)\, t^m,$$

satisfy the recursion

$$(3.30) \qquad (m+2)P_{m+2}(x) = (2m+3)xP_{m+1}(x) - (m+1)P_m(x), \ m \geq 0,$$

where $P_0(x) = 1$ and $P_1(x) = x$.

If in (3.29) we replace x by $\dfrac{x}{y}$ and t by $\dfrac{y\,t^2}{2}$, we then obtain

$$(3.31) \qquad \frac{1}{\sqrt{1 - xt^2 + \frac{y^2}{4}t^4}} = \sum_{m=0}^{\infty} \frac{y^m}{2^m} P_m\left(\frac{x}{y}\right) t^{2m}.$$

If $S_m(x, y) \overset{\text{def}}{=} y^m P_m\left(\dfrac{x}{y}\right)$, $m \geq 0$, then, since $P_m(-x) = (-1)^m P_m(x)$, we easily see that $S_m(x, -y) = S_m(x, y)$, so $S_m(x, y)$ is an even function in y. If in (3.31) we put $x = c_1$ and $y = 2\sqrt{c_2}$, we get

$$\frac{a_1}{\sqrt{1 - c_1 t^2 + c_2 t^4}} = a_1 \sum_{m=0}^{\infty} \frac{S_m(c_1, 2\sqrt{c_2})}{2^m} t^{2m}.$$

Now, if we transform the recursion (3.30) by the same substitutions we used in (3.29), we obtain

$$(m + 2)S_{m+2}(x, y) = (2m + 3)x S_{m+1}(x, y) - (m + 1)y^2 S_m(x, y).$$

Multiplying this recursion through by $\dfrac{a_1}{2^{m+2}}$ and then defining that $U_m = \dfrac{a_1}{2^m} S_m(c_1, 2\sqrt{c_2})$, $m \geq 0$, we discover that the sequence $\{U_m\}_{m=0}^{\infty}$ satisfies the recursion

$$(m + 2)U_{m+2} = \frac{c_1}{2}(2m + 3)U_{m+1} - (m + 1)c_2 U_m, \quad m \geq 0,$$

where $U_0 = a_1$ and $U_1 = \dfrac{a_1 c_1}{2}$. But this is the same recursion as in (3.26). Thus,

$$(3.32) \qquad (2m + 1)a_{2m+1} = b_m = U_m = \frac{a_1}{2^m} S_m(c_1, 2\sqrt{c_2}), \quad m \geq 0,$$

which establishes (3.27). $\quad\square$

Using (3.25), we can find formulas for the first few a's in the expansion of an $f \in \mathcal{F}_1$:

$$(3.33) \qquad a_7 = \frac{9a_3}{7a_1^2}(5a_1 a_5 - 6a_3^2), \quad a_9 = \frac{1}{6a_1^3}(90a_1 a_3^2 a_5 - 189a_3^4 + 25a_1^2 a_5^2),$$

$$a_{11} = \frac{9a_3}{22a_1^4}(125a_1^2 a_5^2 - 81a_3^4 - 150a_1 a_3^2 a_5).$$

In the next corollary we give a formula for the coefficients of an $f \in \mathcal{F}_1$ (cf. (5.21)).

Corollary 3.7. Let $f \in \mathcal{F}_0$. Then $f \in \mathcal{F}_1 \iff$

$$(3.34) \qquad a_{2m+1} = \frac{a_1}{2^m(2m+1)} S_m(c_1, 2\sqrt{c_2}), \ m \geq 0.$$

Proof. The result follows from Theorem 3.5 and (3.32). $\square$

The next corollary gives a recursive relationship between the coefficients of the f_n's for an $f \in \mathcal{F}_1$.

Corollary 3.8. Suppose that $f \in \mathcal{F}_0$. Then for $m \geq 0$ and $0 \leq j \leq m+2$, $f \in \mathcal{F}_1$ if and only if

$$(3.35) \quad \varphi_{2j+1}^{(2m+5)} = (2m+3)^2 c_1 \varphi_{2j+1}^{(2m+3)} + a_1^2 \varphi_{2j-1}^{(2m+3)}$$
$$- 4(m+1)^2(2m+1)(2m+3) c_2 \varphi_{2j+1}^{(2m+1)}$$

and

$$(3.36) \qquad \varphi_{2j+2}^{(2m+6)} = 4(m+2)^2 c_1 \varphi_{2j+2}^{(2m+4)} + a_1^2 \varphi_{2j}^{(2m+4)}$$
$$- 4(m+1)(m+2)(2m+3)^2 c_2 \varphi_{2j+2}^{(2m+2)},$$

where $\varphi_0^{(n)} = \delta_{n,0}$ and $\varphi_n^{(n)} = a_1^n$, $n \geq 0$, $\varphi_1^{(3)} = 6a_3$ and $\varphi_2^{(4)} = 24\, a_1 a_3$. Also, we take $\varphi_r^{(n)} = 0$ when $r < 0$ or $r > n$.

Proof. ($\Longrightarrow$) The initial conditions come from Corollary 1.6.

(3.35) Putting (1.36) into (3.14) gives the stated recursion when the respective coefficients on the two sides are equated.

(3.36) This is established similarly using (1.37) and (3.15).

($\Longleftarrow$) Recursions (3.35) and (3.36) readily imply (3.14) and (3.15), which are equivalent to the condition $f \in \mathcal{F}_1$. $\square$

For $f \in \mathcal{F}_1$, recall that $\{f_n(x)\}_{n=0}^{\infty}$ satisfies a fourth-order recursion of the form given in (3.11), which is much simpler than the general recursion in (1.6). In the next theorem we obtain the condition on Δ, appropriate to this chapter, under which this sequence actually satisfies a second-order recursion.

Corollary 3.9. If $f \in \mathcal{F}_1$, then the polynomials in $\{f_n(x)\}_{n=0}^{\infty}$ satisfy the second-order recursion

$$(3.37) \quad f_{n+2}(x) = a_1 x f_{n+1}(x) + \tfrac{1}{2} n(n+1)c_1 f_n(x), \ n \geq 0, f_0(x) = 1$$
$$\iff f_1(x) = a_1 x \text{ and } \Delta = 0.$$

Proof. ($\Longrightarrow$) Writing equation (3.37) as

$$g_1(x) \stackrel{\text{def}}{=} f_{n+2}(x) - a_1 x f_{n+1}(x) - \frac{1}{2} n(n+1)c_1 f_n(x) = 0,$$

and replacing n in (3.37) by $n+1$ and $n+2$, respectively, we obtain

$$g_2(x) \stackrel{\text{def}}{=} f_{n+3}(x) - a_1 x f_{n+2}(x) - \frac{1}{2}(n+1)(n+2)c_1 f_{n+1}(x) = 0$$

and

$$g_3(x) \overset{\text{def}}{=} f_{n+4}(x) - a_1 x f_{n+3}(x) - \frac{1}{2}(n+2)(n+3)c_1 f_{n+2}(x) = 0.$$

If we then form the linear combination

$$g_3(x) + a_1 x g_2(x) - \frac{1}{2}(n+1)(n+2)\, c_1 g_1(x),$$

we obtain a recursion in which the $f_{n+1}(x)$ and $f_{n+3}(x)$ terms are missing, viz.,

$$(3.38) \quad f_{n+4}(x) - [a_1^2 x^2 + c_1(n+2)^2] f_{n+2}(x) + \frac{1}{4}n(n+1)^2(n+2)c_1^2 f_n(x) = 0.$$

But $f \in \mathcal{F}_1 \subset \mathcal{F}_0$, so the polynomials in $\{f_n(x)\}_{n=0}^{\infty}$ satisfy the recursion (3.11). A comparison of this recursion formula with that in (3.38) gives that $c_2 = \frac{1}{4}c_1^2$, which is that $\Delta = 0$ by (3.3).

($\Longleftarrow$) Let $b = \frac{1}{2}c_1$ and assume $n \geq 0$. Since $\Delta = 0$, then $c_2 = b^2$. Also, from (3.11) we have

$$f_{n+4}(x) - [a_1^2 x^2 + 2b(n+2)^2] f_{n+2}(x) + b^2 n(n+1)^2(n+2) f_n(x) = 0.$$

Rewriting, we obtain the equation

$$(3.39) \quad \begin{aligned} &[f_{n+4}(x) - a_1 x f_{n+3}(x) - b(n+2)(n+3)f_{n+2}(x)] \\ &+ a_1 x[f_{n+3}(x) - a_1 x f_{n+2}(x) - b(n+1)(n+2)f_{n+1}(x)] \\ &- b(n+1)(n+2)[f_{n+2}(x) - a_1 x f_{n+1}(x) - bn(n+1)f_n(x)] = 0. \end{aligned}$$

Now

$$g_n(x) \overset{\text{def}}{=} f_{n+2}(x) - a_1 x f_{n+1}(x) - bn(n+1)f_n(x), \quad n \geq 0.$$

We will show that $g_n(x) = 0$, which will prove the theorem.

Now, (3.39) gives

$$(3.40) \qquad g_{n+2}(x) = -a_1 x g_{n+1}(x) + b(n+1)(n+2)g_n(x).$$

But $g_0(x) = f_2(x) - a_1 x f_1(x) = a_1^2 x^2 - a_1^2 x^2 = 0$ and $g_1(x) = f_3(x) - a_1 x f_2(x) - 2b f_1(x) = a_1^3 x^3 + 6a_3 x - a_1^3 x^3 - 6a_3 x = 0$. Thus, $g_n(x) = 0$. $\square$

It will be convenient in what follows to separate $\mathcal{F}_1$ into classes according to whether the two numbers c_2 and Δ are positive, zero, or negative. Each of these nine classes will be correspondingly designated by $(\epsilon_{c_2}, \epsilon_\Delta)$, where ϵ_{c_2} and ϵ_Δ are $+$, 0, or $-$. By (3.3) and (3.4), we see that $\Delta > 0$ implies that $c_2 > 0$ and $\Delta = 0$ implies that $c_2 \geq 0$, so the three possibilities $(0, +), (-, +)$, and $(-, 0)$ do not occur. The six remaining classes are characterized by the pairs $(+, -), (+, +), (+, 0), (0, 0), (0, -)$, and $(-, -)$.

Since we saw in (3.25) that the three initial coefficients a_1, a_3, and a_5 completely determine an $f \in \mathcal{F}_1$, they also determine the values of c_2 and Δ by (3.3), which is to say they determine which of the six classes a given function f will be in. To show that none of the six classes is empty, we list in Table 3.1 at least one function in each class.

In the first line of Table 3.1, $P_m(y)$ is the mth Legendre polynomial, $0 < k < 1$, and $k' = \sqrt{1 - k^2}$. The third line is a special case of the equation in the first line with $k = 1$. In the second and sixth lines, $\mathrm{sl}^{-1}(t)$ and $\mathrm{slh}^{-1}(t)$ are the lemniscate and hyperbolic lemniscate integrals, respectively [WhW, p. 524].

Table 3.1   Familiar functions in the six classes

(c_2, Δ)	$f(t)$
$(k^2, -\frac{1}{8}(k')^4)$	$\displaystyle\int_0^t \frac{du}{\sqrt{(1-u^2)(1-k^2u^2)}} = \mathrm{sn}^{-1}(t,k)$ $\displaystyle = \sum_{m=0}^{\infty} k^m P_m\left(\tfrac{1}{2}(k+\tfrac{1}{k})\right)\frac{t^{2m+1}}{2m+1}$ $= t + \frac{1}{6}(k^2+1)t^3 + \frac{1}{40}(3k^4+2k^2+3)t^5 + \cdots$
$(1, \frac{1}{2})$	$\displaystyle\int_0^t \frac{du}{\sqrt{1+u^4}} = \mathrm{slh}^{-1}(t) = \sum_{n=0}^{\infty}\frac{(-1)^n}{2^{2n}}\binom{2n}{n}\frac{t^{4n+1}}{4n+1}$ $= t - \frac{1}{10}t^5 + \frac{1}{24}t^9 - \cdots$
$(1, 0)$	$\displaystyle\int_0^t \frac{du}{1-u^2} = \tanh^{-1}(t) = \sum_{n=0}^{\infty}\frac{t^{2n+1}}{2n+1}$ $= t + \frac{1}{3}t^3 + \frac{1}{5}t^5 \cdots$
$(0, 0)$	$\displaystyle\int_0^t du = t$
$(0, -\frac{1}{8})$	$\displaystyle\int_0^t \frac{du}{\sqrt{1-u^2}} = \sin^{-1}(t) = \sum_{n=0}^{\infty}\frac{1}{2^{2n}}\frac{(2n)!}{n!}\frac{t^{2n+1}}{2n+1}$ $= t + \frac{1}{6}t^3 + \frac{3}{40}t^5 + \cdots$
$(-1, -\frac{1}{2})$	$\displaystyle\int_0^t \frac{du}{\sqrt{1-u^4}} = \mathrm{sl}^{-1}(t) = \sum_{n=0}^{\infty}\frac{1}{2^{2n}}\binom{2n}{n}\frac{t^{4n+1}}{4n+1}$ $= t + \frac{1}{10}t^5 + \frac{1}{24}t^9 + \cdots$

Chapter 4

The Moment Polynomials –
$P_n(x, y)$, $Q_n(x, y)$, and $R_n(x, y)$

This is a long chapter, interposed between the chapters discussing the functions in $\mathcal{F}_1$ and $\mathcal{F}_1^{-1}$. It is devoted to the study of the polynomials $P_n(x, y), Q_n(x, y) \in \mathbf{Z}[x, y]$. Their basic properties are given in Theorem 4.2 and their generating functions $g(x, y; t)$ and $h(x, y; t)$ are shown in (4.22) and (4.23) to be elliptic functions respectively satisfying the differential equations

$$(g')^2 = 1 + xg^2 + \frac{y - x^2}{12} g^4 \text{ and } (h')^2 = 2h + 4xh^2 + \frac{2(y - x^2)}{3} h^3.$$

Various other differential equations involving g and h are obtained, as well as a collection of identities satisfied by $P_n(x, y)$ and $Q_n(x, y)$.

Theorem 4.10 establishes the power of x and y that divides these polynomials. This theorem is based on Theorem 4.8, one of the two intricate number-theoretic theorems that precede it. Theorem 4.10 is used later to establish simple factorizations of the polynomial coefficients in the Maclaurin expansions of sn (t, k) and $\text{sn}^2(t, k)$ and other elliptic functions (cf. Theorem 7.11). The expansions of the sinelemniscate function sl (t) and its square, studied by Hurwitz [Hu], as well as other lemniscate functions, also occur. The chapter concludes with the evaluation of two determinants involving P_n and Q_n that are part of a later formula (Theorem 7.11).

An important aspect of these polynomials is that they appear in (5.21) and (5.22) (on the f^{-1}- side) in the next chapter as the coefficients of the

Maclaurin expansions of f^{-1} and $(f^{-1})^2$, evaluated at functions of the first three coefficients in the series of an $f \in \mathcal{F}_1$.

Definition 4.1. The polynomial sequences $\{P_n(x,y)\}_{n=0}^{\infty}$, $\{Q_n(x,y)\}_{n=0}^{\infty}$, and $\{R_n(x,y)\}_{n=0}^{\infty}$ are defined as follows:

(4.1) $P_{n+2}(x,y) =$

$$x\, P_{n+1}(x,y) + (y - x^2) \sum_{j=0}^{n} \binom{2n+2}{2j} P_j(x,y) Q_{n-j}(x,y),$$

where

(4.2) $$Q_n(x,y) = \sum_{j=0}^{n} \binom{2n+1}{2j} P_j(x,y)\, P_{n-j}(x,y),$$

and $P_0(x,y) = 1$, $P_1(x,y) = x$.

Also,

(4.3) $$R_n(x,y) = Q_n\left(\frac{x}{4}, \frac{1}{48}\left(-5x^2 + 8y\right)\right).$$

It is worth pointing out here that utilizing the change of variable $j \to n-j$ produces another form of (4.1) (we will often omit writing the arguments x and y), viz.,

(4.4) $$P_{n+2} = x\, P_{n+1} + (y - x^2) \sum_{j=0}^{n} \binom{2n+2}{2j+2} P_{n-j} Q_j.$$

There are two other forms for Q_n besides (4.2) as well, viz.,

(4.5) $$Q_n = \sum_{j=0}^{n} \binom{2n+1}{2j+1} P_j\, P_{n-j}$$

and

(4.6) $$Q_n = \frac{1}{2} \sum_{j=0}^{n} \binom{2n+2}{2j+1} P_j\, P_{n-j},$$

where (4.5) is obtained from (4.2) by the change of variable $j \to n-j$ and (4.6) is related to (4.2) by (1.31) with $r = 2$ and $s = 0$. There is sometimes an advantage in using (4.6), because it can be put in the form of a Cauchy product, viz.,

(4.7) $$\frac{2Q_n}{(2n+2)!} = \sum_{j=0}^{n} \frac{P_j}{(2j+1)!} \frac{P_{n-j}}{(2n-2j+1)!}, \quad n \geq 0.$$

Note. The polynomial X_{2n+1} in [Du1] is connected with P_{2n+1} by the formula $X_{2n+1}(0,a,b) = abP_n(a^2+b^2, a^4+14a^2b^2+b^4)$. We wish to thank S. Milne for this reference.

The next theorem gives some basic properties of $P_n(x,y)$ and $Q_n(x,y)$.

Theorem 4.2. For $n \geq 0$, we have

$$(4.8) \qquad P_n(\lambda x, \lambda^2 y) = \lambda^n P_n(x,y), \ \lambda \in \mathbb{C} \ (0^0 = 1),$$

$$(4.9) \qquad Q_n(\lambda x, \lambda^2 y) = \lambda^n Q_n(x,y), \ \lambda \in \mathbb{C},$$

$$(4.10) \qquad P_n(1,1) = 1, \ Q_n(1,1) = 2^{2n}, \ R_n(1,1) = 1,$$

$$(4.11) \qquad \begin{aligned} P_n(-1,1) &= (-1)^n, \ Q_n(-1,1) = (-1)^n 2^{2n}, \\ R_n(-1,1) &= (-1)^n, \end{aligned}$$

$$(4.12) \qquad P_n(x,x^2) = x^n, \ Q_n(x,x^2) = (4x)^n,$$

$$(4.13) \qquad P_n(x,y) = \sum_{s=0}^{[\frac{n}{2}]} \alpha_s^{(n)} x^{n-2s} y^s, \ \alpha_s^{(n)} \in \mathbb{Z},$$

$$(4.14) \qquad Q_n(x,y) = \sum_{s=0}^{[\frac{n}{2}]} \beta_s^{(n)} x^{n-2s} y^s, \ \beta_s^{(n)} \in \mathbb{Z}.$$

$$(4.15) \qquad GCD\left(\alpha_0^{(n)}, \alpha_1^{(n)}, \cdots, \alpha_{[\frac{n}{2}]}^{(n)}\right) = 1, \ n \geq 0.$$

Proof. (4.8) True for $n = 0$ and 1 and any real λ. Assume true for $n = 0, 1, \cdots k+1$ for some $k \geq 0$. Then by (4.1) and (4.2)

$$P_{k+2}(\lambda x, \lambda^2 y) = \lambda x P_{k+1}(\lambda x, \lambda^2 y) + \lambda^2(y - x^2) \sum_{j=0}^{k} \binom{2k+2}{2j} P_j(\lambda x, \lambda^2 y)$$

$$\times \sum_{s=0}^{k-j} \binom{2k-2j+1}{2s} P_s(\lambda x, \lambda^2 y) P_{k-j-s}(\lambda x, \lambda^2 y)$$

$$= \lambda^{k+2} x P_{k+1}(x,y) + \lambda^2(y - x^2) \sum_{j=0}^{k} \binom{2k+2}{2j} \lambda^j P_j(x,y)$$

$$\times \sum_{s=0}^{k-j} \binom{2k-2j+1}{2s} \lambda^s P_s(x,y) \lambda^{k-j-s} P_{k-j-s}(x,y)$$

$$= \lambda^{k+2} P_{k+2}(x,y).$$

(4.9) The identity follows from (4.2) and (4.8).

Table 4.1 $P_n(x,y)$, $0 \leq n \leq 11$

n	$P_n(x,y)$
0	1
1	x
2	y
3	$-10x^3 + 11xy$
4	$-80x^4 + 60x^2y + 21y^2$
5	$-920x^3y + 921xy^2$
6	$17600x^6 - 40960x^4y + 20820x^2y^2 + 2541y^3$
7	$418000x^7 - 627000x^5y - 74730x^3y^2 + 283731xy^3$
8	$17780800x^6y - 35210640x^4y^2 + 16406280x^2y^3 + 1023561y^4$
9	$-496672000x^9 + 1756976000x^7y - 1797330000x^5y^2$ $+304366560x^3y^3 + 232659441xy^4$
10	$-23576960000x^{10} + 53048160000x^8y - 6005822400x^6y^2$ $-51842667120x^4y^3 + 27340480440x^2y^4 + 1036809081y^5$
11	$-2487877568000x^9y + 7366536216000x^7y^2 - 6851510928000x^5y^3$ $+1554433150950x^3y^4 + 418419129051xy^5$

Table 4.2 $Q_n(x,y)$, $0 \leq n \leq 11$

n	$Q_n(x,y)$
0	1
1	$4x$
2	$10x^2 + 6y$
3	$-80x^3 + 144xy$
4	$-2000x^4 + 1920x^2y + 336y^2$
5	$-17600x^5 - 5760x^3y + 24384xy^2$
6	$418000x^6 - 1446000x^4y + 954480x^2y^2 + 77616y^3$
7	$25696000x^7 - 49900800x^5y + 11592960x^3y^2 + 12628224xy^3$
8	$496672000x^8 + 18624000x^6y - 1655040000x^4y^2$ $+1088893440x^2y^3 + 50916096y^4$
9	$-23576960000x^9 + 121012224000x^7y - 158521497600x^5y^2$ $+45452206080x^3y^3 + 15634289664xy^4$
10	$-2696809280000x^{10} + 7881496128000x^8y - 5049148032000x^6y^2$ $-2728455552000x^4y^3 + 2516610700800x^2y^4 + 76307083776y^5$
11	$-91442700800000x^{11} + 22968844800000x^9y$ $+533509032960000x^7y^2 - 731328224256000x^5y^3$ $+226734493593600x^3y^4 + 39558557896704xy^5$

Table 4.3 $R_n(x,y)$, $0 \le n \le 11$

n	$R_n(x,y)$
0	1
1	x
2	y
3	$-5x^3 + 6xy$
4	$\frac{1}{3}(-50x^4 + 25x^2y + 28y^2)$
5	$\frac{1}{3}(175x^5 - 680x^3y + 508xy^2)$
6	$\frac{1}{3}(3750x^6 - 7775x^4y + 2950x^2y^2 + 1078y^3)$
7	$\frac{1}{3}(15125x^7 + 8150x^5y - 67120x^3y^2 + 43848xy^3)$
8	$\frac{1}{9}(-1205750x^8 + 5004875x^6y - 6104400x^4y^2 + 1951700x^2y^3 + 353584y^4)$
9	$\frac{1}{9}(-25934375x^9 + 67626500x^7y - 30569100x^5y^2 - 38265880x^3y^3 + 27142864xy^4)$
10	$\frac{1}{9}(10391250x^{10} - 954096125x^8y + 2868890750x^6y^2 - 2829788250x^4y^3 + 816284000x^2y^4 + 88318384y^5)$
11	$\frac{1}{3}(4276365625x^{11} - 20069893750x^9y + 31217670000x^7y^2 - 15517072000x^5y^3 - 3722517200x^3y^4 + 3815447328xy^5)$

Tables 4.1 and 4.2 were recursively computed using (4.1), (4.2), and (4.33). Table 4.3 was computed from Table 4.2 using (4.3).

(4.10) We have that $P_0(1,1) = P_1(1,1) = 1$ by the initial conditions following (4.2). From (4.1), $P_{n+2}(1,1) = P_{n+1}(1,1)$, $n \ge 0$, which gives the first result inductively. From (4.2),

$$Q_n(1,1) = \sum_{j=0}^{n} \binom{2n+1}{2j} P_j(1,1)P_{n-j}(1,1) = \sum_{j=0}^{n} \binom{2n+1}{2j} = 2^{2n}$$

by [Go, (1.93)]. Using (4.9), with $\lambda = \frac{1}{4}$, and (4.10), we have that $R_n(1,1) = Q_n(\frac{1}{4}, \frac{1}{16}) = (\frac{1}{4})^n Q_n(1,1) = 1$.

(4.11) The first two results follow from (4.8) and (4.9) using (4.10). From (4.3), $R_n(-1,1) = Q_n(-\frac{1}{4}, \frac{1}{16}) = (-\frac{1}{4})^n Q_n(1,1) = (-1)^n$ as in the proof of (4.10).

(4.12) Put $x = y = 1$ in (4.8). Put $x = y = 1$ into (4.9) and use (4.10).

(4.13) Let $P_n(x,y) = \sum_{r,s \ge 0} a_{r,s} x^r y^s$ and suppose that $\lambda \ne 0$. Then (4.8) implies

$$\lambda^n \sum_{r,s \ge 0} a_{r,s} x^r y^s = \lambda^n P_n(x,y) = P_n(\lambda x, \lambda^2 y) = \sum_{r,s \ge 0} a_{r,s} \lambda^{r+2s} x^r y^s.$$

If $\alpha_{r,s} \neq 0$, then $\lambda^n = \lambda^{r+2s}$, which implies $n = r + 2s$, so $r = n - 2s \geq 0$, i.e., $2s \leq n$. Write $\alpha_{n-2s,s}$ as $\alpha_s^{(n)}$, where $0 \leq s \leq \left[\frac{n}{2}\right]$. Then

$$P_n(x,y) = \sum_{s=0}^{\left[\frac{n}{2}\right]} \alpha_s^{(n)} x^{n-2s} y^s.$$

That $\alpha_s^{(n)} \in \mathbf{Z}$ follows inductively from the recursion in (4.1), which has integer coefficients.

(4.14) The proof is the same as that in (4.13).

(4.15) By (4.13) and (4.10), we see that a common divisor of all the $\alpha_s^{(n)}$'s must divide $P_n(1,1) = 1$. $\square$

To investigate $P_n(x,y)$ and $Q_n(x,y)$ further, it will be helpful to introduce the following two generating functions.

Definition 4.3. Let

(4.16) $$g = g(t;x,y) \overset{\text{def}}{=} \sum_{n=0}^{\infty} P_n(x,y) \frac{t^{2n+1}}{(2n+1)!}$$

and

(4.17) $$h = h(t;x,y) \overset{\text{def}}{=} \sum_{n=0}^{\infty} Q_n(x,y) \frac{t^{2n+2}}{(2n+2)!}.$$

Theorem 4.4. We have

(4.18) $$g^2 = 2h.$$

Proof. From (4.16), we have $g = \sum_{n=0}^{\infty} \frac{P_n}{(2n+1)!} t^{2n+1}$, so squaring and using (4.6) and (4.17) gives

$$g^2 = \sum_{n=0}^{\infty} \left\{ \sum_{j=0}^{n} \frac{P_j}{(2n+1)!} \frac{P_{n-j}}{(2n-2j+1)!} \right\} t^{2n+2}$$

$$= \sum_{n=0}^{\infty} \left\{ \sum_{j=0}^{n} \binom{2n+2}{2j+1} P_j P_{n-j} \right\} \frac{t^{2n+2}}{(2n+2)!}$$

$$= 2 \sum_{n=0}^{\infty} Q_n \frac{t^{2n+2}}{(2n+2)!} = 2h. \quad \square$$

We next derive an important differential equation for g. In this we assume that x and y are fixed.

Theorem 4.5. We have

$$\text{(4.19)} \qquad g'' = xg + \frac{y - x^2}{6} g^3.$$

Proof. From (4.1), write $\dfrac{P_{n+2} - x\,P_{n+1}}{y - x^2} = \displaystyle\sum_{j=0}^{n} \binom{2n+2}{2j} P_j \, Q_{n-j}$. Summing this equation and differentiating, we obtain

$$D_t \sum_{n=0}^{\infty} \left\{ \frac{P_{n+2} - xP_{n+1}}{y - x^2} \right\} \frac{t^{2n+3}}{(2n+3)!}$$

$$= D_t \sum_{n=0}^{\infty} \left\{ \sum_{j=0}^{n} \binom{2n+2}{2j} P_j \, Q_{n-j} \right\} \frac{t^{2n+3}}{(2n+3)!}$$

$$= \sum_{j=0}^{\infty} \sum_{n=0}^{\infty} \binom{2n+2j+2}{2j} P_j \, Q_n \frac{t^{2n+2j+2}}{(2n+2j+2)!}$$

$$= \sum_{j=0}^{\infty} P_j \frac{t^{2j}}{(2j)!} \sum_{n=0}^{\infty} Q_n \frac{t^{2n+2}}{(2n+2)!} = g'h = \frac{1}{2} g^2 g',$$

where we have used (4.16)–(4.18). Integrating this equation, we find the constant of integration is 0 at $t = 0$, which gives the equation

$$\text{(4.20)} \qquad \sum_{n=0}^{\infty} P_{n+2} \frac{t^{2n+3}}{(2n+3)!} - x \sum_{n=0}^{\infty} P_{n+1} \frac{t^{2n+3}}{(2n+3)!} = \frac{y - x^2}{6} g^3.$$

Next, if we re-index the sum in (4.16) as $g = \displaystyle\sum_{n=-2}^{\infty} P_{n+2} \frac{t^{2n+5}}{(2n+5)!} =$

$P_0 t + \dfrac{P_1 t^3}{6} + \displaystyle\sum_{n=0}^{\infty} P_{n+2} \frac{t^{2n+5}}{(2n+5)!}$, then $\displaystyle\sum_{n=0}^{\infty} P_{n+2} \frac{t^{2n+5}}{(2n+5)!} = g - t - \dfrac{xt^3}{6}$.
Differentiating this equation twice gives

$$\text{(4.21)} \qquad \sum_{n=0}^{\infty} P_{n+2} \frac{t^{2n+3}}{(2n+3)!} = g'' - xt.$$

Also, writing

$$g = \sum_{n=-1}^{\infty} P_{n+1} \frac{t^{2n+3}}{(2n+3)!} = P_0 t + \sum_{n=0}^{\infty} P_{n+1} \frac{t^{2n+3}}{(2n+3)!}$$

gives $\displaystyle\sum_{n=0}^{\infty} P_{n+1} \frac{t^{2n+3}}{(2n+3)!} = g - t$. Substituting this latter result and that from (4.21) into (4.20), we find that

$$g'' - xt - x(g - t) = g'' - xg = \frac{y - x^2}{6} g^3. \quad \square$$

We are now in a position to derive some interesting and useful differential equations in g and h by using (4.18) and (4.19). Equations (4.22) and (4.27) show the elliptic nature of g and h.

Corollary 4.6. We have

(4.22)
$$(g')^2 = 1 + xg^2 + \frac{y - x^2}{12}g^4,$$

(4.23)
$$6hh'' = 3(h')^2 + 12xh^2 + 4(y - x^2)h^3,$$

(4.24)
$$gg'' = -2 - 2xh + 2(g')^2,$$

(4.25)
$$g'' = xg + \frac{y - x^2}{3}gh,$$

(4.26)
$$(g')^2 = 1 + 2xh + \frac{y - x^2}{3}h^2,$$

(4.27)
$$(h')^2 = 2h + 4xh^2 + \frac{2(y - x^2)}{3}h^3,$$

(4.28)
$$h'' = 1 + 4xh + (y - x^2)h^2,$$

(4.29)
$$h'' = -2 - 2xh + 3(g')^2.$$

Proof. (4.22) If we multiply (4.19) by $2g'$ and integrate, we obtain $(g')^2 = c + xg^2 + \frac{y - x^2}{12}g^4$. Since $g(0) = 0$ and $g'(0) = 1$, then $c = 1$.

(4.23) Multiplying (4.19) by g, we find using (4.18) that

(4.30)
$$gg'' = xg^2 + \frac{y - x^2}{6}g^4 = 2xh + \frac{2(y - x^2)}{3}h^2.$$

But $(g^2)'' = 2(gg')' = 2gg'' + 2(g')^2$, so $gg'' = \frac{1}{2}(g^2)'' - (g')^2$. Also, (4.18) gives $g' = \frac{h}{g}$, then

$$gg'' = \frac{1}{2}(2h)'' - \left(\frac{h'}{g}\right)^2 = h'' - \frac{(h')^2}{g^2} = h'' - \frac{(h')^2}{2h}.$$

Combining this result with (4.30) gives the desired formula.

(4.24) Multiplying (4.19) by $\frac{g}{2}$ and using (4.22) we find that

$$\frac{1}{2}gg'' - \frac{1}{2}xg^2 = \frac{y - x^2}{12}g^4 = (g')^2 - 1 - xg^2.$$

The result follows using (4.18).

(4.25) Use (4.18) in the last term of (4.19).

(4.26) Use (4.18) in (4.22).

(4.27) From (4.18), we have $h' = gg'$. Thus, (4.26) and (4.18) imply that

$$(h')^2 = g^2(g')^2 = 2h\left(1 + 2xh + \frac{(y - x^2)}{3}h^2\right)$$
$$= 2h + 4xh^2 + \frac{2(y - x^2)}{3}h^3.$$

(4.28) Differentiating (4.27) and canceling $2h' \neq 0$ gives the result.

(4.29) Use (4.18) in (4.29). $\square$

For the sake of brevity, we will write (4.16) and (4.17), respectively, with $u = t^2$ as

(4.31)
$$g = t \sum_{n=0}^{\infty} \frac{P_n}{(2n+1)!} u^n$$

and

(4.32)
$$h = \sum_{n=0}^{\infty} \frac{Q_n}{(2n+2)!} u^{n+1}.$$

In the next theorem we derive some useful identities involving the P_n's and Q_n's.

Theorem 4.7. For $n \geq 0$, we have

(4.33)
$$Q_{n+2} = 4x\, Q_{n+1} + (y - x^2) \sum_{j=0}^{n} \binom{2n+4}{2j+2} Q_j\, Q_{n-j},$$

(4.34)
$$P_{n+2} = x\, P_{n+1} + \frac{y - x^2}{3} \sum_{j=0}^{n} \binom{2n+3}{2j+1} P_j\, Q_{n-j},$$

(4.35)
$$Q_{n+1} = -2x\, Q_n + 3 \sum_{j=0}^{n+1} \binom{2n+2}{2j} P_j\, P_{n-j+1},$$

(4.36)
$$Q_{n+1} = 4x\, Q_n + 3 \sum_{j=0}^{n+1} \binom{2n+2}{2j} P_j\, P_{n-j+1}$$
$$- 3x \sum_{j=0}^{n} \binom{2n+2}{2j+1} P_j\, P_{n-j},$$

(4.37)
$$Q_{n+2} = 4x\, Q_{n+1} + 6 \sum_{j=0}^{n} \binom{2n+3}{2j+3} \left(P_{j+2} - x P_{j+1} \right) P_{n-j},$$

(4.38)
$$x\, Q_n = -\frac{1}{2n+3} \sum_{j=0}^{n+1} (n-3j) \binom{2n+3}{2j+1} P_j\, P_{n-j+1},$$

(4.39)
$$Q_{n+1} = x\, Q_n + \frac{3}{2} \sum_{j=0}^{n} \binom{2n+2}{2j+1} P_j\, P_{n-j+1},$$

(4.40) $\quad 2x\,Q_{n+1} =$

$$\sum_{j=0}^{n+2} \binom{2n+4}{2j} P_j\, P_{n-j+2} - \frac{y-x^2}{3} \sum_{j=0}^{n} \binom{2n+4}{2j+2} Q_j\, Q_{n-j},$$

(4.41) $\quad Q_{n+1} = \dfrac{1}{(n+1)(2n+5)} \sum_{j=0}^{n} (2n-3j+2)\binom{2n+5}{2j+2} P_{n-j+1}Q_j,$

(4.42) $\qquad \displaystyle\sum_{j=0}^{n}(n-3j)\binom{2n+3}{2j+1} P_j\, Q_{n-j} = 0,$

and

(4.43) $\qquad R_{n+2} = x\, R_{n+1} + \dfrac{1}{6}\,(y-x^2) \displaystyle\sum_{j=0}^{n} \binom{2n+4}{2j+2} R_j\, R_{n-j}.$

Proof. (4.33) Twice differentiating and squaring (4.32) gives, respectively,

(4.44) $$h'' = \sum_{n=0}^{\infty} \frac{Q_n}{(2n)!}\, u^n$$

and

(4.45) $$h^2 = \sum_{n=0}^{\infty}\left\{ \sum_{j=0}^{n} \frac{Q_j}{(2j+2)!}\,\frac{Q_{n-j}}{(2n-2j+2)!}\right\} u^{n+2}$$

$$= \sum_{n=0}^{\infty}\left\{ \sum_{j=0}^{n} \binom{2n+4}{2j+2} Q_j Q_{n-j}\right\} \frac{u^{n+2}}{(2n+4)!}.$$

Substituting these series into (4.28) gives the equation

$$\sum_{n=0}^{\infty} \frac{Q_n}{(2n)!}\, u^n = 1 + 4x \sum_{n=0}^{\infty} \frac{Q_n}{(2n+2)!}\, u^{n+1}$$

$$+ (y-x^2) \sum_{n=0}^{\infty}\left\{ \sum_{j=0}^{n} \binom{2n+4}{2j+2} Q_j\, Q_{n-j}\right\} \frac{u^{n+2}}{(2n+4)!}.$$

For $n \geq 0$, the coefficients of u^{n+2} on the two sides are equal, so we find that

$$\frac{Q_{n+2}}{(2n+4)!} = 4x\frac{Q_{n+1}}{(2n+4)!} + \frac{y-x^2}{(2n+4)!}\sum_{j=0}^{n}\binom{2n+4}{2j+2} Q_j\, Q_{n-j},$$

from which the result follows.

(4.34) From (4.16) we have that

$$(4.46) \qquad g'' = t \sum_{n=1}^{\infty} \frac{P_n}{(2n-1)!} u^{n-1} = t \sum_{n=0}^{\infty} \frac{P_{n+1}}{(2n+1)!} u^n.$$

Also,

$$(4.47) \qquad gh = t \sum_{n=0}^{\infty} \left\{ \sum_{j=0}^{n} \frac{P_j}{(2j+1)!} \frac{Q_{n-j}}{(2n-2j+2)!} \right\} u^{n+1}$$

$$= t \sum_{n=0}^{\infty} \left\{ \sum_{j=0}^{n} \binom{2n+3}{2j+1} P_j Q_{n-j} \right\} \frac{u^{n+1}}{(2n+3)!}.$$

Substituting these results into (4.25) gives

$$t \sum_{n=0}^{\infty} \frac{P_{n+1}}{(2n+1)!} u^n = xt \sum_{n=0}^{\infty} \frac{P_n}{(2n+1)!} u^n$$

$$+ \frac{y-x^2}{3} t \sum_{n=0}^{\infty} \left\{ \sum_{j=0}^{n} \binom{2n+3}{2j+1} P_j Q_{n-j} \right\} \frac{u^{n+1}}{(2n+3)!}.$$

For $n \geq 0$, equating the coefficients of u^{n+1} gives

$$\frac{P_{n+2}}{(2n+3)!} = x \frac{P_{n+1}}{(2n+3)!} + \frac{y-x^2}{3(2n+3)!} \sum_{j=0}^{n} \binom{2n+3}{2j+1} P_j Q_{n-j},$$

which implies the result.

(4.35) From (4.16) we obtain

$$(4.48) \qquad (g')^2 = \sum_{n=0}^{\infty} \left\{ \sum_{j=0}^{n} \frac{P_j}{(2j)!} \frac{P_{n-j}}{(2n-2j)!} \right\} u^n$$

$$= \sum_{n=0}^{\infty} \left\{ \sum_{j=0}^{n} \binom{2n}{2j} P_j P_{n-j} \right\} \frac{u^n}{(2n)!}.$$

Then substituting (4.44) and (4.48) into (4.29) gives

$$\sum_{n=0}^{\infty} \frac{Q_n}{(2n)!} u^n = -2 - 2x \sum_{n=0}^{\infty} \frac{Q_n}{(2n+2)!} u^{n+1}$$

$$+ 3 \sum_{n=0}^{\infty} \left\{ \sum_{j=0}^{n} \binom{2n}{2j} P_j P_{n-j} \right\} \frac{u^n}{(2n)!}.$$

For $n \geq 0$, equating coefficients of u^{n+1} gives

$$\frac{Q_{n+1}}{(2n+2)!} = -2x \frac{Q_n}{(2n+2)!} + \frac{3}{(2n+2)!} \sum_{j=0}^{n+1} \binom{2n+2}{2j} P_j P_{n-j+1}.$$

(4.36) If we add (4.6), written as

$$0 = 6x Q_n - 3x \sum_{j=0}^{n} \binom{2n+2}{2j+1} P_j P_{n-j},$$

to (4.35), we obtain (4.36).

(4.37) We begin by observing that

$$(4.49) \quad \frac{1}{n+2} \sum_{j=0}^{n} (n-2j-2) \binom{2n+4}{2j+4} P_{j+2} P_{n-j} = -P_{n+2}$$

$$- n(2n+3)x P_{n+1} + \frac{1}{n+2} \sum_{j=0}^{n-2} (n-2j-2) \binom{2n+4}{2j+4} P_{j+2} P_{n-j}.$$

Since the re-indexing $j \to n-j-2$ sends the sum on the right side of (4.49) into its negative, that sum is zero. Also, since

$$\frac{n-2j-2}{n+2} \binom{2n+4}{2j+4} = \binom{2n+4}{2j+4} - 2\binom{2n+3}{2j+3},$$

we can rewrite (4.49) as

$$(4.50) \quad \sum_{j=0}^{n} \binom{2n+4}{2j+4} P_{j+2} P_{n-j} =$$

$$2\sum_{j=0}^{n} \binom{2n+3}{2j+3} P_{j+2} P_{n-j} - P_{n+2} - n(2n+3)x P_{n+1}.$$

On the other hand, we also have

$$(4.51) \quad \sum_{j=0}^{n} \binom{2n+4}{2j+4} P_{j+2} P_{n-j} = \sum_{j=2}^{n+2} \binom{2n+4}{2j} P_j P_{n-j+2}$$

$$= \sum_{j=0}^{n+2} \binom{2n+4}{2j} P_j P_{n-j+2} - P_{n+2} - (n+2)(2n+3)x P_{n+1}$$

$$= \frac{1}{3}(Q_{n+2} + 2x Q_{n+1}) - P_{n+2} - (n+2)(2n+3)x P_{n+1},$$

using (4.35). Thus, combining (4.50) and (4.51), we obtain

$$(4.52) \quad Q_{n+2} + 2x Q_{n+1} = 6\sum_{j=0}^{n} \binom{2n+3}{2j+3} P_{j+2}P_{n-j} + 6(2n+3)x P_{n+1}.$$

But by (4.5), we have that

$$Q_{n+1} = \sum_{j=0}^{n+1} \binom{2n+3}{2j+1} P_j\, P_{n-j+1} = (2n+3)P_{n+1} + \sum_{j=1}^{n+1} \binom{2n+3}{2j+1} P_j\, P_{n-j}$$

$$= (2n+3)P_{n+1} + \sum_{j=0}^{n} \binom{2n+3}{2j+3} P_{j+1}\, P_{n-j},$$

so

$$(2n+3)P_{n+1} = Q_{n+1} - \sum_{j=0}^{n} \binom{2n+3}{2j+3} P_{j+1}\, P_{n-j}.$$

Substituting this result into the last term of (4.52) gives (4.37).

(4.38) From (4.31) and (4.46) we obtain

$$(4.53) \qquad gg'' = \sum_{n=0}^{\infty} \left\{ \sum_{j=0}^{n} \frac{P_j}{(2j+1)!} \frac{P_{n-j+1}}{(2n-2j+1)!} \right\} u^{n+1}$$

$$= \sum_{n=0}^{\infty} \left\{ \sum_{j=0}^{n} \binom{2n+2}{2j+1} P_j\, P_{n-j+1} \right\} \frac{u^{n+1}}{(2n+2)!}.$$

Substituting (4.50) and (4.48) into (4.24) then gives

$$\sum_{n=0}^{\infty} \left\{ \sum_{j=0}^{n} \binom{2n+2}{2j+1} P_j\, P_{n-j+1} \right\} \frac{u^{n+1}}{(2n+2)!}$$

$$= -2 - 2x \sum_{n=0}^{\infty} Q_n \frac{u^{n+1}}{(2n+2)!} + 2\sum_{n=0}^{\infty} \left\{ \sum_{j=0}^{n} \binom{2n}{2j} P_j\, P_{n-j} \right\} \frac{u^n}{(2n)!}.$$

Thus, equating coefficients of u^{n+1} gives

$$\frac{1}{(2n+2)!} \sum_{j=0}^{n} \binom{2n+2}{2j+1} P_j\, P_{n-j+1}$$

$$= -2x \frac{Q_n}{(2n+2)!} + \frac{2}{(2n+2)!} \sum_{j=0}^{n+1} \binom{2n+2}{2j} P_j\, P_{n-j+1},$$

so

$$2xQ_n = 2\sum_{j=0}^{n+1} \binom{2n+2}{2j} P_j\, P_{n-j+1} - \sum_{j=0}^{n} \binom{2n+2}{2j+1} P_j\, P_{n-j+1}.$$

But

$$\binom{2n+2}{2j} = \frac{2j+1}{2n+3} \binom{2n+3}{2j+1} \quad \text{and} \quad \binom{2n+2}{2j+1} = \frac{2n-2j+2}{2n+3} \binom{2n+3}{2j+1}.$$

Thus,

$$xQ_n = \sum_{j=0}^{n+1} \frac{2j+1}{2n+3}\binom{2n+3}{2j+1} P_j\, P_{n-j+1}$$

$$- \sum_{j=0}^{n+1} \frac{n-j+1}{2n+3}\binom{2n+3}{2j+1} P_j\, P_{n-j+1}$$

$$= -\frac{1}{2n+3} \sum_{j=0}^{n+1}(n-3j)\binom{2n+3}{2j+1} P_j\, P_{n-j+1}.$$

(4.39) Subtracting xQ_n from both sides of (4.35) and replacing xQ_n on the right by the expression in (4.38) gives

$$Q_{n+1} - xQ_n = \frac{3}{2n+3} \sum_{j=0}^{n+1}(n-3j)\binom{2n+3}{2j+1} P_j P_{n-j+1}$$

$$+ 3\sum_{j=0}^{n+1}\binom{2n+2}{2j} P_j P_{n-j+1}$$

$$= 3\sum_{j=0}^{n+1}\left[\frac{n-3j}{2n+3}\binom{2n+3}{2j+1} + \binom{2n+2}{2j}\right] P_j P_{n-j+1}.$$

Since the term with $j = n+1$ is zero, we obtain

$$Q_{n+1} - xQ_n = 3\sum_{j=0}^{n}\left[\frac{n-3j}{2(n-j+1)}\binom{2n+2}{2j+1}\right.$$

$$\left. + \frac{2j+1}{2(n-j+1)}\binom{2n+2}{2j+1}\right] P_j P_{n-j+1} = \frac{3}{2}\sum_{j=0}^{n}\binom{2n+2}{2j+1} P_j P_{n-j}.$$

(4.40) Substituting (4.48) and (4.45) into (4.26) gives

$$\sum_{n=0}^{\infty}\left\{\sum_{j=0}^{n}\binom{2n}{2j} P_j\, P_{n-j}\right\}\frac{u^n}{(2n)!}$$

$$= 1 + 2x\sum_{n=0}^{\infty} Q_n \frac{u^{n+1}}{(2n+2)!}$$

$$+ \frac{y-x^2}{3}\sum_{n=0}^{\infty}\left\{\sum_{j=0}^{n}\binom{2n+4}{2j+2} Q_j Q_{n-j}\right\}\frac{u^{n+2}}{(2n+4)!}.$$

Equating coefficients of u^{n+2} gives

$$\frac{1}{(2n+4)!}\sum_{j=0}^{n+2}\binom{2n+4}{2j} P_j\, P_{n-j+2}$$

$$= 2x\frac{Q_{n+1}}{(2n+4)!} + \frac{y-x^2}{3(2n+1)!}\sum_{j=0}^{n}\binom{2n+4}{2j+2} Q_j Q_{n-j},$$

which gives the result.

(4.41) Define the sequences $\{a_n\}_{n=0}^{\infty}$ and $\{b_n\}_{n=0}^{\infty}$ by

$$a_n = \frac{P_n}{(2n+1)!} \quad \text{and} \quad b_n = \frac{2Q_n}{(2n+2)!}.$$

Then the equation (4.7) becomes $b_n = \sum_{k=0}^{n} a_k a_{n-k}$, $n \geq 0$. Since $a_0 = 1$, $b_0 = 1$, this equation implies the two formal power series $(1, a_1, a_2, \cdots)$ and $(1, b_1, b_2, \cdots)$ satisfy the equation $(1, b_1, b_2, \cdots) = (1, a_1, a_2, \cdots)^2$. But then using the formula [Po, p. 304, (c)]: $b_{n+1} = \frac{1}{n+1} \sum_{j=0}^{n} (2n - 3j + 2) a_{n-j+1} b_j$, $n \geq 0$, we get (4.41).

(4.42) Subtracting (4.1) from (4.34) gives

$$\sum_{j=0}^{n} \left[\binom{2n+3}{2j+1} - 3\binom{2n+2}{2j} \right] P_j Q_{n-j} = 0.$$

But the equation $\binom{2n+2}{2j} = \frac{2j+1}{2n+3} \binom{2n+3}{2j+1}$ implies that

$$\binom{2n+3}{2j+1} - 3\binom{2n+2}{2j} = \frac{2(n-3j)}{2n+3} \binom{2n+3}{2j+1},$$

from which the result follows.

(4.43) Replace x by $\frac{x}{4}$ and y by $\frac{1}{48}(-5x^2 + 8y)$ in (4.33). $\quad\square$

Note that the relationship in (4.33) permits the calculation of Q_n using only preceding Q_k's but no P_k's.

An examination of the polynomials in Tables 4.1 and 4.2 suggests the following arithmetic properties of the coefficients of the P_n's and Q_n's. Theorems 4.8 and 4.9 are interesting preliminary results on which the proof of the factorizations in Theorem 4.10 is based.

Theorem 4.8. We have

(4.54) $\qquad P_{2n}(x,y) \equiv y^n \pmod{10}$, $n \geq 0$,

(4.55) $\qquad P_{2n+1}(x,y) \equiv xy^n \pmod{10}$, $n \geq 0$,

(4.56) $\qquad Q_{2n}(x,y) \equiv 6y^n \pmod{10}$, $n \geq 1$,

(4.57) $\qquad Q_{2n+1}(x,y) \equiv 4xy^n \pmod{10}$, $n \geq 0$,

(4.58) $\qquad P_{3n}(x,y) \equiv 2x^{3n} + 2x^{3n-2}y \pmod{3}$, $n \geq 1$,

(4.59) $\qquad P_{3n+1}(x,y) \equiv x^{3n+1} \pmod{3}$, $n \geq 0$,

(4.60) $\qquad P_{3n+2}(x,y) \equiv x^{3n}y \pmod{3}$, $n \geq 0$,

and

(4.61) $Q_n(x,y) \equiv x^n \pmod{3}, \ n \geq 0.$

Proof. (4.54)–(4.57) We begin by establishing the four (mod 2) formulas. First, (4.56) follows from (4.37). Congruence (4.57) is true for $n = 0$ because $Q_1(x,y) = 4x$ and, for $n \geq 1$, formula (4.37) holds. Congruences (4.54) and (4.55) are true for $n = 0, 1$. Assume they are true up to n. Then from (4.1), we find that

$$P_{n+2} \equiv xP_{n+1} + (y-x^2)\binom{2n+2}{2n}P_n \equiv xP_{n+1} + (y-x^2)(n+1)P_n \pmod{2},$$

using (4.56) and (4.57). For $n = 2k$, $k \geq 0$, we have

$$P_{2k+2} \equiv xP_{2k+1} + (y-x^2)P_{2k} \equiv x(xy^k) + (y-x^2)y^k = y^{k+1} \pmod{2},$$

while for $n = 2k+1$, $k \geq 0$, we have

$$P_{2k+3} \equiv xP_{2k+2} \equiv xy^{k+1} \pmod{2}.$$

In establishing (4.54)–(4.57) (mod 5), we will need to use the following formula derived from [Go, 1.53]:

(4.62) $$\sum_{j=0}^{[\frac{n-a}{r}]} \binom{n}{rj+a} = \frac{1}{r}\sum_{j=1}^{r} \zeta_r^{-aj}(1+\zeta_r^j)^n,$$

where $\zeta_r = e^{\frac{2\pi i}{r}}$ and $0 \leq a \leq r-1$. The four congruences (mod 5) are true for $n = 0$. Assume they are true up to n. To prove (4.54) (mod 5), we have from (4.1) that

$$P_{2n+2} = xP_{2n+1} + (y-x^2)\sum_{j=0}^{2n}\binom{4n+2}{2j}P_j Q_{2n-j}$$

$$= xP_{2n+1}$$

$$+ (y-x^2)\left\{ \sum_{j=0}^{n}\binom{4n+2}{4j}P_{2j}Q_{2n-2j} + \sum_{j=0}^{n-1}\binom{4n+2}{4j+2}P_{2j+1}Q_{2n-2j+1} \right\}$$

$$\equiv x(xy^n)$$

$$+ (y-x^2)\left\{ \sum_{j=0}^{n}\binom{4n+2}{4j}y^j y^{n-j} + \sum_{j=0}^{n-1}\binom{4n+2}{4j+2}(xy^j)(4xy^{n-j}) \right\}$$

$$= x^2 y^n + (y-x^2)y^n\left\{ \sum_{j=0}^{n}\binom{4n+2}{4j} - x^2\sum_{j=0}^{n-1}\binom{4n+2}{4j+2} \right\} \pmod{5}.$$

From (4.62), we find the congruence

$$\sum_{j=0}^{n} \binom{4n+2}{4j} = 2^{4n} \equiv 1 \pmod 5, \ n \geq 0,$$

and

$$\sum_{j=0}^{n-1} \binom{4n+2}{4j+2} = 2^{4n} - 1 \equiv 0 \pmod 5, \ n \geq 1,$$

so that

$$P_{2n+2} \equiv x^2 y^n + (y - x^2)y^n = y^{n+1} \pmod 5.$$

Next, from (4.1), we have

$$P_{2n+3} = x P_{2n+2} + (y - x^2) \sum_{j=0}^{2n+1} \binom{4n+4}{2j} P_j Q_{2n-j+1}$$

$$= x P_{2n+2}$$

$$+ (y - x^2) \left\{ \sum_{j=0}^{n} \binom{4n+4}{4j} P_{2j} Q_{2n-2j+1} + \sum_{j=0}^{n} \binom{4n+4}{4j+2} P_{2j+1} Q_{2n-2j} \right\}$$

$$\equiv x y^{n+1}$$

$$+ (y - x^2) \left\{ \sum_{j=0}^{n} \binom{4n+4}{4j} y^j (-x y^{n-j}) + \sum_{j=0}^{n} \binom{4n+4}{4j+2} (x y^j) y^{n-j} \right\}$$

$$= x y^{n+1} + x y^n (y - x^2) \left\{ \sum_{j=0}^{n} \binom{4n+4}{4j+2} - \sum_{j=0}^{n} \binom{4n+4}{4j} \right\} \pmod 5.$$

But, from (4.62) we obtain

(4.63) $$\sum_{j=0}^{n} \binom{4n+4}{4j+2} = 4^{2n+1} + 2(-4)^n \equiv 1 \pmod 5$$

and

(4.64) $$\sum_{j=0}^{n} \binom{4n+4}{4j} = 4^{2n+1} - 2(-4)^n - 1 \equiv 1 \pmod 5.$$

Thus, $P_{2n+3} \equiv x y^{n+1} \pmod 5.$

From (4.33), we have

$$Q_{2n+2} = 4xQ_{2n+1} + (y-x^2)\sum_{j=0}^{2n}\binom{4n+4}{2j+2}Q_j Q_{2n-j}$$

$$= 4xQ_{2n+1}$$
$$+ (y-x^2)\left\{\sum_{j=0}^{n}\binom{4n+4}{4j+2}Q_{2j}Q_{2n-2j} + \sum_{j=0}^{n-1}\binom{4n+4}{4j+4}Q_{2j+1}Q_{2n-2j-1}\right\}$$

$$\equiv 4x(4xy^n)$$
$$+ (y-x^2)\left\{\sum_{j=0}^{n}\binom{4n+4}{4j+2}y^j y^{n-j} + \sum_{j=0}^{n-1}\binom{4n+4}{4j+4}(4xy^j)(4xy^{n-j-1})\right\}$$

$$\equiv x^2 y^n + y^{n-1}(y-x^2)\left\{y\sum_{j=0}^{n}\binom{4n+4}{4j+2} + x^2\sum_{j=0}^{n-1}\binom{4n+4}{4j+4}\right\} \pmod 5.$$

The first sum is $\equiv 1 \pmod 5$ by (4.63). Re-indexing the second sum by $j \to j+1$ gives

$$\sum_{j=1}^{n}\binom{4n+4}{4j} = \sum_{j=0}^{n}\binom{4n+4}{4j} - 1 \equiv 0 \pmod 5,$$

using (4.64). Thus,

$$Q_{2n+2} \equiv x^2 y^n + y^n(y-x^2) = y^{n+1} \pmod 5.$$

Finally, from (4.33), we have

$$Q_{2n+3} = 4xQ_{2n+2} + (y-x^2)\sum_{j=0}^{2n+1}\binom{4n+6}{2j+2}Q_j Q_{2n-j+1}$$

$$= 4xQ_{2n+2}$$
$$+ (y-x^2)\left\{\sum_{j=0}^{n}\binom{4n+6}{4j+2}Q_{2j}Q_{2n-2j+1} + \sum_{j=0}^{n}\binom{4n+6}{4j+4}Q_{2j+1}Q_{2n-2j}\right\}$$

$$\equiv 4x(6y^{n+1})$$
$$+ (y-x^2)\left\{\sum_{j=0}^{n}\binom{4n+6}{4j+2}(6y^j)(4xy^{n-j}) + \sum_{j=0}^{n}\binom{4n+6}{4j+4}(4xy^j)(6y^{n-j})\right\}$$

$$\equiv 4xy^{n+1} + 4xy^n(y-x^2)\left\{\sum_{j=0}^{n}\binom{4n+6}{4j+2} + \sum_{j=0}^{n}\binom{4n+6}{4j+4}\right\} \pmod 5.$$

From (4.62), we find that

$$\sum_{j=0}^{n}\binom{4n+6}{4j+2} = 16^{n+1} - 1 \equiv 0 \pmod 5.$$

The map $j \to n - j$ transforms the second sum into the first, so it is also $\equiv 0 \pmod 5$. Thus, $Q_{2n+3} \equiv 4xy^{n+1} \pmod 5$. The formulas (4.54)–(4.57) (mod 10) follow by combining the (mod 2) and (mod 5) results.

(4.58)–(4.60) The three congruences are true for their initial values of n. Assume they are true up to some $n \geq 1$. Then substituting into (4.1), we obtain

$$P_{3n+3} = xP_{3n+2} + (y - x^2) \sum_{j=0}^{3n+1} \binom{6n+4}{2j} P_j \, Q_{3n-j+1}$$

$$= xP_{3n+2} + (y - x^2) \left\{ \sum_{j=0}^{n} \binom{6n+4}{6j} P_{3j} Q_{3n-3j+1} \right.$$

$$\left. + \sum_{j=0}^{n} \binom{6n+4}{6j+2} P_{3j+1} Q_{3n-3j} + \sum_{j=0}^{n-1} \binom{6n+4}{6j+4} P_{3j+2} Q_{3n-3j-1} \right\}$$

$$\equiv x(x^{3n}y) + (y - x^2) \left\{ x^{3n+1} + \sum_{j=1}^{n} \binom{6n+4}{6j} \left(2x^{3j} + 2x^{3j-2}y \right) x^{3n-3j+1} \right.$$

$$\left. + \sum_{j=0}^{n} \binom{6n+4}{6j+2} \left(x^{3j+1} \right) x^{3n-3j} + \sum_{j=0}^{n-1} \binom{6n+4}{6j+4} (x^{3j}y) \, x^{3n-3j-1} \right\}$$

$$= x^{3n+1}y + (y - x^2) \left\{ x^{3n+1} + (2x^{3n+1} + 2x^{3n-1}y) \sum_{j=1}^{n} \binom{6n+4}{6j} \right.$$

$$\left. + x^{3n+1} \sum_{j=0}^{n} \binom{6n+4}{6j+2} + x^{3n-1}y \sum_{j=0}^{n-1} \binom{6n+4}{6j+4} \right\} \pmod 3.$$

But by (4.62), we find that

$$\sum_{j=1}^{n} \binom{6n+4}{6j} = \frac{1}{6} \left(2^{6n+4} - 7 + (-1)^{n+1} 3^{3n+2} \right) \equiv 0 \pmod 3$$

and

$$\sum_{j=0}^{n} \binom{6n+4}{6j+2} = \frac{1}{3} \left(2^{6n+3} + 1 + (-1)^n 3^{3n+2} \right) \equiv 0 \pmod 3.$$

Also, since the third sum equals the first by the map $j \to n - j$, we have that $P_{3n+3} \equiv x^{3n+1}y + (y - x^2)x^{3n+1} = 2x^{3n+3} + 2x^{3n+1}y \pmod 3$.

Next, again from (4.1), we have

$$P_{3n+4} = xP_{3n+3} + (y - x^2) \sum_{j=0}^{3n+2} \binom{6n+6}{2j} P_j Q_{3n-j+2}$$

$$= xP_{3n+3} + (y - x^2) \left\{ \sum_{j=0}^{n} \binom{6n+6}{6j} P_{3j} Q_{3n-3j+2} \right.$$

$$\left. + \sum_{j=0}^{n} \binom{6n+6}{6j+2} P_{3j+1} Q_{3n-3j+1} + \sum_{j=0}^{n} \binom{6n+6}{6j+4} P_{3j+2} Q_{3n-3j} \right\},$$

so

$$P_{3n+4} \equiv x(2x^{3n+3} + 2x^{3n+1}y)$$

$$+ (y - x^2) \left\{ x^{3n+2} + \sum_{j=1}^{n} \binom{6n+6}{6j} (2x^{3j} + 2x^{3j-2}y)x^{3n-3j+2} \right.$$

$$\left. + \sum_{j=0}^{n} \binom{6n+6}{6j+2} (x^{3j+1})(x^{3n-3j+1}) + \sum_{j=0}^{n} \binom{6n+6}{6j+4} (x^{3j}y)(x^{3n-3j}) \right\}$$

$$= 2x^{3n+4} + 2x^{3n+2}y + (y - x^2) \left\{ x^{3n+2} + (2x^{3n+2} + 2x^{3n}y) \sum_{j=1}^{n} \binom{6n+6}{6j} \right.$$

$$\left. + x^{3n+2} \sum_{j=0}^{n} \binom{6n+6}{6j+2} + x^{3n}y \sum_{j=0}^{n} \binom{6n+6}{6j+4} \right\} \pmod{3}.$$

From (4.62), we find that

$$\sum_{j=1}^{n} \binom{6n+6}{6j} = \frac{1}{3}[2^{6n+5} - 5 + (-1)^{n+1}3^{3n+3}] \equiv 0 \pmod{3}$$

and

$$\sum_{j=0}^{n} \binom{6n+6}{6j+2} = \frac{1}{6}[2^{6n+6} - 1 + (-1)^{n}3^{3n+3}] \equiv 0 \pmod{3}.$$

Also, the third sum equals the second sum by the map $j \to n - j$. Thus,

$$P_{3n+4} \equiv 2x^{3n+4} + 2x^{3n+2}y + (y - x^2)x^{3n+2} \equiv x^{3n+4} \pmod{3}.$$

Finally, (4.1) gives

$$P_{3n+5} = xP_{3n+4} + (y - x^2) \sum_{j=0}^{3n+3} \binom{6n+8}{2j} P_j Q_{3n-j+3}$$

$$= xP_{3n+4} + (y - x^2) \left\{ \sum_{j=0}^{n+1} \binom{6n+8}{6j} P_{3j} Q_{3n-3j+3} \right.$$

$$\left. + \sum_{j=0}^{n} \binom{6n+8}{6j+2} P_{3j+1} Q_{3n-3j+2} + \sum_{j=0}^{n} \binom{6n+8}{6j+4} P_{3j+2} Q_{3n-3j+1} \right\},$$

so

$$P_{3n+5} \equiv x(x^{3n+4})$$

$$+ (y - x^2)\left\{ x^{3n+3} + \sum_{j=1}^{n+1} \binom{6n+8}{6j} (2x^{3j} + 2x^{3j-2}y)\, x^{3n-3j+3} \right.$$

$$+ \sum_{j=0}^{n} \binom{6n+8}{6j+2} (x^{3j+1}) x^{3n-3j+2} + \left. \sum_{j=0}^{n} \binom{6n+8}{6j+4} (x^{3j}y)\, x^{3n-3j+1} \right\}$$

$$= x^{3n+5} + (y - x^2)\left\{ x^{3n+3} + (2x^{3n+3} + 2x^{3n+1}y) \sum_{j=1}^{n+1} \binom{6n+8}{6j} \right.$$

$$+ x^{3n+3} \sum_{j=0}^{n} \binom{6n+8}{6j+2} + x^{3n+1}y \left. \sum_{j=0}^{n} \binom{6n+8}{6j+4} \right\} \pmod{3}.$$

From (4.62) we find that

$$\sum_{j=1}^{n+1} \binom{6n+8}{6j} = \frac{1}{6}\left(2^{6n+8} - 7 + (-1)^{n+1}3^{3n+4} \right) \equiv 1 \pmod{3}.$$

But this sum maps into the second by $j \to n + 1 - j$, so the second sum is also $\equiv 1 \pmod{3}$. For the last sum, we have

$$\sum_{j=0}^{n} \binom{6n+8}{6j+4} = \frac{1}{3}\left(2^{6n+7} + 1 + (-1)^{n}3^{3n+4} \right) \equiv 1 \pmod{3}.$$

Thus,

$$P_{3n+5} \equiv x^{3n+5} + (y - x^2)\left\{ x^{3n+3} + 2x^{3n+3} + 2x^{3n+1}y + x^{3n+3} + x^{3n+1}y \right\}$$

$$\equiv x^{3n+3}y \pmod{3}.$$

(4.61) This follows immediately from (4.35) by induction. $\square$

In the next theorem we derive some of the intriguing relationships between the leading coefficients of $P_n(x, y)$ and $Q_n(x, y)$.

Theorem 4.9. For $n \geq 0$,

(4.65) $$a_0^{(3n+1)} = \beta_0^{(3n)},$$

(4.66) $$a_0^{(3n+2)} = 0,$$

and

(4.67) $$a_0^{(3n+3)} = -\beta_0^{(3n+2)}.$$

Proof. For simplicity, write $\alpha_n = P_n(1,0) = \alpha_0^{(n)}$ and $\beta_n = Q_n(1,0) = \beta_0^{(n)}$, $n \geq 0$. From Table 4.1, $\alpha_0 = 1$. The three formulas are true for $n = 0$ and 1. Assume the three formulas are true up to some $n \geq 1$, i.e.,

$$(a) \quad \alpha_{3n-1} = 0, \quad (b) \quad \alpha_{3n} = -\beta_{3n-1}, \quad (c) \quad \alpha_{3n+1} = \beta_{3n}, \ n \geq 1.$$

We will next prove in turn that equations $(a) - (c)$ hold for $n + 1$.

(i)   We first prove that $\alpha_{3n+2} = 0$. Putting $x = 1, y = 0$ into (4.1) gives

$$(4.68) \qquad \alpha_{n+2} = \alpha_{n+1} - \sum_{j=0}^{n} \binom{2n+2}{2j} \alpha_j \beta_{n-j}, \ n \geq 0.$$

Replacing n by $3n$, we get

$$\alpha_{3n+2} = \alpha_{3n+1} - \sum_{j=0}^{3n} \binom{6n+2}{2j} \alpha_j \beta_{3n-j}$$

$$= \alpha_{3n+1} - \left\{ \sum_{j=0}^{n} \binom{6n+2}{6j} \alpha_{3j} \beta_{3n-3j} + \sum_{j=0}^{n-1} \binom{6n+2}{6j+2} \alpha_{3j+1} \beta_{3n-3j-1} \right.$$

$$\left. + \sum_{j=0}^{n-1} \binom{6n+2}{6j+4} \alpha_{3j+2} \beta_{3n-3j-2} \right\}.$$

Since by assumption $\alpha_{3j+2} = 0$ for $0 \leq j \leq n - 1$, the last sum is zero and we have, using (b) and (c), that

$$\alpha_{3n+2} = \alpha_{3n+1} - \beta_{3n} - \sum_{j=1}^{n} \binom{6n+2}{6j} \alpha_{3j} \beta_{3n-3j}$$

$$- \sum_{j=0}^{n-1} \binom{6n+2}{6j+2} \alpha_{3j+1} \beta_{3n-3j-1}$$

$$= \beta_{3n} - \beta_{3n} - \sum_{j=1}^{n} \binom{6n+2}{6j} (-\beta_{3j-1}) \beta_{3n-3j} - \sum_{j=0}^{n-1} \binom{6n+2}{6j+2} \beta_{3j} \beta_{3n-3j-1}$$

$$= \sum_{j=1}^{n} \binom{6n+2}{6j} \beta_{3j-1} \beta_{3n-3j} - \sum_{j=0}^{n-1} \binom{6n+2}{6j+2} \beta_{3j} \beta_{3n-3j-1}.$$

If we re-index the first sum by $j \to n - j$, it transforms into the second sum, so the difference is 0, which proves (i).

(ii)   We next prove that $\alpha_{3n+3} = -\beta_{3n+2}$. We begin by proving the two auxiliary summation results:

$$(4.69) \qquad A_n \stackrel{\text{def}}{=} \sum_{j=0}^{n} \binom{6n+4}{6j+2} \beta_{3j} \beta_{3n-3j} = \frac{1}{3} (2\beta_{3n+1} + \beta_{3n+2}), \ n \geq 0,$$

and for $n \geq 1$ that

$$(4.70) \qquad B_n \stackrel{\text{def}}{=} \sum_{j=0}^{n-1} \binom{6n+4}{6j+4} \beta_{3j+1} \beta_{3n-3j-1} = \frac{1}{3}\left(5\beta_{3n+1} - 2\beta_{3n+2}\right).$$

If we put $x = 1, y = 0$ into (4.35), we obtain

$$(4.71) \qquad \beta_{n+1} = -2\beta_n + 3\sum_{j=0}^{n+1} \binom{2n+2}{2j} \alpha_j \alpha_{n-j+1}, \ n \geq 0.$$

Replacing n by $3n+1$ gives

$$\beta_{3n+2} + 2\beta_{3n+1} = 3\sum_{j=0}^{3n+2} \binom{6n+4}{2j} \alpha_j \alpha_{3n-j+2}$$

$$= 3\left\{ \sum_{j=0}^{n} \binom{6n+4}{6j} \alpha_{3j} \alpha_{3n-3j+2} + \sum_{j=0}^{n} \binom{6n+4}{6j+2} \alpha_{3j+1} \alpha_{3n-3j+1} \right.$$

$$\left. + \sum_{j=0}^{n} \binom{6n+4}{6j+4} \alpha_{3j+2} \alpha_{3n-3j} \right\}.$$

But, in the first sum using (a), we have $\alpha_{3n-3j+2} = 0$ for $0 \leq j \leq n$, and in the third sum using (a), then $\alpha_{3j+2} = 0$, $0 \leq j \leq n$, by the induction assumption. Thus,

$$\beta_{3n+2} + 2\beta_{3n+1} = 3\sum_{j=0}^{n} \binom{6n+4}{6j+2} \beta_{3j} \beta_{3n-3j} = 3A_n,$$

which gives (4.69).

Next, substituting $x = 1, y = 0$ into (4.33) gives

$$(4.72) \qquad \beta_{n+2} = 4\beta_{n+1} - \sum_{j=0}^{n} \binom{2n+4}{2j+2} \beta_j \beta_{n-j}.$$

Replacing n by $3n$ gives

$$\beta_{3n+2} = 4\beta_{3n+1} - \sum_{j=0}^{3n} \binom{6n+4}{2j+2} \beta_j \beta_{3n-j}$$

$$= 4\beta_{3n+1} - \left\{ \sum_{j=0}^{n} \binom{6n+4}{6j+2} \beta_{3j} \beta_{3n-3j} + \sum_{j=0}^{n-1} \binom{6n+4}{6j+4} \beta_{3j+1} \beta_{3n-3j-1} \right.$$

$$\left. + \sum_{j=0}^{n-1} \binom{6n+4}{6j+6} \beta_{3j+2} \beta_{3n-3j-2} \right\}.$$

But, re-indexing the third sum by $j \to n - j - 1$ shows it equals the second sum, so we have $\beta_{3n+2} = 4\beta_{3n+1} - (A_n + 2B_n)$. Substituting from (4.69), we obtain (4.70).

We next replace n by $3n + 1$ in (4.68), so

$$\alpha_{3n+3} = \alpha_{3n+2} - \sum_{j=0}^{3n+1} \binom{6n+4}{2j} \alpha_j \beta_{3n-j+1}$$

$$= \alpha_{3n+2} - \left\{ \sum_{j=0}^{n} \binom{6n+4}{6j} \alpha_{3j} \beta_{3n-3j+1} \right.$$

$$+ \sum_{j=0}^{n} \binom{6n+4}{6j+2} \alpha_{3j+1} \beta_{3n-3j} + \left. \sum_{j=0}^{n-1} \binom{6n+4}{6j+4} \alpha_{3j+2} \beta_{3n-3j-1} \right\}$$

$$= -\beta_{3n+1} - \sum_{j=1}^{n} \binom{6n+4}{6j} (-\beta_{3j-1}) \beta_{3n-3j+1} - \sum_{j=0}^{n} \binom{6n+4}{6j+2} \beta_{3j} \beta_{3n-3j}$$

$$= -\beta_{3n+1} + B_n - A_n = -\beta_{3n+2},$$

where we have used $\alpha_{3j+2} = 0$, $0 \le j \le n$, and values (b) and (c). We also re-indexed the first sum by $j \to n - j$ and used (4.69) and (4.70).

(iii)   We must prove that $\alpha_{3n+4} = \beta_{3n+3}$.

We begin, as in (ii), by first proving the two summation results:

$$(4.73) \qquad C_n \overset{\text{def}}{=} \sum_{j=0}^{n} \binom{6n+6}{6j+2} \beta_{3j} \beta_{3n-3j+1} = \frac{2}{3} (\beta_{3n+2} - \beta_{3n+3}), \quad n \ge 0,$$

and for $n \ge 1$,

$$(4.74) \qquad D_n \overset{\text{def}}{=} \sum_{j=1}^{n} \binom{6n+6}{6j} \beta_{3j-1} \beta_{3n-3j+2} = \frac{1}{3} (8\beta_{3n+2} + \beta_{3n+3}).$$

We evaluate D_n first. Substituting $3n + 2$ for n in (4.71) and using $\alpha_0 = 1$ and $\alpha_{3j+2} = 0$, $0 \le j \le n$, and (b), we obtain the equation

$$\beta_{3n+3} + 2\beta_{3n+2} = 3 \sum_{j=0}^{3n+3} \binom{6n+6}{2j} \alpha_j \alpha_{3n-j+3}$$

$$= 3 \left\{ \sum_{j=0}^{n+1} \binom{6n+6}{6j} \alpha_{3j} \alpha_{3n-3j+3} + \sum_{j=0}^{n} \binom{6n+6}{6j+2} \alpha_{3j+1} \alpha_{3n-3j+2} \right.$$

$$+ \left. \sum_{j=0}^{n} \binom{6n+6}{6j+4} \alpha_{3j+2} \alpha_{3n-3j+1} \right\}$$

$$= 3 \left\{ 2\alpha_{3n+3} + \sum_{j=1}^{n} \binom{6n+6}{6j} \alpha_{3j} \alpha_{3n-3j+3} \right\} = -6\beta_{3n+2} + 3D_n,$$

which implies (4.74).

Next, if we replace n by $3n+1$ in (4.72), we obtain

$$\beta_{3n+3} = 4\beta_{3n+2} - \sum_{j=0}^{3n+1} \binom{6n+6}{2j+2} \beta_j\, \beta_{3n-j+1}$$

$$= 4\beta_{3n+2} - \left\{ \sum_{j=0}^{n} \binom{6n+6}{6j+2} \beta_{3j}\, \beta_{3n-3j+1} + \sum_{j=0}^{n} \binom{6n+6}{6j+4} \beta_{3j+1}\, \beta_{3n-3j} \right.$$

$$\left. + \sum_{j=0}^{n-1} \binom{6n+6}{6j+6} \beta_{3j+2}\, \beta_{3n-3j-1} \right\}.$$

The first sum is C_n, and if we re-index the second sum by $j \to n-j$, it becomes C_n. Re-indexing the third sum by $j \to j-1$ transforms that sum into D_n, Thus, $\beta_{3n+3} = 4\beta_{3n+2} - 2C_n - D_n$, from which we obtain (4.73) using (4.74).

To finish the proof, if we replace n by $3n+2$ in (4.68), we obtain

$$\alpha_{3n+4} = \alpha_{3n+3} - \sum_{j=0}^{3n+2} \binom{6n+6}{2j} \alpha_j\, \alpha_{3n-j+2}$$

$$= \alpha_{3n+3} - \left\{ \sum_{j=0}^{n} \binom{6n+6}{6j} \alpha_{3j}\, \beta_{3n-3j+2} + \sum_{j=0}^{n} \binom{6n+6}{6j+2} \alpha_{3j+1}\, \beta_{3n-3j+1} \right.$$

$$\left. + \sum_{j=0}^{n} \binom{6n+6}{6j+4} \alpha_{3j+2}\, \beta_{3n-3j} \right\}.$$

But the third sum is zero by $\alpha_{3j+2} = 0$, so replacing the α's by the respective β's on the right, we find using (4.73) and (4.74) that

$$\alpha_{3n+4} = -2\beta_{3n+2} + \sum_{j=1}^{n} \binom{6n+6}{6j} \beta_{3j-1}\, \beta_{3n-3j+2}$$

$$- \sum_{j=0}^{n} \binom{6n+6}{6j+2} \beta_{3j}\, \beta_{3n-3j+1} = -2\beta_{3n+2} + D_n - C_n = \beta_{3n+3},$$

which completes the proof. $\square$

The next theorem was developed originally to account for the observed factorizations listed in Theorem 8.11.

Theorem 4.10. For $n \geq 0$,

(4.75) x is not a factor of $P_{2n}(x,y)$ and $x \parallel P_{2n+1}(x,y)$,

(4.76) y does not divide $P_n(x,y)$, $n \not\equiv 2 \pmod 3$, and $y \parallel P_{3n+2}(x,y)$,

$$(4.77) \quad \begin{aligned} deg_x\{P_n(x,y)\} = n, \ n \not\equiv 2 \pmod 3, \ \text{and} \\ deg_x\{P_{3n+2}(x,y)\} = 3n, \end{aligned}$$

$$(4.78) \quad deg_y\{P_n(x,y)\} = \left[\frac{n}{2}\right],$$

(4.79) x is not a factor of $Q_{2n}(x,y)$ and $x \,\|\, Q_{2n+1}(x,y)$,

(4.80) y is not a factor of $Q_n(x,y)$,

and

$$(4.81) \quad deg_x\{Q_n(x,y)\} = n \ \text{and} \ deg_y\{Q_n(x,y)\} = \left[\frac{n}{2}\right].$$

Proof. (4.75) From (4.13), we have

$$P_{2n}(x,y) = \sum_{s=0}^{n} \alpha_s^{(2n)} x^{2n-2s} y^s = \cdots + \alpha_n^{(2n)} y^n.$$

Since $\alpha_n^{(2n)} \neq 0$ by (4.54), x does not divide $P_{2n}(x,y)$. On the other hand,

$$P_{2n+1}(x,y) = \sum_{s=0}^{n} \alpha_s^{(2n+1)} x^{2n+1-2s} y^s = x \left(\cdots + \alpha_n^{(2n+1)} y^n \right).$$

Since $\alpha_n^{(2n+1)} \neq 0$ by (4.55), only the first power of x divides $P_{2n+1}(x,y)$.

(4.76) From (4.13), (4.58), and (4.59), we have $\alpha_0^{(3n)} \alpha_0^{(3n+1)} \neq 0$, $n \geq 0$, so y does not divide P_{3n} or P_{3n+1}. Also, by (4.13) and (4.60), y divides P_{3n+2}. But by (4.60), $\alpha_1^{3n+2} \neq 0$, $n \geq 2$, so y^2 does not divide P_{3n+2}.

(4.77) The first part follows from (4.13), (4.58), and (4.59), while the second part follows from (4.13) and (4.60).

(4.78) This follows from (4.13) and that $\alpha_{\{\frac{n}{2}\}}^{(n)} \neq 0$ by (4.54) and (4.55).

(4.79) The first part follows from (4.14) and (4.56), while the second part follows from (4.14) and (4.57).

(4.80) This follows from (4.14) and (4.61).

(4.81) These follow from (4.61), (4.56), and (4.57), respectively. □

We can prove the divisibility results in Theorem 4.10 in another way, as the following theorem illustrates (cf. (4.75)).

Theorem 4.11. For $n \geq 0$, $x \mid P_{2n+1}(x,y)$.

Proof. We will show that $P_{2n+1}(0,y) = 0$, where y is fixed. From (4.16), we see for $n \geq 0$ that $\rho_n \stackrel{\text{def}}{=} P_n(0,y) = \left[D_t^{2n+1} g\right]_{t=0}$, where $g = g(t;0,y)$.

From (4.19), $D_t^2 g = \frac{y}{6} g^3$, so $\rho_{2n+1} = \left[D_t^{4n+3} g \right]_{t=0} = \frac{y}{6} \left[D_t^{4n+1} g^3 \right]_{t=0}$.

Now, Leibniz's formula for the product of three functions is:

$$D^n(fgh) = \sum_{\substack{i,j,k \geq 0 \\ i+j+k=n}} \frac{n!}{i!\,j!\,k!} D_t^i f \, D_t^j g \, D_t^k h.$$

Thus,

(4.82) $\qquad \rho_{2n+1} = \frac{y}{6} \sum_{\substack{i,j,k \geq 0 \\ i+j+k=4n+1}} \frac{(4n+1)!}{i!\,j!\,k!} \left[D_t^i g \, D_t^j g \, D_t^k g \right]_{t=0}.$

Since g is an odd function, $[D_t^{2s} g]_{t=0} = 0$, $s \geq 0$. Thus, the only possible non-zero terms in (4.82) are those in which all of the indices i, j, and k are odd, i.e., $i = 2p+1$, $j = 2q+1$, and $k = 2r+1$. Thus,

$$\rho_{2n+1} = \frac{y}{6} \sum_{\substack{p,q,r \geq 0 \\ p+q+r=2n-1}} \frac{(4n+1)!}{(2p+1)!\,(2q+1)!\,(2r+1)!} \rho_p \, \rho_q \, \rho_r.$$

Next, note that $\rho_1 = 0$, so if we assume inductively that $\rho_{2k+1} = 0$ for $0 \leq k \leq n-1$ for some $n \geq 1$, then the sum on the right side of the previous equation will be zero, because at least one of p, q, or r must be odd, so at least one ρ in each term will inductively be 0. $\square$

Remarks.

1. We can also use induction to prove that x divides P_{2n+1} and Q_{2n+1}, $n \geq 0$. For $P_1 = x$ and $Q_1 = 4x$, so assume that x divides P_{2k+1} and Q_{2k+1} for some $k \geq 0$. Then from (4.1) and (4.2) we get

$$P_{2n+3} \equiv y \sum_{j=0}^{2n+1} \binom{4n+4}{2j} P_j \, Q_{2n+1-j} \pmod{x}$$

and

$$Q_{2n+3} \equiv \sum_{j=0}^{2n+1} \binom{4n+3}{2j} P_j \, P_{2n+1-j} \pmod{x}.$$

But x divides one or the other of the two factors in each term of the sums, because the subscripts j and $2n+1-j$ differ in parity, so at least one of them is odd.

2. It is reasonable to raise the question whether $P_n(x,y)$ and $Q_n(x,y)$ are irreducible in $Z[x,y]$ when any factors x or y have been removed. We have not investigated this question.

Definition 4.12. For $n \geq 0$, let

(4.83) $$P_n^*(z) = P_n(-z - 1, z^2 + 14z + 1)$$

and

(4.84) $$Q_n^*(z) = Q_n(-z - 1, z^2 + 14z + 1).$$

Theorem 4.13. For $n \geq 0$,

(4.85) $$P_n^*(z) \text{ and } Q_n^*(z) \text{ are palindromic}$$

and

(4.86) $$deg\{P_n^*(z)\} = deg\{Q_n^*(z)\} = n.$$

Proof. (4.85) That z does not divide $P_n^*(z)$ follows from (4.8), since $P_n^*(0) = P_n(-1,1) = (-1)^n \neq 0$. Thus, $\deg\left\{z^n P_n^*\left(\frac{1}{z}\right)\right\} = \deg\{P_n^*(z)\}$ and by (4.3),

$$
\begin{aligned}
z^n P_n^*\left(\frac{1}{z}\right) &= z^n P_n\left(-\frac{1}{z} - 1, \frac{1}{z^2} + \frac{14}{z} + 1\right) \\
&= P_n\left(z\left(-\frac{1}{z} - 1\right), z^2\left(\frac{1}{z^2} + \frac{14}{z} + 1\right)\right) \\
&= P_n(-1 - z, 1 + 14z + z^2) = P_n^*(z).
\end{aligned}
$$

The proof for $Q_n^*(z)$ is similar.

(4.86) From (4.13) we get $P_n^*(z) = \sum_{s=0}^{[\frac{n}{2}]} a_s^{(n)}(-z - 1)^{n-2s}(z^2 + 14z + 1)^s$, each non-zero term of which has degree n. Since $a_{[\frac{n}{2}]}^{(n)} \neq 0$ by (4.54) and (4.55), then $\deg\{P_n^*(z)\} = n$. The proof for $Q_n^*(z)$ is similar. □

In the next theorem we evaluate two interesting determinants. We will defer the proof of these results until after Lemma 7.9.

Theorem 4.14. For $m \geq 1$, we have that

$$
D_P(x, y) \overset{\text{def}}{=}
\begin{vmatrix}
P_0(x, y) & \cdots & P_{m-1}(x, y) \\
P_1(x, y) & \cdots & P_m(x, y) \\
\vdots & & \vdots \\
P_{m-1}(x, y) & \cdots & P_{2m-2}(x, y)
\end{vmatrix}
$$

(4.87) $$= \left(\frac{y - x^2}{12}\right)^{\frac{m(m-1)}{2}} \prod_{j=1}^{2m-1} j!$$

and

(4.88) $$
D_Q(x, y) \overset{\text{def}}{=}
\begin{vmatrix}
Q_0(x, y) & \cdots & Q_{m-1}(x, y) \\
Q_1(x, y) & \cdots & Q_m(x, y) \\
\vdots & & \vdots \\
Q_{m-1}(x, y) & \cdots & Q_{2m-2}(x, y)
\end{vmatrix}
$$

$$= \frac{1}{2^m}\left(\frac{y - x^2}{12}\right)^{\frac{m(m-1)}{2}} \prod_{j=1}^{2m} j!.$$

Chapter 5

The Functions $\mathcal{F}_1^{-1}$ – Elliptic Polynomials of the Second Kind

Recall from the Introduction that $\mathcal{F}_1^{-1} = \{f^{-1} : f \in \mathcal{F}_1\}$. We begin this chapter by showing that $\mathcal{F}_1 \cap \mathcal{F}_1^{-1} = \{a_1 t, a_1 \neq 0\}$. We then deduce the f^{-1}-side version of the two $f_n(x)$ recursions (3.11) (Theorem 5.4) and (3.37) (Corollary 5.6). Then follows the important Theorem 5.7 and formulas for the Maclaurin expansions of f^{-1} and $(f^{-1})^2$ for $f \in \mathcal{F}_1$. This theorem is then used to obtain the Maclaurin expansions of $\operatorname{sn}(t, k)$, $\operatorname{sn}^2(t, k)$, and other functions, including the sinelemniscate and its square.

Lemma 5.1. If f and f^{-1} are both in $\mathcal{F}_1$, then $c_1 = 0$.

Proof. We begin by deriving some useful formulas:

(5.1)
$$a_3 = \frac{a_1 c_1}{6}, \quad a_5 = \frac{a_1}{40}(3c_1^2 - 4c_2),$$
$$\bar{c}_1 = -\frac{c_1}{a_1^2}, \quad \bar{c}_2 = \frac{1}{3a_1^4}(2c_1^2 - 3c_2).$$

The first formula comes from (3.3), as does the second, using the first. Since f and f^{-1} are both in $\mathcal{F}_1$, then inverse barring holds. Thus, (3.3) implies

$$\bar{c}_1 = \frac{6\bar{a}_3}{\bar{a}_1} = -\frac{6a_3}{a_1^3} = -\frac{c_1}{a_1^2},$$

using Table 2.2 and (3.3) again. In the same way, using (3.3), Table 2.2, and (5.1), we find that

$$\bar{c}_2 = \frac{1}{\bar{a}_1^2}(27\bar{a}_3^2 - 10\bar{a}_1\bar{a}_5) = \frac{1}{a_1^6}(10a_1a_5 - 3a_3^2) = \frac{1}{3a_1^4}(2c_1^2 - 3c_2).$$

Also, since $f \in \mathcal{F}_1$, we obtain from (3.33) the formulas

(5.2) $a_7 = \dfrac{a_1c_1}{112}(5c_1^2 - 12c_2)$ and $a_9 = \dfrac{a_1}{1152}(35c_1^4 - 120c_1^2c_2 + 48c_2^2).$

Now suppose that $c_1 \neq 0$. Then substituting the results for a_3 and a_5 from (5.1) and a_7 from (5.2) into Table 2.2, we get

(5.3) $\bar{a}_7 = -\dfrac{c_1}{5040a_1^7}(c_1^2 + 132c_2).$

Furthermore, barring (5.2) and using Table 2.2 and (5.1), then

$$\bar{a}_7 = \frac{\bar{a}_1\bar{c}_1}{112}(5\bar{c}_1^2 - 12\bar{c}_2) = \frac{3c_1}{112a_1^7}(c_1^2 - 4c_2).$$

Equating this result with that in (5.3) and canceling $c_1 \neq 0$ gives

(5.4) $c_1^2 = 3c_2.$

We now use (5.4) to simplify the formulas for a_5, a_7, and a_9 in (5.1) and (5.2), obtaining

(5.5) $a_5 = \dfrac{a_1c_2}{8}, \quad a_7 = \dfrac{3a_1c_1c_2}{112}, \quad$ and $\quad a_9 = \dfrac{a_1c_2^2}{384}.$

Substituting these results into Table 2.2 and using (5.5) gives

(5.6) $\bar{a}_9 = \dfrac{149c_2^2}{8064a_1^9}.$

Also, barring a_9 in (5.5) and using (5.1) and (5.5), we obtain

$$\bar{a}_9 = \frac{\bar{a}_1\bar{c}_2^2}{384} = \frac{c_2^2}{384a_1^9}.$$

Equating this result with that in (5.6) gives that $c_2 = 0$, which by (5.4) implies $c_1 = 0$, a contradiction. $\square$

Theorem 5.2. The functions f and f^{-1} are both in $\mathcal{F}_1 \iff f(t) = a_1t$, $a_1 \neq 0$.

Proof. ($\Longleftarrow$) Since $a_3 = a_5 = 0$, then $c_1 = c_2 = 0$. But then (3.8) gives $A(t) = 1$. It can then be readily verified that both f and f^{-1} satisfy (3.12), so by Theorem 3.5 both functions are in $\mathcal{F}_1$.

($\Longrightarrow$) By Lemma 5.1, we have that $c_1 = 0$. It then follows from (5.1) and (5.2) that

$$(5.7) \qquad a_3 = a_7 = 0, \quad a_5 = -\frac{a_1 c_2}{10}, \quad a_9 = \frac{a_1 c_2^2}{24}, \quad \text{and} \quad \bar{c}_2 = -\frac{c_2}{a_1^4}.$$

Using inverse barring on a_9 in (5.7) and again using (5.7), we have that

$$(5.8) \qquad \bar{a}_9 = \frac{\bar{a}_1 \bar{c}_2^2}{24} = \frac{c_2^2}{24 a_1^9}.$$

From Table 2.2 and (5.7), we find that

$$\bar{a}_9 = \frac{1}{a_1^{11}}(5a_5^2 - a_1 a_9) = \frac{c_2^2}{120 a_1^9}.$$

Combining this result with that in (5.8) gives $c_2 = 0$. The theorem then follows from (3.24). $\quad\square$

Definition 5.3. If $f \in \mathcal{F}_1^{-1}$, then we respectively call $\bar{f}_n(x)$ and $\bar{G}_m(z)$, $\bar{H}_m(z)$ "primary" and "secondary" elliptic polynomials of the "second kind."

When $f \in \mathcal{F}_1$, we know by Theorem 3.5 and Theorem 3.2 that the terms of $\{f_n(x)\}_{n=0}^{\infty}$ satisfy the fourth-order recursion (3.11). It seems reasonable then to expect the terms of the corresponding sequence $\{\bar{f}_n(x)\}_{n=0}^{\infty}$ to satisfy some kind of recursion on its side. The next theorem shows this is true, the recursion being of the second order and involving up to fourth derivatives.

Theorem 5.4. Suppose that $f \in \mathcal{F}_1$ and also that we write $R_1(D) = \frac{x}{2a_1^2}\{2x A(D) + A'(D)\}$, where $A(t)$ is defined in (3.8). Then for $n \geq 0$,

$$(5.9) \qquad \bar{f}_{n+2} = R_1(D)\, \bar{f}_n,$$

where $\bar{f}_0(x) = 1$ and $\bar{f}_1(x) = \frac{x}{a_1}$.

Proof. Differentiating (3.13) gives

$$(5.10) \qquad D_t^2 f^{-1}(t) = \frac{1}{2a_1^2} A'(f^{-1}(t))$$

when the non-zero factor $D_t f^{-1}(t)$ is canceled.

Next, consider

$$(5.11) \qquad \bar{G} = \bar{G}(x,t) = e^{x f^{-1}(t)} = \sum_{n=0}^{\infty} \bar{f}_n(x)\frac{t^n}{n!}.$$

It then follows that

$$(5.12) \qquad D_t^2 \tilde{G} = \sum_{n=0}^{\infty} \tilde{f}_{n+2}(x) \frac{t^n}{n!}.$$

On the other hand,

$$D_t \tilde{G} = D_t e^{x f^{-1}(t)} = x\big(D_t f^{-1}(t)\big) \tilde{G}$$

and

$$D_t^2 \tilde{G} = x^2 \big(D_t f^{-1}(t)\big)^2 \tilde{G} + x\big(D_t^2 f^{-1}(t)\big)\tilde{G} = x^2\left(\frac{1}{a_1^2} A(f^{-1}(t))\right)\tilde{G}$$
$$+ x\left(\frac{1}{2a_1^2} A'(f^{-1}(t))\right)\tilde{G} = \frac{x}{a_1^2}\left(xA(f^{-1}(t)) + \frac{1}{2}A'(f^{-1}(t))\right)\tilde{G},$$

using (3.13) and (5.10). It is also clear from (5.11) that

$$A(D_x)\tilde{G} = A(f^{-1}(t))\tilde{G} \quad \text{and} \quad A'(D_x)\tilde{G} = A'(f^{-1}(t))\,\tilde{G}.$$

Thus,

$$D_t^2\tilde{G} = \frac{x}{a_1^2}\left(xA(D_x) + \frac{1}{2}A'(D_x)\right)\tilde{G} = \frac{x}{a_1^2}\left(xA(D_x) + \frac{1}{2}A'(D_x)\right)\sum_{n=0}^{\infty}\tilde{f}_n(x)\frac{t^n}{n!},$$

so finally

$$D_t^2\tilde{G} = \sum_{n=0}^{\infty}\frac{x}{2a_1^2}\big(2xA(D) + A'(D)\big)\tilde{f}_n(x)\frac{t^n}{n!}.$$

Combining this equation with (5.12) and equating corresponding coefficients gives (5.9). $\square$

Corollary 5.5. If $f \in \mathcal{F}_1$, then we have

$$(5.13) \qquad \bar{\varphi}_n^{(n)} = \bar{a}_1^n, \ n \geq 0,$$

$$(5.14) \qquad \bar{\varphi}_n^{(n+2)} = n(n+1)(n+2)\bar{a}_1^{n-1}\bar{a}_3, \ n \geq 0,$$

$$(5.15) \qquad \bar{\varphi}_0^{(n)} = \delta_{n,0}, \ n \geq 0,$$

$$(5.16) \qquad \bar{\varphi}_1^{(2n+1)} = (2n+1)!\,\bar{a}_{2n+1}, \ n \geq 0,$$

$$(5.17) \qquad \bar{\varphi}_k^{(n)} = 0, \ 0 \leq k < n, \ n - k \ \text{odd}, \ n \geq 1,$$

$$(5.18) \qquad a_1^2 \bar{\varphi}_1^{(2n+1)} = 12 c_2 \bar{\varphi}_3^{(2n-1)} - c_1 \bar{\varphi}_1^{(2n-1)}, \ n \geq 2, \ (\text{cf. } (1.13))$$

and, for $2 \leq k \leq n-2$ and $n \geq 4$,

$$(5.19) \qquad a_1^2 \bar{\varphi}_k^{(n+2)} = k(k+1)^2(k+2) c_2 \, \bar{\varphi}_{k+2}^{(n)} - k^2 c_1 \, \bar{\varphi}_k^{(n)} + \bar{\varphi}_{k-2}^{(n)}.$$

Proof. Formulas (5.13)–(5.17) come from Corollary 1.5 using inverse barring (and the fact that $\bar{a}_{2k} = 0$). If (2.4) is substituted into (5.9) and the corresponding coefficients are equated, we obtain results for $n = 0, 1, 2, 3$ that are covered by the above formula as well as the formula for general n given in (5.19). □

The next corollary is the f^{-1}-side version of Corollary 3.9. Here the recursion is of order one for $\bar{f}_n$ and involves a second derivative. The comparable relationship on the f-side is a linear recursion for f_n of the second order.

Corollary 5.6. If $f \in \mathcal{F}_1$ and $R_2(D) = \dfrac{x}{a_1}\left(1 - \dfrac{c_1}{2}D^2\right)$, then for $n \geq 0$,

$$(5.20) \qquad \bar{f}_{n+1}(x) = R_2(D)\,\bar{f}_n(x) \iff \Delta = 0.$$

Proof. ($\Longrightarrow$) Substituting $\bar{f}_4(x)$ and $\bar{f}_5(x)$ from Table 2.3 into (5.20) and simplifying gives $\Delta = 0$.

($\Longleftarrow$) Since $\Delta = 0$, we have from (3.3) that $c_2 = \dfrac{c_1^2}{4}$. Thus, from (3.8), $A(t) = \left(1 - \dfrac{c_1 t^2}{2}\right)^2$. After some routine calculation, it follows then that $R_1(D) = R_2^2(D)$, so from (5.9) we have $R_2^2(D)\bar{f}_n = \bar{f}_{n+2}$, $n \geq 0$. If we combine this result with the fact that $\bar{f}_1 = \dfrac{x}{a_1} = R_2(D)\bar{f}_0$, we find that $R_2(D)\bar{f}_1 = R_2^2(D)\bar{f}_0 = \bar{f}_2$. Continuing in this way, we find inductively that $R_2(D)\bar{f}_n = \bar{f}_{n+1}$. □

The next theorem is one of the main results of this work. For any $f \in \mathcal{F}_1$, it respectively gives the Maclaurin expansions of f^{-1} and $(f^{-1})^2$ in terms of $P_n(x,y)$ and $Q_n(x,y)$ that are evaluated at points that depend only on a_1, a_3, and a_5.

Theorem 5.7. If $f \in \mathcal{F}_0$, then

$$(5.21) \quad f \in \mathcal{F}_1 \iff f^{-1}(t) = \frac{1}{a_1} \sum_{n=0}^{\infty} P_n\left(-\frac{c_1}{a_1^2}, \frac{c_1^2 + 12 c_2}{a_1^4}\right) \frac{t^{2n+1}}{(2n+1)!}.$$

If $f \in \mathcal{F}_1$, then

$$(5.22) \qquad [f^{-1}(t)]^2 = \frac{2}{a_1^2} \sum_{n=0}^{\infty} Q_n\left(-\frac{c_1}{a_1^2}, \frac{c_1^2 + 12 c_2}{a_1^4}\right) \frac{t^{2n+2}}{(2n+2)!}.$$

Proof. (5.21) ($\Longrightarrow$) From (3.13), we have that

$$(5.23) \qquad [a_1 (f^{-1})'(t)]^2 = 1 - \frac{c_1}{a_1^2}[a_1 f^{-1}(t)]^2 + \frac{c_2}{a_1^4}[a_1 f^{-1}(t)]^4.$$

But the function $g(t)$, defined in (4.16), satisfies (4.22). Putting $x = -\dfrac{c_1}{a_1^2}$ and $y = \dfrac{1}{a_1^4}(c_1^2 + 12c_2)$ in (4.22) gives an equation the same as (5.23).

Next, we obtain from (5.23) that

$$(5.24) \qquad a_1(f^{-1})'(t) = \sqrt{1 - \frac{c_1}{a_1^2}[a_1 f^{-1}(t)]^2 + \frac{c_2}{a_1^4}[a_1 f^{-1}(t)]^4},$$

since $f^{-1}(0) = 0$ and $a_1(f^{-1})'(0) = 1$. Also, from (4.22), we obtain that

$$(5.25) \qquad g'(t) = \sqrt{1 - \frac{c_1}{a_1^2}[g(t)]^2 + \frac{c_2}{a_1^4}[g(t)]^4},$$

since $g(0) = 0$ and $g'(0) = 1$. It now readily follows from Picard's existence and uniqueness theorem for ordinary differential equations [Cro, p. 14] that the initial value problem

$$y' = \sqrt{1 - c_1 y^2 + c_2 y^4}, \ y(0) = 0,$$

has a unique solution for small $|t|$. Thus, $g(t) = a_1 f^{-1}(t)$, which implies (5.21) using (4.16).

($\Longleftarrow$) From (5.21) and (4.16), we find that $g(t) = a_1 f^{-1}(t)$. But then from (4.22), it follows that (3.13) is satisfied, which implies $f \in \mathcal{F}_1$ by Theorem 3.5.

(5.22) By the first part, we have that $f^{-1}(t) = \dfrac{1}{a_1} g(t)$, so using (4.18) and (4.17), we obtain

$$[f^{-1}(t)]^2 = \frac{1}{a_1^2} g^2(t) = \frac{2}{a_1^2} h(t)$$

$$= \frac{2}{a_1^2} \sum_{n=0}^{\infty} Q_n\left(-\frac{c_1}{a_1^2}, \frac{c_1^2 + 12c_2}{a_1^4}\right) \frac{t^{2n+2}}{(2n+2)!}. \qquad \square$$

Corollary 5.8. The functions $\mathcal{F}_1^{-1}$ can be parameterized by the three-parameter set $S_1 = \{(a_1, x, y) \in \mathbb{R}^3 : a_1 \neq 0\}$.

Proof. It is clear from (5.21) that the set of functions $\mathcal{F}_1^{-1}$ is parameterized by $\{(a_1, c_1, c_2) \in \mathbb{R}^3, a_1 \neq 0\}$, since given any $a_1 \neq 0$, there is a $c_1 \in \mathbb{R}$ such that the first argument $x = -\dfrac{c_1}{a_1^2}$ can be any real number. For each pair a_1, c_1, and so for $y = \dfrac{c_1^2 + 12c_2}{a_1^4}$, there is a $c_2 \in \mathbb{R}$ such that y is any real number. $\square$

We end this chapter with a collection of expansions obtained from Theorem 5.6. For this purpose, we will need the expansions of the following functions in $\mathcal{F}_1$ (see [WhW, Example, p. 494]).

Definition 5.9. For $0 \le k \le 1$,

$$(5.26) \quad \mathrm{sn}^{-1}(t, k) \stackrel{\text{def}}{=} \int_0^t \frac{du}{\sqrt{(1 - u^2)(1 - k^2 u^2)}}$$
$$= t + \frac{1}{6}(k^2 + 1) t^3 + \frac{1}{40}(3k^4 + 2k^2 + 3) t^5 + \cdots,$$

$$(5.27) \quad \mathrm{sc}^{-1}(t, k) \stackrel{\text{def}}{=} \int_0^t \frac{du}{\sqrt{[1 + u^2][1 + (1 - k^2)u^2]}}$$
$$= t + \frac{1}{6}(k^2 - 2) t^3 + \frac{1}{40}(3k^4 - 8k^2 + 8) t^5 + \cdots,$$

$$(5.28) \quad \mathrm{sd}^{-1}(t, k) \stackrel{\text{def}}{=} \int_0^t \frac{du}{\sqrt{[1 + k^2 u^2][1 - (1 - k^2)u^2]}}$$
$$= t - \frac{1}{6}(2k^2 - 1) t^3 + \frac{1}{40}(8k^4 - 8k^2 + 3) t^5 + \cdots,$$

and (cf. (8.49))

$$(5.29) \quad f^*(t, k) \stackrel{\text{def}}{=} \int_0^t \frac{du}{\sqrt{1 - 2(1 - 2k^2)u^2 + u^4}}$$
$$= t + \frac{1}{3}(1 - 2k^2) t^3 + \frac{1}{5}(6k^4 - 6k^2 + 1) t^5 + \cdots.$$

The next theorem expresses $f^*(t, k)$ in terms of the sn^{-1} function.

Theorem 5.10. If $0 < k < 1$ and $k' = \sqrt{1 - k^2}$, then

$$(5.30) \qquad f^*(t, k) = \frac{1}{2} \mathrm{sn}^{-1}\left(\frac{2t}{1 + t^2}, k'\right).$$

Proof. Let

$$g(t, k) = \frac{1}{2} \mathrm{sn}^{-1}\left(\frac{2t}{1 + t^2}, k'\right).$$

Since $f^*(0, k) = g(0, k) = 0$, it is only necessary to prove that $(f^*)'(t, k) = g'(t, k)$. But

$$(f^*)'(t, k) = \frac{1}{\sqrt{1 - 2(1 - 2k^2)t^2 + t^4}}.$$

Also, by [GrR, p. 917, 8.159], we have that

$$\frac{d}{du}(\mathrm{sn}\, u) = \sqrt{(1 - \mathrm{sn}^2 u)(1 - k^2 \mathrm{sn}^2 u)},$$

so, from the formula $(h^{-1})'(y) = \dfrac{1}{h'(h^{-1}(y))}$, we find that

$$g'(t,k) = (sn^{-1})'\left(\frac{2t}{1+t^2}, k'\right) \cdot \frac{1-t^2}{(1+t^2)^2}$$

$$= \frac{1-t^2}{(1+t^2)^2} \frac{1}{\sqrt{[1-(\frac{2t}{1+t^2})^2][1-(k')^2(\frac{2t}{1+t^2})^2]}}$$

$$= \frac{1}{\sqrt{1-2(1-2k^2)t^2+t^4}}. \quad \square$$

Corollary 5.11.

$$(5.31) \qquad (f^*)^{-1}(t,k) = \frac{sn\,(t,k')\,dn\,(t,k')}{cn\,(t,k')} = -\frac{d}{dt}\log\,(cn\,(t,k')).$$

Proof. Setting $y = f^*(t,k)$, we obtain from equation (5.30) the result [AbS, p. 574, 16.118.1]

$$(5.32) \qquad \frac{2t}{1+t^2} = sn\,(2y) = \frac{2sn\,(y)\,cn\,(y)\,dn\,(y)}{cn^2(y)+sn^2(y)\,dn^2(y)}.$$

Solving for t gives

$$t = \frac{cn^2(y)+sn^2(y)\,dn^2(y) \pm [cn^2(y)-sn^2(y)\,dn^2(y)]}{2\,sn\,(y)\,cn\,(y)\,dn\,(y)},$$

where the minus sign must be taken, since $y = 0$ when $t = 0$. Hence, we find [AbS, p. 574, 16.16.2] that

$$t = \frac{sn\,(y)\,dn\,(y)}{cn\,(y)} = -\frac{d}{dt}\log\,(cn\,(y)). \quad \square$$

The next theorem presents the Maclaurin expansions of certain functions using Theorem 5.7. Because of the similarity of the formula in (5.22) to that in (5.21), we will omit writing out the expansions of the squares of these functions except in the first case, which illustrates the formula.

Theorem 5.12. For $0 < k < 1$,

$$(5.33) \qquad sn\,(t,k) = \sum_{n=0}^{\infty}(-1)^n P_n(k^2+1, k^4+14k^2+1)\frac{t^{2n+1}}{(2n+1)!},$$

$$(5.34) \qquad sn^2(t,k) = 2\sum_{n=0}^{\infty}(-1)^n Q_n(k^2+1, k^4+14k^2+1)\frac{t^{2n+2}}{(2n+2)!},$$

$$(5.35) \quad \log\left(\operatorname{cn}(t,k)\right) = -\sum_{n=0}^{\infty} 2^n P_n(1 - 2k^2, 4(k^4 - k^2 + 1)) \frac{t^{2n+2}}{(2n+2)!}$$

$(5.36) \quad \operatorname{cn}(t,k) =$

$$1 - \sum_{n=0}^{\infty} \frac{(-1)^n}{2^{n+1}} \left\{ \sum_{j=0}^{n} 2^j \binom{2n+2}{2j+1} P_j(k^2 + 1, k^4 + 14k^2 + 1) \right.$$

$$\left. \times P_{n-j}(2k^2 - 1, 4(k^4 - k^2 + 1)) \right\} \frac{t^{2n+2}}{(2n+2)!},$$

$(5.37) \quad \operatorname{dn}(t,k) =$

$$1 - k^2 \sum_{n=0}^{\infty} \frac{(-1)^n}{2^{n+1}} \left\{ \sum_{j=0}^{n} 2^j \binom{2n+2}{2j+1} P_j(k^2 + 1, k^4 + 14k^2 + 1) \right.$$

$$\left. \times P_{n-j}(k^2 - 2, 4(k^4 - k^2 + 1)) \right\} \frac{t^{2n+2}}{(2n+2)!},$$

$$(5.38) \quad \log\left(\operatorname{dn}(t,k)\right) = -k^2 \sum_{n=0}^{\infty} 2^n P_n(k^2 - 2, 4(k^4 - k^2 + 1)) \frac{t^{2n+2}}{(2n+2)!},$$

$(5.39) \quad \dfrac{1}{\operatorname{cn}(t,k)} =$

$$1 + \sum_{n=0}^{\infty} \frac{(-1)^n}{2^{n+1}} \left\{ \sum_{j=0}^{n} 2^j \binom{2n+2}{2j+1} P_j(k^2 - 2, k^4 - 16k^2 + 16) \right.$$

$$\left. \times P_{n-j}(2k^2 - 1, 4(k^4 - k^2 + 1)) \right\} \frac{t^{2n+2}}{(2n+2)!},$$

$(5.40) \quad \dfrac{1}{\operatorname{dn}(t,k)} =$

$$1 + k^2 \sum_{n=0}^{\infty} \frac{1}{2^{n+1}} \left\{ \sum_{j=0}^{n} 2^j \binom{2n+2}{2j+1} P_j(2k^2 - 1, 16k^4 - 16k^2 + 1) \right.$$

$$\left. \times P_{n-j}(k^2 - 2, 4(k^4 - k^2 + 1)) \right\} \frac{t^{2n+2}}{(2n+2)!},$$

$$(5.41) \quad \operatorname{sc}(t,k) = \sum_{n=0}^{\infty} P_n(2 - k^2, k^4 - 16k^2 + 16) \frac{t^{2n+1}}{(2n+1)!},$$

$$(5.42) \qquad \mathrm{sd}\,(t,k) = \sum_{n=0}^{\infty} P_n(2k^2 - 1, 16k^4 - 16k^2 + 1)\,\frac{t^{2n+1}}{(2n+1)!},$$

and

$$(5.43) \qquad (f^*)^{-1}(t,k) = \sum_{n=0}^{\infty} 2^n\, P_n(2k^2 - 1, 4(k^4 - k^2 + 1))\,\frac{t^{2n+1}}{(2n+1)!}.$$

Proof. (5.33), (5.34) From (5.26), we find that $a_1 = 1$, $a_3 = \frac{1}{6}(k^2 + 1)$, $a_5 = \frac{1}{40}(3k^4 + 2k^2 + 3)$, $c_1 = k^2 + 1$, and $c_2 = k^2$. The results follow from (5.21), (5.22), (4.8), and (4.9).

(5.35) From (5.29), we have $a_1 = 1$, $a_3 = \frac{1}{3}(1-2k^2)$, $a_5 = \frac{1}{5}(6k^4 - 6k^2 + 1)$, $c_1 = 2(1 - 2k^2)$, and $c_2 = 1$, so combining (5.31) with (5.21) we obtain

$$(5.44) \qquad -\frac{d}{dt}\,\log\,(\mathrm{cn}\,(t,k'))$$

$$= \sum_{n=0}^{\infty} P_n\!\left(2\,(2k^2 - 1), 16\,(k^4 - k^2 + 1)\right)\frac{t^{2n+1}}{(2n+1)!}.$$

Integrating (the constant of integration is 0 at $t = 0$) gives

$$\log\,(\mathrm{cn}\,(t,k')) = -\sum_{n=0}^{\infty} P_n\!\left(2\,(2k^2 - 1), 16\,(k^4 - k^2 + 1)\right)\frac{t^{2n+2}}{(2n+2)!}.$$

Interchanging k with k' and using (4.8) gives the result.

(5.36) From [AbS, p. 574, 16.18.2], we have

$$(5.45) \qquad \mathrm{cn}\,(2y) = \frac{\mathrm{cn}^2(y) - \mathrm{sn}^2(y)\,\mathrm{dn}^2(y)}{\mathrm{cn}^2(y) + \mathrm{sn}^2(y)\,\mathrm{dn}^2(y)},$$

so using (5.45), (5.33), and (5.31), it follows that

$$(5.46) \qquad \frac{1 - \mathrm{cn}\,(2y)}{\mathrm{sn}\,(2y)} = \frac{\mathrm{sn}\,(y)\,\mathrm{dn}\,(y)}{\mathrm{cn}\,(y)} = -\frac{d}{dy}\,\log\,(\mathrm{cn}\,(y)).$$

Thus, from (5.33), (5.35), and (5.46), we have

$$\mathrm{cn}\,(2t) = 1 + \mathrm{sn}\,(2t)\frac{d}{dt}\,\log\,(\mathrm{cn}\,(t))$$

$$= 1 + \left\{\sum_{n=0}^{\infty} 2^{2n+1}\,P_n(-k^2 - 1, k^4 + 14k^2 + 1)\frac{t^{2n+1}}{(2n+1)!}\right\}$$

$$\times\left\{-\sum_{n=0}^{\infty} 2^n P_n(1 - 2k^2, 4(k^4 - k^2 + 1))\frac{t^{2n+1}}{(2n+1)!}\right\},$$

which gives the result using (4.8) and replacing $2t$ by t.

(5.37) From [Han, p. 252] or [GrR, p. 916], we have

$$(5.47) \qquad \mathrm{dn}\,(t,k) = \mathrm{cn}\left(kt,\frac{1}{k}\right).$$

The result follows from (5.36).

(5.38) Since by (5.47) we have that $\log\,(\mathrm{dn}\,(t,k)) = \log\left(\mathrm{cn}\,(kt,\frac{1}{k})\right)$, the result follows from (5.35) using (4.8).

(5.39) From [GrR, p. 916], $\dfrac{1}{\mathrm{cn}\,(t,k)} = \mathrm{cn}\,(it,k')$, which implies the result using (5.36).

(5.40) The result follows from (5.39) using (5.47) and (4.8).

(5.41) From (5.27), we find that $a_1 = 1$, $a_3 = \frac{1}{6}(k^2 - 2)$, $a_5 = \frac{1}{40}(3k^4 - 8k^2 + 8)$, $c_1 = k^2 - 2$, and $c_2 = 1 - k^2$. The result follows from (5.21).

(5.42) From (5.28), we find that $a_1 = 1$, $a_3 = \frac{1}{6}(1 - 2k^2)$, $a_5 = \frac{1}{40}(6k^4 - 6k^2 + 1)$, $c_1 = 1 - 2k^2$, and $c_2 = k^2(k^2 - 1)$, the result follows from (5.21).

(5.43) Combine (5.31) and (5.44) and use (5.8). □

Remark. The divisibility patterns for the coefficients in expansions of the functions in (5.33), (5.41), and (5.42) mentioned in the Preface, as well as for the coefficients in the expansions of their squares, follow directly from applying the factorizations in Theorem 4.10 to the coefficients in the expansions in (5.21) and (5.22).

We next derive the Maclaurin expansions of the sinelemniscate function $\mathrm{sl}\,(t)$, the hyperbolic sinelemniscate function $\mathrm{slh}\,(t)$ and their squares, and the cosinelemniscate function $\mathrm{cl}\,(t)$ [WhW, p. 524]. (The function slh^{-1} is also dealt with later in Sec. 9(b).) In what follows, we will write α_n for $\alpha_n^{(2n)}$ and β_n for $\beta_n^{(2n)}$.

Before we consider these expansions, we will derive some useful results. From (4.13) and (4.14) we have for $n \geq 0$ that

$$(5.48) \qquad P_{2n+1}(0,y) = Q_{2n+1}(0,y) = 0$$

and

$$(5.49) \qquad P_{2n}(0,\pm 3) = (\pm 3)^n \alpha_n \text{ and } Q_{2n}(0,\pm 3) = (\pm 3)^n \beta_n.$$

These evaluations will be used in the proofs that follow without reference. If we specialize identities (4.33), (4.34), (4.37), and (4.42) at $(0, -3)$, we obtain the respective formulas

$$(5.50) \qquad \beta_{n+1} = \sum_{j=0}^{n} \binom{4n+4}{4j+2} \beta_j\,\beta_{n-j}, \ \beta_0 = 1, \ n \geq 0,$$

(5.51) $$a_{n+1} = \frac{1}{3} \sum_{j=0}^{n} \binom{4n+3}{4j+1} a_j \beta_{n-j}, \quad a_0 = \beta_0 = 1, \ n \geq 0,$$

(5.52) $$\beta_{n+1} = 6 \sum_{j=0}^{n} \binom{4n+3}{4j+3} a_{j+1} a_{n-j}, \quad a_0 = 1, \ n \geq 0.$$

and, for $\alpha_1 = \beta_0 = 1$ and $n \geq 1$,

(5.53) $$\beta_n = \frac{-1}{n(4n+3)} \sum_{j=1}^{n} (n-3j) \binom{4n+3}{4j+1} a_j \beta_{n-j}.$$

No simple formula seems to exist which expresses a_n only in terms of the β's.

Corollary 5.13.

(5.54) $$\mathrm{sl}\,(t) = \sum_{n=0}^{\infty} (-12)^n a_n \frac{t^{4n+1}}{(4n+1)!},$$

(5.55) $$\mathrm{sl}^2(t) = 2 \sum_{n=0}^{\infty} (-12)^n \beta_n \frac{t^{4n+2}}{(4n+2)!},$$

(5.56) $$\mathrm{slh}\,(t) = \sum_{n=0}^{\infty} 12^n a_n \frac{t^{4n+1}}{(4n+1)!},$$

(5.57) $$\mathrm{slh}^2(t) = 2 \sum_{n=0}^{\infty} 12^n \beta_n \frac{t^{4n+2}}{(4n+2)!},$$

(5.58) $$\mathrm{cl}\,(t) = 1 - \sum_{n=0}^{\infty} (-1)^n \left\{ \sum_{j=0}^{\left[\frac{n}{2}\right]} \binom{2n+2}{4j+1} 3^j a_j P_{n-2j}(3,33) \right\} \frac{t^{2n+2}}{(2n+2)!},$$

where $a_0 = 1$ and for $n \geq 0$,

(5.59) $$a_{n+1} = \sum_{j=0}^{n} \binom{4n+2}{4j+2} a_{n-j} \left\{ \sum_{k=0}^{j} \binom{4j+1}{4k} a_k a_{j-k} \right\},$$

and β_n is computed from (5.50).

Proof. (5.54) Using the formula $\mathrm{sl}\,(t) = \dfrac{1}{\sqrt{2}}\,\mathrm{sd}\left(t\sqrt{2},\dfrac{1}{\sqrt{2}}\right)$ from [WhW, p. 524], we obtain from (5.42) that

$$\mathrm{sl}\,(t) = \frac{1}{\sqrt{2}}\sum_{n=0}^{\infty} P_n(0,-3)\frac{(t\sqrt{2})^{2n+1}}{(2n+1)!} = \sum_{n=0}^{\infty} 2^{2n}P_{2n}(0,-3)\frac{t^{4n+1}}{(4n+1)!}$$

$$= \sum_{n=0}^{\infty}(-12)^n\alpha_n\frac{t^{4n+1}}{(4n+1)!}.$$

(5.55) Using (5.42) and (5.22), we find that

$$\mathrm{sd}^2(t,k) = 2\sum_{n=0}^{\infty} Q_n(2k^2-1,16k^4-16k^2+1)\frac{t^{2n+2}}{(2n+2)!}.$$

Thus,

$$\mathrm{sl}^2(t) = \frac{1}{2}\mathrm{sd}^2(t\sqrt{2},\tfrac{1}{\sqrt{2}}) = \sum_{n=0}^{\infty} Q_n(0,-3)\frac{(t\sqrt{2})^{2n+2}}{(2n+2)!}$$

$$= \sum_{n=0}^{\infty} 2^{2n+1}Q_{2n}(0,-3)\frac{t^{4n+2}}{(4n+2)!} = 2\sum_{n=0}^{\infty}(-12)^n\beta_n\frac{t^{4n+2}}{(4n+2)!}.$$

(5.56) From (5.29), we find that $\mathrm{slh}^{-1}(t) = \displaystyle\int_0^t \frac{du}{\sqrt{1+u^4}} = f^*\left(t,\frac{1}{\sqrt{2}}\right)$. Thus, from (5.43) we have

$$\mathrm{slh}\,(t) = (f^*)^{-1}\left(t,\frac{1}{\sqrt{2}}\right) = \sum_{n=0}^{\infty} 2^n P_n(0,3)\frac{t^{2n+1}}{(2n+1)!}$$

$$= \sum_{n=0}^{\infty} 2^{2n} P_{2n}(0,3)\frac{t^{4n+1}}{(4n+1)!} = \sum_{n=0}^{\infty} 12^n\alpha_n\frac{t^{4n+1}}{(4n+1)!}.$$

(5.57) Using (5.22) and (4.8), we have

$$\mathrm{slh}^2(t) = \left((f^*)^{-1}(t,\tfrac{1}{\sqrt{2}})\right)^2 = 2\sum_{n=0}^{\infty} 2^n Q_n(0,3)\frac{t^{2n+2}}{(2n+2)!}$$

$$= 2\sum_{n=0}^{\infty} 2^{2n}Q_{2n}(0,3)\frac{t^{4n+2}}{(4n+2)!} = 2\sum_{n=0}^{\infty} 12^n\beta_n\frac{t^{4n+2}}{(4n+2)!}.$$

(5.58) Using the formula $\mathrm{cl}\,(t) = \mathrm{cn}\left(\sqrt{2}\,t,\dfrac{1}{\sqrt{2}}\right)$ from [WhW, p. 524], we obtain from (5.36), (5.8), replacing j by $n-j$, (5.48), and (5.49) that

$$\text{cl}(t) = 1 - \sum_{n=0}^{\infty} \frac{(-1)^n}{2^{n+1}} \left\{ \sum_{j=0}^{n} 2^j \binom{2n+2}{2j+1} P_j\left(\frac{3}{2},\frac{33}{4}\right) P_{n-j}(0,3) \right\} \frac{(\sqrt{2}t)^{2n+2}}{(2n+2)!}$$

$$= 1 - \sum_{n=0}^{\infty} (-1)^n \left\{ \sum_{j=0}^{n} \binom{2n+2}{2j+1} P_j(3,33) P_{n-j}(0,3) \right\} \frac{t^{2n+2}}{(2n+2)!}$$

$$= 1 - \sum_{n=0}^{\infty} (-1)^n \left\{ \sum_{j=0}^{[\frac{n}{2}]} \binom{2n+2}{4j+1} 3^j \alpha_j P_{n-2j}(3,33) \right\} \frac{t^{2n+2}}{(2n+2)!}.$$

(5.59) We find that $P_{2n}(0,1) = \alpha_n$ and $Q_{2n}(0,1) = \beta_n$ using (4.13) and (4.14). Then from (4.1), we have

$$\alpha_{n+1} = P_{2n+2}(0,1) = \sum_{j=0}^{n} \binom{4n+2}{4j} P_{2j}(0,1) Q_{2n-2j}(0,1)$$

$$= \sum_{j=0}^{n} \binom{4n+2}{4j+2} P_{2n-2j}(0,1) Q_{2j}(0,1) = \sum_{j=0}^{n} \binom{4n+2}{4j+2} \alpha_{n-j}\beta_j.$$

From (4.2) we find that

$$\beta_j = Q_{2j}(0,1) = \sum_{k=0}^{2j} \binom{4j+1}{2k} P_k(0,1) P_{2j-k}(0,1)$$

$$= \sum_{k=0}^{j} \binom{4j+1}{4k} P_{2k}(0,1) P_{2j-2k}(0,1) = \sum_{k=0}^{j} \binom{4j+1}{4k} \alpha_k \alpha_{j-k}.$$

Combining these results gives the desired formula. □

Theorem 5.14.

(5.60) $$\alpha_n \equiv 1 \pmod{20}, \ n \geq 0,$$

(5.61) $$\alpha_n \equiv 0 \pmod{3}, \ n \geq 2,$$

and

(5.62) $$\beta_n \equiv 6 \pmod{30}, \ n \geq 1.$$

Proof. (5.60) For $n \geq 0$, (4.54) gives $\alpha_n = P_{2n}(0,1) \equiv 1 \pmod{10}$. Thus, we need only show that $\alpha_n \equiv 1 \pmod 4$ for $n \geq 0$. This is true for $n = 0$ and 1. Assume for some $n \geq 1$ that $\alpha_1 \equiv \alpha_2 \equiv \cdots \equiv \alpha_n \equiv 1 \pmod 4$. Also, it is true that $\beta_0 = 1$, $\beta_1 = 6$, and $\beta_2 = 336$. We first show that $\beta_n \equiv 0 \pmod 4$, $n \geq 2$.

From (4.2) we obtain

$$\beta_n = Q_{2n}(0,1) = \sum_{j=0}^{n}\binom{4n+1}{4j}\alpha_j\,\alpha_{n-j} \equiv \sum_{j=0}^{n}\binom{4n+1}{4j} \pmod 4.$$

From (4.62) we find that

$$\sum_{j=0}^{n}\binom{4n+1}{4j} = \frac{1}{4}\sum_{j=1}^{4}(1+i^j)^{4n+1}$$

$$= \frac{1}{4}\left((1+i)^{4n+1} + (1-i)^{4n+1} + 2^{4n+1}\right) = \frac{1}{2}\left((-4)^n + 2^{4n}\right),$$

from which the result follows.

For $n \geq 1$, equation (4.1) gives

$$\alpha_{n+1} = P_{2n+2}(0,1) = \sum_{j=0}^{n}\binom{4n+2}{4j}P_{2j}(0,1)\,Q_{2n-2j}(0,1)$$

$$= \sum_{j=0}^{n}\binom{4n+2}{4j+2}\alpha_{n-j}\beta_j \equiv \sum_{j=0}^{1}\binom{4n+2}{4j+2}\beta_j$$

$$= \binom{4n+2}{2}\beta_0 + \binom{4n+2}{6}\beta_1 \equiv 2n+1+(2n+1)2n \equiv 1 \pmod 4.$$

(5.61) This congruence follows from combining (4.58)–(4.60).

(5.62) For $n \geq 1$, congruence (4.56) gives $\beta_n = Q_{2n}(0,1) \equiv 6 \pmod{10}$, while (4.61) gives $\beta_n = Q_{2n}(0,1) \equiv 0 \pmod 3$. $\square$

Comments.

1. Congruence (5.62) implies that $6\,|\,\beta_n$, $n \geq 1$. This is also clear from (5.52). Note that we can also compute the β's more efficiently for $m \geq 1$ using the formulas

(5.63)
$$\beta_{2m} = 2\sum_{j=0}^{m-1}\binom{8m}{4j+2}\beta_j\beta_{2m-1-j}$$

and

(5.64)
$$\beta_{2m+1} = 2\sum_{j=0}^{m-1}\binom{8m+4}{4j+2}\beta_j\beta_{2m-j} + \binom{8m+4}{4m+2}\beta_m^2.$$

To prove (5.63), replace n by $2m-1$ in (5.50) and use $j \to 2m-1-j$, so that

$$\beta_{2m} = \sum_{j=0}^{2m-1}\binom{8m}{4j+2}\beta_j\beta_{2m-1-j} = \sum_{j=0}^{m-1}\binom{8m}{4j+2}\beta_j\beta_{2m-1-j}$$

$$+ \sum_{j=m}^{2m-1}\binom{8m}{4j+2}\beta_j\beta_{2m-1-j} = 2\sum_{j=0}^{m-1}\binom{8m}{4j+2}\beta_j\beta_{2m-1-j}.$$

The second result is obtained in a similar way replacing n by $2m$.

2. Congruence (5.60) implies that the prime divisors of α_n are odd.

The next theorem shows that (since 3 divides α_2) each prime $p \equiv 3 \pmod 4$ divides the terms of the α and β sequences.

Theorem 5.15. If $p = 4n - 1$ is a prime, then

$$(5.65) \qquad p \text{ divides } \alpha_n, \ n \geq 2,$$

and

$$(5.66) \qquad p \text{ divides } \beta_n, \ n \geq 1.$$

Proof. (5.65) Note in (5.51) that $\alpha_1 = 1$ for $n = 0$. For $n \geq 1$, if $4n + 3$ is a prime, then p divides each binomial coefficient in the sum in (5.51), so p divides α_{n+1}.

(5.66) For $n \geq 0$, the same argument gives the result using (5.50). □

The next two tables are readily computed using (5.59) and (5.50).

Table 5.1 $\alpha_n^{(2n)}$, $0 \leq n \leq 13$

n	$\alpha_n^{(2n)}$
0	1
1	1
2	$21 = 3 \cdot 7$
3	$2541 = 3 \cdot 7 \cdot 11^2$
4	$1023561 = 3^2 \cdot 7^2 \cdot 11 \cdot 211$
5	$1036809081 = 3^2 \cdot 7^3 \cdot 11 \cdot 19 \cdot 1607$
6	$2219782435101 = 3^4 \cdot 7^3 \cdot 11^2 \cdot 19 \cdot 23 \cdot 1511$
7	$8923051855107621 = 3^3 \cdot 7^4 \cdot 11^2 \cdot 19 \cdot 23 \cdot 2603099$
8	$61797392100611962641 = 3^4 \cdot 7^4 \cdot 11^3 \cdot 19 \cdot 23 \cdot 31 \cdot 3931 \cdot 4483$
9	$690766390156657904866161 = 3^4 \cdot 7^5 \cdot 11^3 \cdot 19 \cdot 23 \cdot 31$ $\cdot 28140693319$
10	$11839493254591562294152214181 = 3^5 \cdot 7^5 \cdot 11^3 \cdot 19^2 \cdot 23 \cdot 31$ $\cdot 179 \cdot 47272527433$
11	$298556076626963858753929987732701 = 3^5 \cdot 7^6 \cdot 11^4 \cdot 19^2 \cdot 23 \cdot 31$ $\cdot 43 \cdot 64445926746077$
12	$10706038142052878970311146962646277721 = 3^7 \cdot 7^7 \cdot 11^4 \cdot 19^2$ $\cdot 23^2 \cdot 31 \cdot 43 \cdot 47 \cdot 280199 \cdot 121105961$
13	$530588758323899225681861502684757146635241 = 3^6 \cdot 7^7 \cdot 11^4$ $\cdot 19^2 \cdot 23^2 \cdot 31 \cdot 43 \cdot 47 \cdot 383 \cdot 13172974161370579$

Remark. For a discussion of the remarkable way in which the β's in the next table factor, see [Ca3], [Ca5], and the wonderful paper of Hurwitz [Hu].

Table 5.2 $\beta_n^{(2n)}$, $0 \leq n \leq 13$

n	$\beta_n^{(2n)}$
0	1
1	$6 = 2 \cdot 3$
2	$336 = 2^4 \cdot 3 \cdot 7$
3	$77616 = 2^4 \cdot 3^2 \cdot 7^2 \cdot 11$
4	$50916096 = 2^8 \cdot 3^2 \cdot 7^2 \cdot 11 \cdot 41$
5	$76307083776 = 2^9 \cdot 3^3 \cdot 7^4 \cdot 11^2 \cdot 19$
6	$226653840838656 = 2^{12} \cdot 3^3 \cdot 7^3 \cdot 11^2 \cdot 19 \cdot 23 \cdot 113$
7	$1207012936807028736 = 2^{11} \cdot 3^4 \cdot 7^4 \cdot 11^2 \cdot 19 \cdot 23 \cdot 223 \cdot 257$
8	$10696277678308486742016 = 2^{16} \cdot 3^4 \cdot 7^5 \cdot 11^3 \cdot 19 \cdot 23 \cdot 31$ $\cdot 61 \cdot 109$
9	$148900090457044541209706496 = 2^{17} \cdot 3^5 \cdot 7^5 \cdot 11^4 \cdot 19^2 \cdot 23 \cdot$ $\cdot 31^2 \cdot 2381$
10	$3110043187741674836967136690176 = 2^{20} \cdot 3^5 \cdot 7^6 \cdot 11^3 \cdot 19^4 \cdot 23$ $\cdot 31 \cdot 397 \cdot 2113$
11	$93885206124269301790338015801901056 = 2^{20} \cdot 3^6 \cdot 7^6 \cdot 11^4$ $\cdot 19^2 \cdot 23^2 \cdot 31 \cdot 43 \cdot 241 \cdot 1162253$
12	$3970859549814416912519992571903015387136 = 2^{24} \cdot 3^6$ $\cdot 7^7 \cdot 11^4 \cdot 19^2 \cdot 23^2 \cdot 31 \cdot 43 \cdot 47 \cdot 157 \cdot 1613 \cdot 8887$
13	$229208148540736805825727314928056159925436416 = 2^{25} \cdot 3^7$ $\cdot 7^7 \cdot 11^5 \cdot 19^2 \cdot 23^2 \cdot 31 \cdot 43^2 \cdot 47 \cdot 61 \cdot 113 \cdot 127 \cdot 52288$

Chapter 6

Inner Products, Integrals, and Moments – Favard's Theorem

This chapter reviews the basic definitions and theorems that form a general mathematical framework for discussing the orthogonality of a sequence of real polynomials. This makes the book somewhat more self-contained and perhaps more helpful to readers who are not familiar with some of the basic material about orthogonal polynomials. The important theorem of Favard, which is the theoretical basis for orthogonality proofs in this work, is included. There is also a simple proof that a primary sequence is not orthogonal, as well as a formula that expresses a polynomial in an orthogonal sequence as a determinant involving the moments of that sequence.

Definition 6.1. Let P be the real vector space $\mathbb{R}[x]$ and P' be its algebraic dual. We will say a sequence $\{P_n\}_{n=0}^{\infty} \subset P$ is "simple" if and only if $P_0 \neq 0$ and $\deg(P_n) = n$, $n \geq 0$. (Observe that a simple sequence in P is a Hamel basis of P.) An inner product on $P, (*,*) : P \times P \longrightarrow \mathbb{R}$, is called "proper" if and only if $(f, g) = (fg, 1)$ for all $f, g \in P$. **Definition 6.2.** An "orthogonal sequence" in P is a sequence $\{P_n\}_{n=0}^{\infty}$ which has the following two properties:

(i) $\{P_n\}_{n=0}^{\infty}$ is simple;

(ii) There exists a proper inner product on P with the property that

$(P_m, P_n) = 0$ for all $m, n \geq 0$ and $m \neq n$.

Remark. Let $\{P_n\}_{n=0}^{\infty} \subset P$ be an orthogonal sequence. Then there exists a unique, proper, inner product on P such that $(P_m, P_n) = 0$, for all $m, n \geq 0$ and $m \neq n$, and $(1, 1) = 1$.

Definition 6.3. Let $\mu : \underline{B}(\mathbb{R}) \longrightarrow [0, \infty]$ be a Borel measure and let $x_0 \in \mathbb{R}$. Then μ is "locally null at x_0" if and only if there exist $a, b \in \mathbb{R}$ such that $a < x_0 < b$ and $\mu((a, b)) = 0$. Also, let $\sigma(\mu) = \{x \in \mathbb{R} : \mu$ is not locally null at $x\}$.

Remark. The next theorem is "equivalent" to the later Theorem 6.7 by the theorem in [HeS, p. 331]. Also, Theorem 6.7 is "essentially" the same as the result given in [Ch, p. 58].

Theorem 6.4. (a) Let $(*, *) : P \times P \longrightarrow \mathbb{R}$ be a proper inner product on P such that $(1, 1) = 1$. Then there exists a (not necessarily unique) Borel probability measure $\mu : \underline{B}(\mathbb{R}) \longrightarrow [0, 1]$ such that

(i) $P \subset L_1(\mathbb{R}, \underline{B}(\mathbb{R}), \mu)$,

(ii) $(f, g) = \int_{\mathbb{R}} f g \, d\mu$, for all $f, g \in P$,

(iii) $\sigma(\mu)$ is infinite.

(b) Let $\mu : \underline{B}(\mathbb{R}) \longrightarrow [0, 1]$ be a Borel probability measure such that

(i) $P \subset L_1(\mathbb{R}, \underline{B}(\mathbb{R}), \mu)$

(ii) $\sigma(\mu)$ is infinite,

and let $(*, *) : P \times P \longrightarrow \mathbb{R}$ be defined by $(f, g) = \int_{\mathbb{R}} f g \, d\mu$, for all $f, g \in P$. Then $(*, *)$ is a proper inner product on P such that $(1, 1) = 1$.

Definition 6.5. Let X be a non-empty topological space, x_0 a point in X, Y a non-empty set, and $h : X \longrightarrow Y$. Then h is "locally constant at x_0" if and only if there exists an open neighborhood $U \subset X$ of x_0 such that $h|U$ is constant (i.e., $h(x) = h(x_0)$ for all $x \in U$). Also, $\sigma(h) = \{x \in X : h$ is not locally constant at $x\}$.

Definition 6.6. Let $h : \mathbb{R} \longrightarrow \mathbb{R}$ be a non-decreasing function. Then $\mu_h : \underline{B}(\mathbb{R}) \longrightarrow [0, \infty]$ is the Borel-Stieltjes measure defined by h (i.e., $\mu_h : \underline{B}(\mathbb{R}) \longrightarrow [0, \infty]$ is the unique Borel measure such that $\mu_h([a, b)) = h(b-) - h(a-)$, for all $a, b \in \mathbb{R}$ such that $a < b$).

Theorem 6.7. (a) Let $(*, *) : P \times P \longrightarrow \mathbb{R}$ be a proper inner product such that $(1, 1) = 1$. Then there exists a (not necessarily unique) bounded, non-decreasing function $h : \mathbb{R} \longrightarrow \mathbb{R}$ such that

(i) $\lim\limits_{x \to \infty} h(x) - \lim\limits_{x \to -\infty} h(x) = 1$,

(ii) $P \subset L_1(\mathbb{R}, \underline{B}(\mathbb{R}), \mu_h)$,

(iii) $(f, g) = \int_{\mathbb{R}} f g \, d\mu_h$, for all $f, g \in P$,

(iv) $\sigma(h)$ is infinite.

(b) Let $h : \mathbb{R} \longrightarrow \mathbb{R}$ be a bounded, non-decreasing function such that

  (i) $\lim\limits_{x \to \infty} h(x) - \lim\limits_{x \to -\infty} h(x) = 1$,

  (ii) $\mathcal{P} \subset L_1(\mathbb{R}, \underline{\mathcal{B}}(\mathbb{R}), \mu_h)$,

  (iii) $\sigma(h)$ is infinite,

and let $(*,*) : \mathcal{P} \times \mathcal{P} \longrightarrow \mathbb{R}$ be defined by $(f, g) = \int_{\mathbb{R}} f g \, d\mu_h$ for all $f, g \in \mathcal{P}$. Then $(*,*)$ is a proper inner product on $\mathcal{P}$ such that $(1, 1) = 1$.

Definition 6.8. Let $(*,*) : \mathcal{P} \times \mathcal{P} \longrightarrow \mathbb{R}$ be a proper inner product, and let $\{p_n\}_{n=0}^{\infty} \subset \mathcal{P}$ be defined by $p_n(x) = x^n$. The sequence $\{\mu_n\}_{n=0}^{\infty}$ of moments, using the above inner product, is defined by $\mu_n = (p_n, 1)$, $n \geq 0$.

The next theorem is the primary theorem for determining orthogonal sequences and is fundamental in the development that follows.

Theorem 6.9. Favard (see also [AsI, p. 7], [Ch, p. 21], [Na, p. 167], [Ne, p. 4], [Su, p. 481], [Vi, I-4].) Let $\{S_n\}_{n=0}^{\infty} \subset \mathcal{P}$ be a simple sequence of real polynomials and define $S_{-1} = 0$. Then the following two statements are equivalent:

(1) $\{S_n\}_{n=0}^{\infty}$ is an orthogonal sequence.

(2) There exist three sequences $\{\alpha_n\}_{n=0}^{\infty}$, $\{\beta_n\}_{n=0}^{\infty}$, and $\{\gamma_n\}_{n=0}^{\infty} \subset \mathbb{R}$ such that:

(6.1) $\qquad S_{n+1}(x) = (\alpha_n x + \beta_n) S_n(x) - \gamma_n S_{n-1}(x)$, $n \geq 0$,

and

(6.2) $\qquad\qquad\qquad \alpha_{n-1}\alpha_n\gamma_n > 0$, $n \geq 1$.

Furthermore, if $\{S_n\}_{n=0}^{\infty}$ has either of the above properties and $(*,*)_S$: $\mathcal{P} \times \mathcal{P} \longrightarrow \mathbb{R}$ is a proper inner product such that $(S_m, S_n)_S = 0$ for $m, n \geq 0$ and $m \neq n$, then

(6.3) $\qquad\qquad (S_n, S_n)_S = \dfrac{\alpha_0 \gamma_1 \gamma_2 \cdots \gamma_n}{\alpha_n} S_0^2 (1, 1)_S$, $n \geq 1$.

Using Theorem 6.9, we can now investigate the orthogonality of the sequence $\{f_n(x)\}_{n=0}^{\infty}$, generated using an $f \in \mathcal{F}$.

Theorem 6.10. If $f \in \mathcal{F}$, then the polynomial sequence $\{f_n(x)\}_{n=0}^{\infty}$ is not orthogonal.

Proof. Suppose that $\{f_n(x)\}_{n=0}^{\infty}$ is orthogonal and that $(*, *)$ is the associated inner product. Since we have $a_1 \neq 0$ and $f_0(x) = 1$, $f_1(x) = a_1 x$, $f_2(x) = a_1^2 x^2 + 2a_2 x$, so $f_2 = f_1^2 + \dfrac{2a_2}{a_1} f_1$, we find that

$$0 = \left(f_2, f_0\right) = \left(f_1^2 + \frac{2a_2}{a_1} f_1, 1\right) = \left(f_1^2, 1\right) + \frac{2a_2}{a_1}\left(f_1, 1\right) = (f_1, f_1) > 0,$$

a contradiction. □

Remark. The proof of the non-orthogonality of the sequence in Theorem 6.10 depends only on the first few terms of $\{f_n(x)\}_{n=0}^{\infty}$, which is typical of such arguments.

We next give a formula which expresses a polynomial in a *monic* orthogonal sequence as a determinant involving moments.

Theorem 6.11. Let $(*,*) : \mathcal{P} \times \mathcal{P} \longrightarrow \mathbb{R}$ be a proper inner product with moments $\mu_n = \mu(n) = (p_n, 1)$, $n \geq 0$. Also, let $\{S_m\}_{m=0}^{\infty} \subset \mathcal{P}$ be a simple sequence of monic polynomials, which is orthogonal with respect to $(*,*)$. Then, for $m \geq 1$,

$$(6.4) \qquad S_m(y) = \frac{1}{D_m} \begin{vmatrix} \mu_0 & \mu_1 & \cdots & \mu_m \\ \mu_1 & \mu_2 & \cdots & \mu_{m+1} \\ \vdots & \vdots & \ddots & \vdots \\ \mu_{m-1} & \mu_m & \cdots & \mu_{2m-1} \\ 1 & y & \cdots & y^m \end{vmatrix}_{(m+1)\times(m+1)} ,$$

where $D_1 = \mu_0$ and, for $m \geq 2$,

$$(6.5) \quad D_m = \begin{vmatrix} \mu_0 & \mu_1 & \cdots & \mu_{m-1} \\ \mu_1 & \mu_2 & \cdots & \mu_m \\ \vdots & \vdots & \ddots & \vdots \\ \mu_{m-1} & \mu_m & \cdots & \mu_{2m-2} \end{vmatrix}_{m\times m}$$

$$= \mu_0 \prod_{j=1}^{m-1} (S_j, S_j) = \mu_0^m \prod_{j=1}^{m-1} \gamma_j^{m-j},$$

where the γ_j is the positive coefficient in (6.1).

Proof. From [Ak2, p. 65, (3)], we have that (6.4) holds, where D_m is some number. Since $S_m(y)$ is monic, then D_m is equal to the coefficient of y^m in the expansion of the determinant in (6.4), i.e., the determinant given in (6.5). To see that $D_m > 0$, we can rewrite this determinant as

$$D_m = \begin{vmatrix} (1,1) & (y,1) & \cdots & (y^{m-1},1) \\ (y,1) & (y^2,1) & \cdots & (y^m,1) \\ \vdots & \vdots & \ddots & \vdots \\ (y^{m-1},1) & (y^m,1) & \cdots & (y^{2m-2},1) \end{vmatrix}$$

$$= \begin{vmatrix} (1,1) & (1,y) & \cdots & (1,y^{m-1}) \\ (y,1) & (y,y) & \cdots & (y,y^{m-1}) \\ \vdots & \vdots & \ddots & \vdots \\ (y^{m-1},1) & (y^{m-1},y) & \cdots & (y^{m-1},y^{m-1}) \end{vmatrix} ,$$

because the inner product is proper. The latter determinant, however, is a Gram determinant [Gr, p. 170], so it is positive, since the sequence $\{1, p_1, \cdots, p_{m-1}\}$ is linearly independent in $\mathcal{P}$.

Normalizing the polynomial sequence, we next have that

$$(6.6) \qquad \tilde{S}_m(y) = \frac{1}{\sqrt{(S_m, S_m)}} S_m(y).$$

On the other hand, from [Ak2, p. 67, (3′)], we find using (6.4) that

$$(6.7) \quad \tilde{S}_m(y) = \frac{1}{\sqrt{D_m D_{m+1}}} \begin{vmatrix} \mu_0 & \mu_1 & \cdots & \mu_m \\ \mu_1 & \mu_2 & \cdots & \mu_{m+1} \\ \vdots & \vdots & \ddots & \vdots \\ \mu_{m-1} & \mu_m & \cdots & \mu_{2m-1} \\ 1 & y & \cdots & y^m \end{vmatrix}$$

$$= \sqrt{\frac{D_m}{D_{m+1}}} S_m(y).$$

Combining equations (6.6) and (6.7), we obtain the recursion $D_{m+1} = (S_m, S_m) D_m$, $m \geq 1$, which implies in turn that

$$(6.8) \qquad D_{m+1} = D_1 \prod_{j=1}^{m} (S_j, S_j) = \mu_0 \prod_{j=1}^{m} (S_j, S_j).$$

For $m \geq 2$, reducing the index by one gives the first product in (6.5).

Now, since $S_m(y)$, $m \geq 0$ is monic, then $S_0(y) = 1$ and, from (6.1), $\alpha_m = 1$ for $m \geq 0$. Thus, (6.3) becomes

$$(S_j, S_j) = \gamma_1 \gamma_2 \cdots \gamma_j (1, 1) = \mu_0 \gamma_1 \gamma_2 \cdots \gamma_j.$$

Thus, from (6.5) we obtain

$$D_m = \mu_0 \prod_{j=1}^{m-1} (\mu_0 \gamma_1 \gamma_2 \cdots \gamma_j),$$

which gives the final product in (6.5). Equation (6.5) is trivial if we have $m = 1$. $\square$

We conclude this chapter by giving a linear functional characterization of a class of functions in $\mathcal{F}_1$.

Theorem 6.12. Let $f \in \mathcal{F}_0$ with $a_1 = 1$ and $c_2 \neq 0$ and let $\{G_m(x)\}_{m=0}^\infty$ and $\{H_m(x)\}_{m=0}^\infty$ be the secondary sequences derived from f. Then $f \in \mathcal{F}_1 \iff$ there exists a unique, non-zero linear functional $\Psi_G \in \mathcal{P}'$ with the three properties:

   (a) $\Psi_G(1) = 1$,

   (b) $\Psi_G(G_m G_n) = 0$, for all $m, n \geq 0$, $m \neq n$,

   (c) For each $n \geq 0$,

$$\begin{vmatrix} \Psi_G(x^0) & \Psi_G(x^1) & \cdots & \Psi_G(x^n) \\ \Psi_G(x^1) & \Psi_G(x^2) & \cdots & \Psi_G(x^{n+1}) \\ \vdots & \vdots & \ddots & \vdots \\ \Psi_G(x^n) & \Psi_G(x^{n+1}) & \cdots & \Psi_G(x^{2n}) \end{vmatrix}_{(n+1)\times(n+1)} \neq 0.$$

The same theorem holds with G replaced by H throughout.

Proof. ($\Longrightarrow$) From Table 1.2, we see that $G_0(x) = 1$ and $G_1(x) = x + c_1$. Since $f \in \mathcal{F}_1$, Theorem 3.5 and (3.14) imply for $n \geq 2$ that

(6.9) $G_n(x) =$
$$[x + (2n-1)^2 c_1] G_{n-1}(x) - 4(n-1)^2(2n-3)(2n-1) c_2 G_{n-2}(x).$$

If we put $\gamma_n = -(2n-1)^2 c_1$, $n \geq 1$, $\lambda_1 = 1$, $\lambda_n = 4(n-1)^2(2n-3)(2n-1) c_2$, $n \geq 2$ and $G_{-1}(x) = 0$, then (6.9) becomes

$$G_n(x) = (x - \gamma_n) G_{n-1}(x) - \lambda_n G_{n-2}(x), \ n \geq 1.$$

By the generalized Favard theorem [Ch, pp. 21–22], there exists a unique linear functional Ψ on the complex vector space $\mathcal{P}^{(\mathbb{C})} = \mathbb{C}[x]$ with properties (a)–(c). Define $\Psi_G = \Psi|\mathcal{P}$. Then, certainly Ψ_G has properties (a)–(c). It remains to show that $\Psi_G \in \mathcal{P}'$, i.e., $\Psi_G : \mathcal{P} \longrightarrow \mathbb{R}$.

Since $x^n = \sum_{j=0}^{n} \bar{\wp}_{2j+1}^{(2n+1)} G_j(x)$ by (2.9), we have that

$$\Psi_G(x^n) = \sum_{j=0}^{n} \bar{\wp}_{2j+1}^{(2n+1)} \Psi(G_j(x)) = \bar{\wp}_1^{(2n+1)} \in \mathbb{R},$$

since by condition (b), we see that $\Psi(G_0) = \Psi(1) = 1$ and $\Psi(G_n) = \Psi(G_n G_0) = 0$, $n \geq 1$. It follows by linearity that range(Ψ_G) $\subset \mathbb{R}$, so $\Psi_G \in \mathcal{P}'$.

The proof for Ψ_H is similar and uses (3.15) and (2.9).

($\Longleftarrow$) Suppose that $\Psi_G \in \mathcal{P}'$ has properties (a)–(c) and let $\Psi \in (\mathcal{P}^{(\mathbb{C})})'$ be such that $\Psi|\mathcal{P} = \Psi_G$. That such a Ψ exists is clear, since for $P \in \mathcal{P}^{(\mathbb{C})}$, we can write $P = Q + iR$, where $Q, R \in \mathcal{P}$. Then define $\Psi(P) = \Psi_G(Q) + i\Psi_G(R)$. The checking that Ψ is a complex linear functional is routine.

It follows from [Ch] (Definition 2.2, the Corollary on p. 9, and Exercise 3.3 on p. 17) that $\{G_m\}_{m=0}^{\infty}$ is the unique, monic orthogonal polynomial sequence (OPS) for Ψ (note that Exercise 3.3 implies that $\Psi_G(G_n^2) \neq 0$). Thus, by [Ch, Theorem 4.1, p. 18], there exist $\{\gamma_m\}_{m=1}^{\infty} \subset \mathbb{C}$ and $\{\lambda_m\}_{m=1}^{\infty} \subset \mathbb{C} \setminus \{0\}$ such that for $m \geq -1$,

$$G_{m+2}(x) = (x - \gamma_{m+2}) G_{m+1}(x) - \lambda_{m+2} G_m(x),$$

with $G_{-1}(x) = 0$ and $G_0(x) = 1$. If we substitute (1.36) into this recursion, by equating corresponding coefficients of x^j for $0 \leq j \leq m + 2$, we obtain the following relations: For $y = m + 1$,

(6.10) $$\varphi_{2n+3}^{(2n+5)} = \varphi_{2m+1}^{(2m+3)} - \gamma_{m+2}\,\varphi_{2m+3}^{(2m+3)},$$

and for $0 \leq j \leq m$, with $\varphi_{-1}^{(2m+3)} = 0$,

(6.11) $$\varphi_{2j+1}^{(2m+5)} = \varphi_{2j-1}^{(2m+3)} - \gamma_{n+2}\,\varphi_{2j+1}^{(2m+3)} - \lambda_{m+2}\varphi_{2j+1}^{(2m+1)}.$$

Since the φ's in these equations have been evaluated in Chapter 2, we can solve for γ_{m+2} and λ_{m+2}. Thus, in (6.10), since $\varphi_{2m+3}^{(2m+3)} = 1$ by (1.9), we have using (1.11) that

$$\gamma_{m+2} = \varphi_{2m+1}^{(2m+3)} - \varphi_{2m+3}^{(2m+5)} = -6(2m+3)^2\, a_3 = -(2m+3)^2\, c_1.$$

From (6.11), we have for $j = m$ that

$$\lambda_{m+2} = \varphi_{2m-1}^{(2m+3)} - \gamma_{m+2}\,\varphi_{2m+1}^{(2m+3)} - \varphi_{2m+1}^{(2m+5)}.$$

Evaluating these φ's by (1.11) and (1.30), we find that the result is $\lambda_{m+2} = 4(m+1)^2(2m+1)(2m+3)\, c_2$. Thus, for $m \geq 0$,

$$G_{m+2}(x) = [x+(2m+3)^2\, c_1]\, G_{m+1}(x) - 4(m+1)^2(2m+1)(2m+3)\, c_2\, G_m(x),$$

so $f \in \mathcal{F}_1$ by (3.14) and Theorem 3.5. The result for H is established in a similar way. $\square$

Chapter 7

The Functions $\mathcal{F}_2$ –
Orthogonal Sequences $\{G_m(z)\}$, $\{H_m(z)\}$

In this chapter we determine the subset of functions $\mathcal{F}_2$ in $\mathcal{F}_1$ for which the secondary sequences $\{G_m(y)\}_{m=0}^{\infty}$ and $\{H_m(y)\}_{m=0}^{\infty}$ are both orthogonal. Also, we show that $\mathcal{F}_2 \cap \mathcal{F}_2^{-1} = \emptyset$. Formulas are given for the inner product of two polynomials in these sequences and for the nth moments, particularly in Theorem 7.7, which expresses these moments as values of the $P_n(x, y)$ and $Q(x, y)$ polynomials. Theorem 7.8 gives formulas for the Maclaurin expansions of f^{-1} and $(f^{-1})^2$, where the coefficients are, respectively, G and H moments of the secondary sequences.

In Chapter 3, the set $\mathcal{F}_1$ was split into six sub-classes. The condition $c_2 > 0$, which defines $\mathcal{F}_2$ within $\mathcal{F}_1$, does not hold in three of these classes, so the remaining three sub-classes partition $\mathcal{F}_2$ into three parts, labeled Classes I, II, and III. At the end of the chapter, the functions in each of these three classes are represented by certain elliptic integrals.

Definition 7.1. With c_2 as defined in (3.3), let

$$(7.1) \qquad \qquad \mathcal{F}_2 = \{f \in \mathcal{F}_1 : c_2 > 0\}.$$

The next result deals with f^{-1}, when $f \in \mathcal{F}_2$.

Theorem 7.2. If $f \in \mathcal{F}_2$, then $f^{-1} \notin \mathcal{F}_2$.

Proof. Since $\mathcal{F}_2 \subset \mathcal{F}_1$, it follows from Theorem 3.8 that $f(t) = a_1 t$ is the

CHAPTER 7

only possible function with the required property. However, $c_2 = 0$ for this
function, so $a_1 t \notin \mathcal{F}_2$ by (7.1). $\square$

Theorem 7.3. Let $f \in \mathcal{F}_1$ and $a_1, c_1, c_2 \in \mathbb{R}$. Then the following statements are equivalent:

$$(7.2) \quad f \in \mathcal{F}_2 \stackrel{\text{def}}{=} \left\{ f : f(t) = a_1 \int_0^t \frac{du}{\sqrt{1 - c_1 u^2 + c_2 u^4}}, \ a_1 \neq 0, \ c_2 > 0 \right\},$$

$$(7.3) \qquad\qquad \{G_m(z)\}_{m=0}^\infty \text{ is orthogonal,}$$

$$(7.4) \qquad\qquad \{H_m(z)\}_{m=0}^\infty \text{ is orthogonal.}$$

Proof. $(7.2) \Longrightarrow (7.3), (7.4)$ Since $f \in \mathcal{F}_2$, then $f \in \mathcal{F}_1$, so by Theorem 3.5 the G and H sequences satisfy (3.14) and (3.15), respectively. Since $c_2 > 0$ by (7.1), Theorem 6.9, implies the two sequences are orthogonal.

$(7.3) \Longrightarrow (7.2)$ Since $\{G_m(z)\}_{m=0}^\infty$ is orthogonal, then by Theorem 6.9 we have for $n \geq -1$ and $G_{-1}(z) = 0$ that

$$(7.5) \qquad G_{n+2}(z) = (\alpha_{n+1} z + \beta_{n+1})G_{n+1}(z) - \gamma_{n+1}G_n(z),$$

for certain sequences $\{\alpha_{n+1}\}_{n=-1}^\infty$, $\{\beta_{n+1}\}_{n=-1}^\infty$, and $\{\gamma_{n+1}\}_{n=-1}^\infty$, where $\alpha_n \alpha_{n+1} \gamma_{n+1} > 0$, $n \geq 0$. To find what these sequences are, we substitute

$$G_n(z) = \sum_{j=0}^n \varphi_{2j+1}^{(2n+1)} z^j.$$

from (1.36) into (7.5). Equating the coefficients of z^{n+2}, z^{n+1}, and z^n on the two sides of the equation for $n \geq 0$ gives

$$(7.6) \qquad \varphi_{2n+5}^{(2n+5)} = \alpha_{n+1} \varphi_{2n+3}^{(2n+3)},$$

$$(7.7) \qquad \varphi_{2n+3}^{(2n+5)} = \alpha_{n+1} \varphi_{2n+1}^{(2n+3)} + \beta_{n+1} \beta_{2n+3}^{(2n+3)},$$

and

$$(7.8) \qquad \varphi_{2n+1}^{(2n+5)} = \alpha_{n+1} \varphi_{2n-1}^{(2n+3)} + \beta_{n+1} \varphi_{2n+1}^{(2n+3)} - \gamma_{n+1} \varphi_{2n+1}^{(2n+1)}.$$

Using (1.9) in (7.6) gives that $\alpha_n = a_1^2$, $n \geq 0$. Note, if we put (1.24) into (1.11), we obtain the auxiliary formula

$$(7.9) \qquad \varphi_n^{(n+2)} = n(n+1)(n+2) a_1^{n-1} a_3.$$

Next, using $\alpha_{n+1} = a_1^2$, (1.9), (7.9), and (3.3), we find that

$$(7.10) \qquad \beta_{n+1} = c_1 (2n+3)^2, \ n \geq 0.$$

Finally, substituting the values for α_{n+1} and β_{n+1} into (7.8) and using (1.30), (7.9), and (3.3), we find after some simplification that

$$(7.11) \qquad \gamma_{n+1} = 4\,(n+1)^2(2n+1)(2n+3)\,c_2, \ n \geq 0,$$

i.e., $G_n(z)$ satisfies (3.14) for $n \geq 0$. The formulas also hold for $n = -1$. Thus, $f \in \mathcal{F}_1$ by Theorem 3.5, and since $\gamma_{n+1} > 0$, then $c_2 > 0$, so $f \in \mathcal{F}_2$ by (7.1).

(7.4) $\implies$ (7.2) This implication is established in the same way as the previous one using (1.37) and (3.15). $\square$

Remarks.

1. Since $f^{-1} \notin \mathcal{F}_2$ when $f \in \mathcal{F}_2$ by Theorem 7.2, inverse barring does not hold in $\mathcal{F}_2$. It also follows from Theorem 7.3 that the sequences of polynomials $\{\tilde{G}_m(z)\}_{m=0}^{\infty}$ and $\{\bar{H}_m(z)\}_{m=0}^{\infty}$ that are derived from f^{-1} are not orthogonal. It is interesting to observe, nonetheless, that the polynomials in some of these non-orthogonal sequences have some of the properties of orthogonal sequences, such as having all real zeros and the separation property of orthogonal polynomials. There are many unanswered questions about the f^{-1} side, not the least of which is whether there is some property comparable to orthogonality that holds on that side (see Chapter 18, Questions 4 and 5).

2. In another direction, it is worth noting that the "rigidity" of f, i.e., that its coefficients are recursively determined from just the initial coefficients a_1, a_3, and a_5 as in Corollary 3.6, depends only on the fact that $f \in \mathcal{F}_1$, which is to say that the associated sequences $\{G_m(z)\}_1^{\infty}$ and $\{H_m(z)\}_1^{\infty}$, respectively, satisfy second-order recursions (3.14) and (3.15) for any c_2. The orthogonality of these two sequences is thus not a consequence of this rigidity, because orthogonality requires f to be in $\mathcal{F}_2$ with the condition $c_2 > 0$ holding in (3.14) and (3.15).

In the next theorem we compute the inner products of the G's and H's.

Theorem 7.4. If $f \in \mathcal{F}_2$, then for $m, n \geq 0$

$$(7.12) \qquad (G_m, G_n)_G = (m+n)!\,(m+n+1)!\,a_1^2\,c_2^m(1,1)_G\,\delta_{m,n}$$

and

$$(7.13) \qquad (H_m, H_n)_H = \frac{1}{2}\,(m+n+1)!\,(m+n+2)!\,a_1^4 c_2^m(1,1)_H\,\delta_{m,n}.$$

Proof. These results follow from (6.3), (3.14), and (3.15). $\square$

Note. In what follows we will take $(1,1)_G = (1,1)_H = 1$.

The next theorem shows an important relationship that exists between the G and H moments and the coefficients of $f^{-1}(t)$. Of particular note is (7.14) which expresses $\bar{a}_{2n+1}$ in terms of $\mu_G(n)$.

Theorem 7.5. If $f \in \mathcal{F}_2$, then for $n \geq 0$,

$$(7.14) \qquad \mu_G(n) = a_1\bar{\varphi}_1^{(2n+1)} = (2n+1)!\,a_1\bar{a}_{2n+1}$$

and

$$(7.15) \quad \mu_H(n) = a_1^2 \bar{\varphi}_2^{(2n+2)} = \frac{1}{2}(2n+2)! \, a_1^2 \sum_{j=0}^{n} \bar{a}_{2j+1} \, \bar{a}_{2n+1-2j}$$

$$= \frac{1}{2} \sum_{j=0}^{n} \binom{2n+2}{2j+1} \mu_G(j) \, \mu_G(n-j) = \sum_{j=0}^{n} \binom{2n+1}{2j} \mu_G(j) \, \mu_G(n-j).$$

Proof. (7.14) Using (2.9), the orthogonality of $\{G_m(z)\}_{m=0}^{\infty}$, Table 1.2, and (1.13), we have

$$\mu_G(n) = (z^n, 1)_G = \left(\sum_{j=0}^{n} \bar{\varphi}_{2j+1}^{(2n+1)} G_j(z), 1 \right)_G = \sum_{j=0}^{n} \bar{\varphi}_{2j+1}^{(2n+1)} (G_j(z), 1)_G$$

$$= \bar{\varphi}_1^{(2n+1)} (G_0(z), 1)_G = a_1 \bar{\varphi}_1^{(2n+1)} = (2n+1)! \, a_1 \bar{a}_{2n+1}.$$

(7.15) Similarly, using (2.9), the orthogonality of $\{H_m(z)\}_{m=0}^{\infty}$, (1.14), and Table 1.3, we have

$$\mu_H(n) = (z^n, 1)_H = \left(\sum_{j=0}^{n} \bar{\varphi}_{2j+2}^{(2n+2)} H_j(z), 1 \right)_H = \sum_{j=0}^{n} \bar{\varphi}_{2j+2}^{(2n+2)} (H_j(z), 1)_H$$

$$= \bar{\varphi}_2^{(2n+2)} (H_0(z), 1)_H = \frac{1}{2}(2n+2)! \, a_1^2 \sum_{j=0}^{n} \bar{a}_{2j+1} \, \bar{a}_{2n+1-2j}.$$

The third equality is established by using (7.14) in the preceding sum. The last equality follows from (1.31) with $r = 2$ and $s = 0$. $\quad\Box$

Note using (7.14), (7.15), and Table 2.2, that $\mu_G(0) = (1,1)_G = a_1 \bar{a}_1 = 1$ and $\mu_H(0) = (1,1)_H = a_1^2 \bar{a}_1^2 = 1$, as they should be. We also have the useful relations

$$(7.16) \qquad \mu_G(1) = -\frac{6a_3}{a_1^3} \text{ and } \mu_G(2) = \frac{120}{a_1^6}(3a_3^2 - a_1 a_5),$$

and for (3.3) and (3.8), $c_1 = -a_1^2 \mu_G(1)$, $c_2 = -\frac{a_1^4}{12}\left[\mu_G^2(1) - \mu_G(2) \right]$, and

$$A(t) = \frac{1}{12}\left[\mu_G(2) - \mu_G^2(1) \right](a_1 t)^4 + \mu_G(1)(a_1 t)^2 + 1.$$

From (7.15) we also have

$$(7.17) \qquad \mu_H(1) = 4\mu_G(1) = -\frac{24a_3}{a_1^3} \text{ and } \mu_H(2) = \frac{360}{a_1^6}(7a_3^2 - 2a_1 a_5).$$

The next theorem gives a recursive method for computing the moments $\mu_G(n)$ and $\mu_H(n)$.

Theorem 7.6. If $f \in \mathcal{F}_2$, then for $n \geq 0$,

$$(7.18) \qquad \mu_G(n+1) = -\frac{1}{a_1^{2n+3}} \sum_{j=0}^{n} \varphi_{2j+1}^{(2n+3)} \mu_G(j)$$

and

$$(7.19) \qquad \mu_H(n+1) = -\frac{1}{a_1^{2n+4}} \sum_{j=0}^{n} \varphi_{2j+2}^{(2n+4)} \mu_H(j),$$

where $\mu_G(0) = \mu_H(0) = 1$.

Proof. Using (7.14) in (2.11) and replacing n by $n+1$ gives

$$\sum_{j=0}^{n+1} \varphi_{2j+1}^{(2n+3)} \mu_G(j) = 0,$$

from which (7.18) follows by solving this equation for $\mu_G(n+1)$, using (1.9). Equation (7.19) is proved in a similar way using (2.12) and (1.9). $\square$

Note. Equations (7.18) and (7.19) give us an effective method for computing $\mu_G(n)$ and $\mu_H(n)$ without needing to know the weight functions or interval of orthogonality. We can compute the coefficients $\varphi_k^{(n)}$ in these recursions using (3.35). Also observe that the recursion in (7.19) allows us to compute $\mu_H(n)$ from only the previous μ_H's, in contrast to the recursion in (7.15).

The following table expresses $\mu_G(n)$ in terms of the ratios $r_n = \dfrac{a_n}{a_1^n}$, $n \geq 1$, using (7.14).

Table 7.1 $\mu_G(n)$, $0 \leq n \leq 5$

n	$\mu_G(n)$
0	1
1	$-6r_3$
2	$120(3r_3^2 - r_5)$
3	$-720r_3(30r_3^2 - 11r_5)$
4	$8640(393r_3^4 - 240r_3^2 r_5 + 35r_5^2)$
5	$-259200r_3(3r_3^2 - r_5)(829r_3^2 - 307r_5)$

The next theorem gives a second recursive method for computing $\mu_G(n)$ and $\mu_H(n)$ using the respective recursions for $P_n(x,y)$ and $Q_n(x,y)$ in (4.1) and (4.33).

Theorem 7.7. If $f \in \mathcal{F}_2$, then for $n \geq 0$,

$$(7.20) \quad \mu_G(n) = P_n\Big(\mu_G(1), \mu_G(2)\Big) = P_n\Big(-\frac{6a_3}{a_1^3}, \frac{120}{a_1^6}(3a_3^2 - a_1a_5)\Big),$$

$$(7.21) \quad \mu_H(n) = Q_n\Big(\mu_G(1), \mu_G(2)\Big) = Q_n\Big(-\frac{6a_3}{a_1^3}, \frac{120}{a_1^6}(3a_3^2 - a_1a_5)\Big),$$

and

$$(7.22) \quad \mu_H(n) = R_n\Big(\mu_H(1), \mu_H(2)\Big) = R_n\Big(-\frac{24a_3}{a_1^3}, \frac{360}{a_1^6}(7a_3^2 - 2a_1a_5)\Big).$$

Proof. (7.20) Write $\mu_n = \mu_G(n)$. From (7.14) we have $\bar{\varphi}_1^{(2n+1)} = \frac{\mu_n}{a_1} = (2n+1), \bar{a}_{2n+1}, \ n \geq 0$. From (7.14), Table 2.2, and (3.3) we find $\mu_1 = 6a_1\bar{a}_3 = 6a_1\big(-\frac{a_3}{a_1^4}\big) = -\frac{c_1}{a_1^2}$ and $\mu_2 = 120\,a_1\bar{a}_5 = \frac{120}{a_1^8}(-a_1^3a_5 + 3a_1^2a_3^2) = \frac{120}{a_1^6}(3a_3^2 - a_1a_5)$, so $c_1 = -a_1^2\,\mu_1$ and $c_2 = \frac{a_1^4}{12}\,(\mu_2 - \mu_1^2)$.

By using the barred version of (1.15), we find for $n \geq 2$ that

$$\bar{\varphi}_3^{(2n-1)} = \frac{(2n-2)!}{2}\sum_{j=2}^{2n-2}(2n-j-1)\bar{a}_{2n-j-1}\sum_{k=1}^{j-1}\bar{a}_k\,\bar{a}_{j-k}$$

$$= \frac{(2n-2)!}{2}\sum_{j=1}^{n-1}(2n-2j-1)\bar{a}_{2n-2j-1}\sum_{k=1}^{j}\bar{a}_{2k-1}\,\bar{a}_{2j-2k+1}.$$

But since $\bar{a}_{2n+1} = \frac{\mu_n}{a_1(2n+1)!}$, the latter equation becomes

$$\bar{\varphi}_3^{(2n-1)} = \frac{(2n-2)!}{2a_1^3}$$

$$\times \sum_{j=1}^{n-1}\frac{(2n-2j-1)\,\mu_{n-j-1}}{(2n-2j-1)!}\sum_{k=1}^{j}\frac{\mu_{k-1}}{(2k-1)!}\frac{\mu_{j-k}}{(2j-2k+1)!}$$

$$= \frac{1}{2a_1^3}\sum_{j=1}^{n-1}\binom{2n-2}{2j}\mu_{n-j-1}\sum_{k=1}^{j}\binom{2j}{2k-1}\mu_{k-1}\,\mu_{j-k}.$$

Now substitute the above results into (5.18), obtaining for $n \geq 2$ that

$$\mu_n = \mu_1\mu_{n-1} + \frac{\mu_2 - \mu_1^2}{2}\sum_{j=1}^{n-1}\binom{2n-2}{2j}\mu_{n-j-1}\sum_{k=1}^{j}\binom{2j}{2k-1}\mu_{k-1}\,\mu_{j-k}.$$

Now replace n by $n+2$ and make the change of variable $j \to n - j + 1$, so the equation becomes

$$\mu_{n+2} = \mu_1 \mu_{n+1}$$

$$+ \frac{\mu_2 - \mu_1^2}{2} \sum_{j=0}^{n} \binom{2n+2}{2j} \mu_j \sum_{k=1}^{n-j+1} \binom{2n-2j+2}{2k-1} \mu_{k-1} \mu_{n-j-k+1}$$

$$= \mu_1 \mu_{n+1} + (\mu_2 - \mu_1^2) \sum_{j=0}^{n} \binom{2n+2}{2j} \mu_j \left\{ \frac{1}{2} \sum_{k=0}^{n-j} \binom{2n-2j+2}{2k+1} \mu_k \mu_{n-j-k} \right\}.$$

Transforming the inner sum by (1.31) and letting $k \to n - j - k$, we have

(7.23) $\mu_{n+2} = \mu_1 \mu_{n+1}$

$$+ (\mu_2 - \mu_1^2) \sum_{j=0}^{n} \binom{2n+2}{2j} \mu_j \sum_{k=0}^{n-j} \binom{2n-2j+1}{2k} \mu_k \mu_{n-j-k}.$$

Now (7.20) holds for $n = 0, 1,$ and 2. Also, $\mu_G(n)$ satisfies (7.23), which is the same recursion that $P_n(x, y)$ satisfies in equation (4.1). Thus, $\mu_G(n) = P_n(\mu_G(1), \mu_G(2))$ with $x = \mu_G(1)$ and $y = \mu_G(2)$. The second form comes from (7.16).

(7.21) From (7.15), (7.20), and (4.2) we have that

$$\mu_H(n) = \sum_{j=0}^{n} \binom{2n+1}{2j} \mu_G(j) \mu_G(n-j) =$$

$$\sum_{j=0}^{n} \binom{2n+1}{2j} P_j(\mu_G(1), \mu_G(2)) P_{n-j}(\mu_G(1), \mu_G(2)) = Q_n(\mu_G(1), \mu_G(2)).$$

The second form comes from (7.16).

(7.22) From equation (7.21) and Table 4.2, we obtain the values $\mu_H(1) = Q_1(\mu_G(1), \mu_G(2)) = 4\mu_G(1)$. Thus, $\mu_G(1) = \frac{1}{4}\mu_H(1)$. Also, we have that $\mu_H(2) = Q_2(\mu_G(1), \mu_G(2)) = 10[\mu_G(1)]^2 + 6\mu_G(2)$ and $\mu_G(2) = \frac{1}{48}(-5\mu_H^2(1) + 8\mu_H(2))$. Finally, using (4.2), we have

$$\mu_H(n) = Q_n\left(\frac{\mu_H(1)}{4}, \frac{1}{48}(-5\mu_H^2(1) + 8\mu_H(2)) \right) = R_n(\mu_H(1), \mu_H(2)).$$

The second form comes from (7.17). □

If $f \in \mathcal{F}_2$, then the following theorem gives a formula for the respective Maclaurin expansions of $f^{-1}(t)$ and $(f^{-1}(t))^2$ in terms of the moments $\mu_G(n)$ and $\mu_H(n)$. It is worth noting in this case that $f^{-1} \in \mathcal{F}_0 - \mathcal{F}_1$ by Corollary 2.2 and Theorem 5.2.

Theorem 7.8. If $f \in \mathcal{F}_2$, then

$$(7.24) \quad f^{-1}(t) = \frac{1}{a_1} \sum_{n=0}^{\infty} \mu_G(n) \frac{t^{2n+1}}{(2n+1)!}$$

$$= \frac{1}{a_1} \sum_{n=0}^{\infty} P_n\left(-\frac{6a_3}{a_1^3}, \frac{120}{a_1^6}(3a_3^2 - a_1 a_5)\right) \frac{t^{2n+1}}{(2n+1)!}$$

and

$$(7.25) \quad (f^{-1}(t))^2 = \frac{2}{a_1^2} \sum_{n=0}^{\infty} \mu_H(n) \frac{t^{2n+2}}{(2n+2)!}$$

$$= \frac{2}{a_1^2} \sum_{n=0}^{\infty} Q_n\left(-\frac{6a_3}{a_1^3}, \frac{120}{a_1^6}(3a_3^2 - a_1 a_5)\right) \frac{t^{2n+2}}{(2n+2)!}.$$

Proof. These results follow immediately from Theorem 5.7 and Theorem 7.7. $\square$

We will now give the proof of Theorem 4.14 which we previously deferred. In this proof we will need the following simple lemma, which we give without proof.

Lemma 7.9. A polynomial $P(x,y) \in \mathbb{R}[x,y]$ that vanishes for each pair $(x,y) \in \mathbb{R}^2$ for which $y > x^2$ also vanishes for all $(x,y) \in \mathbb{R}^2$.

Proof of Theorem 4.14. First assume that $x, y \in \mathbb{R}$ such that $y > x^2$ and consider the function $f = f(t; x, y) \in \mathcal{F}_1$ with $a_1 = 1$, $a_3 = -\frac{x}{6}$, and $a_5 = \frac{10x^2 - y}{120}$. Then by (3.3), $c_1 = -x$ and $c_2 = \frac{y - x^2}{12} > 0$, so $f \in \mathcal{F}_2$ by (7.1). Let the sequences of orthogonal polynomials associated with f be $\{G_m(z; x, y)\}_{m=0}^{\infty}$ and $\{H_m(z; x, y)\}_{m=0}^{\infty}$. Then from the second equations in (7.20) and (7.21) we have for $n \geq 0$ that

$$(7.26) \qquad \mu_G(n; x, y) = P_n(x, y) \text{ and } \mu_H(n; x, y) = Q_n(x, y).$$

It follows from (7.12) and (7.13) that

$$(7.27) \qquad (G_j, G_j)_G = (2j)!\,(2j+1)!\left(\frac{y-x^2}{12}\right)^j$$

and

$$(7.28) \qquad (H_j, H_j)_H = \frac{1}{2}(2j+1)!\,(2j+2)!\left(\frac{y-x^2}{12}\right)^j.$$

(Observe the Note following the proof of Theorem 7.4.) Substituting the results from (7.26), (7.27), and (7.28) into (6.5) with $\mu_0 = 1$ (and still assuming that $y > x^2$) gives

$$(7.29) \quad D_P(x,y) = \prod_{j=1}^{m-1} (2j)!\,(2j+1)!\left(\frac{y-x^2}{12}\right)^j$$

$$= \left(\frac{y-x^2}{12}\right)^{\frac{m(m-1)}{2}} \prod_{j=1}^{2m-1} j!$$

and

$$(7.29) \quad D_Q(x,y) = \prod_{j=1}^{m-1} \frac{1}{2}\,(2j+1)!\,(2j+2)!\left(\frac{y-x^2}{12}\right)^j$$

$$= \frac{1}{2^m}\left(\frac{y-x^2}{12}\right)^{\frac{m(m-1)}{2}} \prod_{j=1}^{2m} j!.$$

By Lemma 7.9, equations (7.29) and (7.29) are valid for all $(x,y) \in \mathbb{R}^2$. □

Corollary 7.10. The functions $\mathcal{F}_2^{-1}$ can be parameterized by the set $S_2 = \{(a_1, x, y) \in \mathbb{R}^3 : y > x^2 \text{ and } a_1 \neq 0\}$.

Proof. Since $\mathcal{F}_2$ is the subset of $\mathcal{F}_1$ defined in (7.1), we see by Corollary 5.8 that $\mathcal{F}_2$ is parameterized by the set S_1 with the further condition that $c_2 > 0$. But, with $x = -\dfrac{c_1}{a_1^2}$ and $y = \dfrac{c_1^2 + 12c_2}{a_1^4}$ as in the proof of Corollary 5.8, we find that $y - x^2 = \dfrac{12c_2}{a_1^4} > 0$ if and only if $c_2 > 0$. □

In the above proof of Theorem 4.14, a function $f \in \mathcal{F}_2$ was introduced which depends on the two parameters $(x, y) \in \mathbb{R}^2$ such that $y > x^2$. The power series for this function was selected so that its associated moments would satisfy the equations $\mu_G(n; x, y) = P_n(x, y)$ and $\mu_H(n; x, y) = Q_n(x, y)$, $n \geq 0$. Since $f \in \mathcal{F}_1$, the function f is completely determined since by (3.25) the coefficients a_{2n+1} of its series depend recursively only on a_1, a_3, and a_5. This function appears in the next theorem.

Theorem 7.11. Let $x, y \in \mathbb{R}$ be such that the condition $y > x^2$ holds, and let $\{G_m(z; x, y)\}_{m=0}^{\infty}$ and $\{H_m(z; x, y)\}_{m=0}^{\infty}$ be the sequences of orthogonal polynomials associated with the function

$$f(t; x, y) = t - \frac{x}{6}t^3 + \frac{10x^2 - y}{120}t^5 - \cdots$$

in $\mathcal{F}_1$. Then for $m \geq 1$,

$$(7.31) \qquad G_m(z; x,y) = \frac{1}{D_P(x,y)} \begin{vmatrix} P_0(x,y) & \cdots & P_m(x,y) \\ P_1(x,y) & \cdots & P_{m+1}(x,y) \\ \vdots & & \vdots \\ P_{m-1}(x,y) & \cdots & P_{2m-1}(x,y) \\ 1 & \cdots & z^m \end{vmatrix}$$

and

$$(7.32) \quad H_m(z; x, y) = \frac{1}{D_Q(x,y)} \begin{vmatrix} Q_0(x,y) & \cdots & Q_m(x,y) \\ Q_1(x,y) & \cdots & Q_{m+1}(x,y) \\ \vdots & & \vdots \\ Q_{m-1}(x,y) & \cdots & Q_{2m-1}(x,y) \\ 1 & \cdots & z^m \end{vmatrix},$$

where we have the formulas

$$D_P(x,y) = \left(\frac{y-x^2}{12}\right)^{\frac{m(m-1)}{2}} \prod_{j=1}^{2m-1} j!$$

and

$$D_Q(x,y) = \frac{1}{2^m}\left(\frac{y-x^2}{12}\right)^{\frac{m(m-1)}{2}} \prod_{j=1}^{2m} j!.$$

Proof. Using (3.3), we have that $c_2 = \frac{1}{12}(y - x^2) > 0$, so it follows that $f \in \mathcal{F}_2$. The sequences $\{G_m(z; x,y)\}_{m=0}^\infty$ and $\{H_m(z; x,y)\}_{m=0}^\infty$ are orthogonal, using Theorem 7.3. Equation (7.31) follows by specializing (6.4) using that $\mu_G(n; x, y) = P_n(x,y)$ from (7.26) and the product form of $D_P(x,y)$ in (4.87). That the formula is true for all $x, y \in \mathbb{R}$ follows from Lemma 7.9. The proof of (7.32) is similar. $\square$

Because the integral in Definition 3.3 is an elliptic integral of the first kind, we will introduce a basic function of this type.

Definition 7.12. For $0 \le k \le 1$, let

$$(7.33) \quad F(t,k) = \operatorname{sn}^{-1}(t,k) = \int_0^t \frac{du}{\sqrt{(1-u^2)(1-k^2u^2)}}.$$

We next obtain an elliptic integral representation for functions in the three classes of functions in $\mathcal{F}_2$.

Since $c_2 > 0$, we can use (3.3) and rewrite the integral in (3.24) as

$$(7.34) \quad f(t) = \frac{a_1}{\sqrt{c_2}} \int_0^t \left[\left(u^2 - \frac{c_1}{2c_2}\right)^2 + \frac{2a_1^2}{c^2}\Delta\right]^{-\frac{1}{2}} du$$

and accordingly characterize the three Classes I, II, and III by the symbols $(c_2, \Delta) = (+, -), (+, +), (+, 0)$, respectively. The integrals in (7.37), (7.38), (7.39), (7.40), (7.41), (7.43), and (7.44) give a parametric form for the members of each class.

Class I $(+,-)$ $\Delta < 0$.
Here we have that $a_3 \ne 0$, since if $a_3 = 0$, then it would follow from (3.24) that $\Delta = \frac{1}{2}c > 0$, a contradiction. In this case we can rewrite (7.34) as

$$f(t) = \frac{a_1}{\sqrt{c_2}} \int_0^t \left[\left(u^2 - \frac{c_1}{2c_2}\right)^2 - \left(\frac{a_1\sqrt{2|\Delta|}}{c}\right)^2\right]^{-\frac{1}{2}} du$$

or

$$(7.35) \qquad f(t) = \frac{a_1}{\sqrt{c_2}} \int_0^t \frac{du}{\sqrt{(u^2 + \alpha_-)(u^2 + \alpha_+)}},$$

where

$$\alpha_\pm = -\frac{a_1}{c}(3a_3 \pm \sqrt{2|\Delta|}).$$

Since $\alpha_- \cdot \alpha_+ = \frac{a_1^2}{c} > 0$ by (3.4), it follows that $\alpha_- \cdot \alpha_+ \neq 0$ and that α_- and α_+ have the same sign. Thus, we distinguish the two cases: $\alpha_\pm < 0$ and $\alpha_\pm > 0$.

I (a) $\alpha_\pm < 0$.
There are two sub-cases here: $a_1 > 0$, $a_3 > \frac{1}{3}\sqrt{2|\Delta|}$ and $a_1 < 0$, $a_3 < -\frac{1}{3}\sqrt{2|\Delta|}$.

I(a)(i) $a_1 > 0$ and $a_3 > \frac{1}{3}\sqrt{2|\Delta|}$.
Here $\alpha_- < \alpha_+ < 0$. Then (7.35) becomes

$$(7.36) \qquad f(t) = \frac{a_1}{\sqrt{c_2}} \int_0^t \frac{du}{\sqrt{(|\alpha_-| - u^2)(|\alpha_+| - u^2)}}.$$

Put $u = \sqrt{|\alpha_-|}\,v$. Then, with $k^2 = \frac{\alpha_-}{\alpha_+}$, so $0 < k < 1$, we have

$$f(t) = \frac{a_1}{\sqrt{c_2|\alpha_+|}} \int_0^{\frac{t}{\sqrt{\alpha_-}}} \frac{dv}{\sqrt{(1 - v^2)(1 - k^2 v^2)}},$$

or by (7.33),

$$(7.37) \qquad f(t) = \frac{a_1}{\sqrt{c_2|\alpha_+|}} F\left(\frac{t}{\sqrt{|\alpha_-|}}, k\right).$$

I (a) (ii) $a_1 < 0$ and $a_3 < -\frac{1}{3}\sqrt{2|\Delta|}$.
Here $|\alpha_-| > |\alpha_+|$. The integral in this case is the same as in II(a)(i), because the integrand in (7.36) is symmetric in α_- and α_+. Thus, taking $k^2 = \frac{\alpha_+}{\alpha_-}$, we find accordingly that

$$(7.38) \qquad f(t) = \frac{a_1}{\sqrt{c_2|\alpha_-|}} F\left(\frac{t}{\sqrt{|\alpha_+|}}, k\right), \quad 0 < k < 1.$$

I(b) $\alpha_\pm > 0$.
There are also two sub-cases here: $a_1 > 0$, $a_3 < -\frac{1}{3}\sqrt{2|\Delta|}$ and $a_1 < 0$, $a_3 > \frac{1}{3}\sqrt{2|\Delta|}$.

I(b)(i) $a_1 > 0$ and $a_3 < -\frac{1}{3}\sqrt{2|\Delta|}$.

Here $\alpha_- > \alpha_+ > 0$, so (7.35) becomes

$$f(t) = \frac{a_1}{\sqrt{c_2}} \int_0^t \frac{du}{\sqrt{(u^2+\alpha_-)(u^2+\alpha_+)}}.$$

Putting $u = \sqrt{\alpha_+}\dfrac{v}{\sqrt{1-v^2}}$ and setting $k^2 = \dfrac{\alpha_+}{\alpha_-}$ and $k' = \sqrt{1-k^2}$, we obtain by routine calculation that

$$f(t) = \frac{a_1}{\sqrt{c_2\alpha_-}} \int_0^{\frac{t}{\sqrt{t^2+\alpha_+}}} \frac{dv}{\sqrt{[1-v^2][1-(k')^2v^2]}}, \quad 0 < k < 1,$$

so by (7.33)

(7.39) $$f(t) = \frac{a_1}{\sqrt{c_2\alpha_-}}F\left(\frac{t}{\sqrt{t^2+\alpha_+}}, k'\right).$$

I (b) (ii) $a_1 < 0$ and $a_3 > \frac{1}{3}\sqrt{2|\Delta|}$.

Here $\alpha_+ > \alpha_- > 0$, so with $k^2 = \dfrac{\alpha_-}{\alpha_+}$ and again using the symmetry of the integrand in the alphas, we obtain

(7.40) $$f(t) = \frac{a_1}{\sqrt{c_2\alpha_+}}F\left(\frac{t}{\sqrt{t^2+\alpha_-}}, k'\right).$$

Class II $(+,+)$ $\Delta > 0$.

Here $c > 9a_3^2$. Let $\gamma \stackrel{\text{def}}{=} \sqrt{\dfrac{|a_1|}{\sqrt{c}}}$ and $\lambda_\pm \stackrel{\text{def}}{=} \dfrac{1}{2}\left(1 \pm \dfrac{1}{\sqrt{2}}\sqrt{1+\dfrac{3a_1a_3}{|a_1|\sqrt{c}}}\right)$. It follows easily that $\lambda_+ > \lambda_- > 0$. From (7.34), we then have that

$$f(t) = \frac{a_1}{\sqrt{c_2}} \int_0^t \frac{du}{\sqrt{[\lambda_+(\gamma-u)^2+\lambda_-(u+\gamma)^2][\lambda_+(u+\gamma)^2+\lambda_-(\gamma-u)^2]}}.$$

Putting $u = \gamma\dfrac{1-x}{1+x}$, so that $x = \dfrac{\gamma-u}{\gamma+u}$, $\gamma-u = x(u+\gamma)$, and $u+\gamma = \dfrac{2\gamma}{1+x}$, we get that

$$f(t) = \frac{a_1\gamma}{2\sqrt{\lambda_-\lambda_+}} \int_r^1 \frac{dx}{\sqrt{(x^2+k^2)(x^2+\frac{1}{k^2})}},$$

where $r = \dfrac{\gamma-t}{\gamma+t}$, $k^2 = \dfrac{\lambda_-}{\lambda_+}$ for $0 < k < 1$.

Now set $x = \dfrac{kv}{\sqrt{1-v^2}}$. We then have that $v = \dfrac{x}{\sqrt{x^2+k^2}}$ and

$$f(t) = \frac{a_1 \gamma}{2\sqrt{\lambda_- \lambda_+}} \int_r^s \frac{dv}{\sqrt{(1-v^2)(1-(1-k^2)v^2)}},$$

where $r = \dfrac{\gamma - t}{\sqrt{(\gamma-t)^2 + k^2(\gamma+t)^2}}$ and $s = \dfrac{1}{\sqrt{k^2+1}}$, or

(7.41) $\quad f(t) = \dfrac{a_1 \gamma}{2\sqrt{\lambda_- \lambda_+}} \times$

$$\left[F\left(\frac{1}{\sqrt{k^2+1}}, k' \right) - F\left(\frac{\gamma-t}{\sqrt{(\gamma-t)^2 + k^2(\gamma+t)^2}}, k' \right) \right].$$

Class III $(+, 0)$ $\Delta = 0$.
Here we have $c_2 = \frac{1}{4}c_1^2$ and by (3.4) and (3.3) that

$$9a_3^2 = c > 0 \Longrightarrow a_3 \neq 0 \Longrightarrow c_1 \neq 0.$$

We distinguish the two sub-cases: $c_1 > 0$ and $c_1 < 0$.
III(a) $c_1 > 0$.
In this case (7.34) becomes

(7.42) $$f(t) = \frac{2a_1}{c_1} \int_0^t \frac{du}{\left(\sqrt{\frac{2}{c_1}}\right)^2 - u^2},$$

which by standard integration yields

(7.43) $$f(t) = a_1 \sqrt{\frac{2}{c_1}} \tanh^{-1}\left(\sqrt{\frac{c_1}{2}} t \right) = \frac{a_1}{\sqrt{2c_1}} \log\left(\frac{1 + \sqrt{\frac{c_1}{2}} t}{1 - \sqrt{\frac{c_1}{2}} t} \right).$$

III(b) $c_1 < 0$.
In this case (7.34) becomes

(7.44) $$f(t) = \frac{2a_1}{|c_1|} \int_0^t \frac{du}{u^2 + \left(\sqrt{\frac{2}{|c_1|}}\right)^2} = a_1 \sqrt{\frac{2}{|c_1|}} \tan^{-1}\left(\sqrt{\frac{|c_1|}{2}} t \right).$$

Note, however, that the classical inverse sinelemniscate function [WhW, p. 524]

(7.45) $$\mathrm{sl}^{-1}(t) \overset{\text{def}}{=} \int_0^t \frac{du}{\sqrt{1-u^4}}$$

is obtained from (3.24) by setting $c_1 = 0$ and $c_2 = -1$. However, since $c_2 < 0$ in this case, it follows that $\mathrm{sl}^{-1}(t) \in \mathcal{F}_1 - \mathcal{F}_2$, so this function does not generate sequences of orthogonal polynomials.

Chapter 8

The Functions in Classes I and II

In this chapter we discuss two families of functions that lie, respectively, in Classes I and II. The family in Class I, discussed in part (a), is $f(t, k) = \mathrm{sn}^{-1}(t, k)$, $0 < k < 1$. Besides a discussion of some of the properties of the primary sequence and determinant representations for the polynomials in the secondary sequences, there are in this section the Maclaurin expansions for $\mathrm{sn}\,(t, k)$ and $\mathrm{sn}^2(t, k)$. Theorem 8.11 shows when the factors $k^2 + 1$ and $k^4 + 14k^2 + 1$ divide the coefficients in these expansions. The section ends with Carlitz's results on discrete measures for the orthogonal secondary sequences.

The family in Class II, discussed in part (b), is defined by the integral $f^*(t, k)$ in (8.49). This family contains the inverse hyperbolic sinelemniscate function $\mathrm{slh}^{-1}(t)$ in (8.56). The main result of this section is the construction of measures for the two secondary sequences, modeled along the lines of Carlitz's construction mentioned in part (a).

8 (a) Class I functions – The elliptic functions sn (t, k)

In this section we investigate the sequences of polynomials derived from the Class I functions $\mathrm{sn}^{-1}(t, k)$. Although these functions were defined for $0 \leq k \leq 1$, in this section we consider $f(t, k) = \mathrm{sn}^{-1}(t, k)$ for only

$0 < k < 1$, where

$$\text{sn}^{-1}(t,k) = F(t,k) = \int_0^t \frac{du}{\sqrt{(1-u^2)(1-k^2 u^2)}}$$

(8.1)
$$= \sum_{m=0}^{\infty} k^m P_m \left(\frac{1}{2}\left(k + \frac{1}{k}\right)\right) \frac{t^{2m+1}}{2m+1} = t + \frac{1}{6}(k^2+1)t^3$$

$$+ \frac{1}{40}(3k^4 + 2k^2 + 3)\, t^5 + \frac{1}{112}(5k^6 + 3k^4 + 3k^2 + 5)\, t^7 + \cdots$$

and $P_n(x)$ is the nth Legendre polynomial (cf. [Ke, p. 48]). Here we have $a_1 = 1$, $a_3 = \frac{1}{6}(k^2+1)$, and $a_5 = \frac{1}{40}(3k^4 + 2k^2 + 3)$, so $c_1 = k^2 + 1$, $c_2 = k^2$, $\alpha_- = -1$ and $\alpha_+ = -\frac{1}{k^2}$ by (7.37), and $\Delta = -\frac{1}{8}(k')^4 < 0$. Note that the series expansion in (8.1) can be obtained by integrating equation (3.31), using $x = k^2 + 1$ and $y = 2k$.

Table 8.1 $\hat{P}_{2n}(k)$, $0 \le n \le 9$

n	$\hat{P}_{2n}(k)$
0	1
1	$\frac{1}{2}(k^2+1)$
2	$\frac{1}{8}(3k^4 + 2k^2 + 3)$
3	$\frac{1}{16}(k^2+1)(5k^4 - 2k^2 + 5)$
4	$\frac{1}{128}(35k^8 + 20k^6 + 18k^4 + 20k^2 + 35)$
5	$\frac{1}{256}(k^2+1)(63k^8 - 28k^6 + 58k^4 - 28k^2 + 63)$
6	$\frac{1}{1024}(231k^{12} + 126k^{10} + 105k^8 + 100k^6 + 105k^4 + 126k^2 + 231)$
7	$\frac{1}{2048}(k^2+1)(429k^{12} - 198k^{10} + 387k^8 - 212k^6 + 387k^4$ $-198k^2 + 429)$
8	$\frac{1}{32768}(6435k^{16} + 3432k^{14} + 2772k^{12} + 2520k^{10} + 2450k^8$ $+2520k^6 + 2272k^4 + 3432k^2 + 6435)$
9	$\frac{1}{65536}(k^2+1)(12155k^{16} - 5720k^{14} + 10868k^{12} - 6248k^{10}$ $+10658k^8 - 6248k^6 + 10868k^4 - 5720k^2 + 12155)$

The most direct way to compute the coefficients in the power series (8.1) is to modify the familiar recursion formula for $P_n(x)$ [GrR, p. 1026, 9.914, 1]:

(8.2) $(n+2)P_{n+2}(x) = (2n+3)xP_{n+1}(x) - (n+1)P_n(x),\ n \ge 0,$

where $P_0(x) = 1$ and $P_1(x) = x$. Setting

(8.3) $$\hat{P}_{2n}(k) \stackrel{\text{def}}{=} k^n P_n\left(\frac{1}{2}\left(k + \frac{1}{k}\right)\right),$$

then for $n \geq 1$ recursion (8.2) becomes

(8.4) $\quad (2n + 4)\hat{P}_{2n+4}(k) = (2n + 3)(k^2 + 1)\hat{P}_{2n+2}(k) - (2n + 2)k^2\hat{P}_{2n}(k),$

where $\hat{P}_0(k) = 1$ and $\hat{P}_2(k) = \frac{1}{2}(k^2 + 1)$. (cf. [Wr, p. 562])
There is also a formula for $\hat{P}_{2n}(k)$ [Ke, p. 480, (2)]:

(8.5) $$\hat{P}_{2n}(k) = \frac{1}{2^{2n}} \sum_{j=0}^{n} \binom{2j}{j}\binom{2n - 2j}{n - j} k^{2j}, \; n \geq 0.$$

The polynomial sequences we derive from $f(t, k) = \text{sn}^{-1}(t, k)$ will be denoted here by $\delta_n(x, k)$, $A_m(z, k)$, and $B_m(z, k)$, respectively, instead of the more general $f_n(x)$, $G_m(z; x, y)$, and $H_m(z; x, y)$. Also, the corresponding moments will be written as $\mu_{A(k)}(n)$ and $\mu_{B(k)}(n)$, respectively, instead of the general $\mu_G(n)$ and $\mu_H(n)$.

From (3.11), we find for $n \geq 0$ that the recursion for $\delta_n(x, k)$ is

(8.6) $\quad \delta_{n+4}(x, k) =$
$$\left(x^2 + (n + 2)^2(k^2 + 1)\right)\delta_{n+2}(x, k) - n(n + 1)^2(n + 2)\, k^2\, \delta_n(x, k),$$

where $\delta_0(x, k) = 1$, $\delta_1(x, k) = x$, $\delta_2(x, k) = x^2$, $\delta_3(x, k) = x^3 + (k^2 + 1)\, x$. Using this recursion (or Table 1.1), we obtain Table 8.2.

It appears in Table 8.2 that each polynomial coefficient of the powers of x is palindromic. This and other properties of $\delta_n(x, k)$ are established in the following theorems.

Theorem 8.1. For $n \geq 0$, we have

(8.7) $$k^n \delta_n\left(\frac{x}{k}, \frac{1}{k}\right) = \delta_n(x, k).$$

Proof. The result is readily checked from Table 8.2 for $n = 0, 1, 2, 3$. Assume that (8.7) is true for all $n < s + 4$ for some $s \geq 0$. Then from (8.6) we find that

$$k^{s+4}\delta_{s+4}\left(\frac{x}{k}, \frac{1}{k}\right) = \left(x^2 + (n + 2)^2(k^2 + 1)\right)k^{s+2}\delta_{s+2}\left(\frac{x}{k}, \frac{1}{k}\right)$$
$$- n(n + 1)^2(n + 2)k^2 \cdot k^s\delta_s\left(\frac{x}{k}, \frac{1}{k}\right)$$
$$= \left(x^2 + (n + 2)^2(k^2 + 1)\right)\delta_{s+2}(x, k)$$
$$- n(n + 1)^2(n + 2)\, k^2\delta_s(x, k) = \delta_{s+4}(x, k),$$

which proves the result by induction. $\square$

Next let

(8.8) $$\delta_n(x,k) = \sum_{j=0}^{n} d_j^{(n)}(k)\,x^j,\; n \geq 0.$$

Corollary 8.2. For $0 \leq j \leq n$,

(8.9) $$k^{n-j} d_j^{(n)}\left(\frac{1}{k}\right) = d_j^{(n)}(k).$$

Proof. From (8.8) and (8.7), we find that

$$\sum_{j=0}^{n} d_j^{(n)}(k)\,x^j = \delta_n(x,k) = k^n \delta_n\left(\frac{x}{k},\frac{1}{k}\right)$$

$$= k^n \sum_{j=0}^{n} d_j^{(n)}\left(\frac{1}{k}\right)\left(\frac{x}{k}\right)^j = \sum_{j=0}^{n}\left\{k^{n-j} d_j^{(n)}\left(\frac{1}{k}\right)\right\} x^j,$$

from which the result follows by equating the corresponding coefficients of x^j. $\square$

Table 8.2 $\delta_n(x,k)$, $0 \leq n \leq 9$

n	$\delta_n(x,k)$
0	1
1	x
2	x^2
3	$x(x^2 + k^2 + 1)$
4	$x^2[x^2 + 4(k^2 + 1)]$
5	$x[x^4 + 10(k^2 + 1)x^2 + 3(3k^4 + 2k^2 + 3)]$
6	$x^2[x^4 + 20(k^2 + 1)x^2 + 8(8k^4 + 7k^2 + 8]$
7	$x[x^6 + 35(k^2 + 1)x^4 + 7(37k^4 + 38k^2 + 37)x^2$ $+45(k^2 + 1)(5k^4 - 2k^2 + 5)]$
8	$x^2[x^6 + 56(k^2 + 1)x^4 + 112(7k^4 + 8k^2 + 7)x^2$ $+384(k^2 + 1)(6k^4 - k^2 + 6)]$
9	$x[x^8 + 84(k^2 + 1)x^6 + 42(47k^4 + 58k^2 + 47)x^4$ $+4(k^2 + 1)(3229k^4 + 86k^2 + 3229)x^2$ $+315(35k^8 + 20k^6 + 18k^4 + 20k^2 + 35)]$

Theorem 8.3. We have for $n \geq 0$ that

(8.10) $$d_n^{(n)}(k) = 1,$$

(8.11) $$d_n^{(n+2)}(k) = \frac{1}{6}n(n+1)(n+2)(k^2+1),$$

(8.12) $$d_0^{(n)}(k) = \delta_{n,0},$$

(8.13) $$d_1^{(2n+1)}(k) = \frac{(2n)!}{2^{2n}}\sum_{j=0}^{n}\binom{2j}{j}\binom{2n-2j}{n-j}k^{2j},$$

and

(8.14) $$d_2^{(2n+2)}(k) = \sum_{j=0}^{n}\binom{2n+1}{2j+1}d_1^{(2j+1)}(k)\,d_1^{(2n+1-2j)}(k)$$

$$= \frac{1}{2}\sum_{j=0}^{n}\binom{2n+2}{2j+1}d_1^{(2j+1)}(k)\,d_1^{(2n+1-2j)}(k).$$

Also,

(8.15) $$d_{r+1}^{(n+1)}(k) = \sum_{j=r}^{n}\binom{n}{j}d_r^{(j)}(k)\,d_1^{(n+1-j)}(k),\ 0\le r\le n,$$

(8.16) $$d_j^{(n)}(k) = 0,\ 0\le j< n,\ n-j\ \text{odd},$$

and for $n \ge 1$,

(8.17) $$d_1^{(n+4)}(k) = (n+2)^2(k^2+1)\,d_1^{(n+2)}(k) - n(n+1)^2(n+2)\,k^2 d_1^{(n)}(k).$$

For $2 \le j \le n$, we have

(8.18) $$d_j^{(n+4)}(k) = d_{j-2}^{(n+2)}(k) + (n+2)^2(k^2+1)\,d_j^{(n+2)}(k)$$
$$- n(n+1)^2(n+2)\,k^2 d_j^{(n)}(k),$$

(8.19) $$d_j^{(n)}(k) = \sum_{s=0}^{\frac{n-j}{2}}a_s(n,j)\,k^{2s},\ a_s(n,j)\in Z^+,\ n-j\ \text{even},$$

and

(8.20) $$d_j^{(n)}(k)\ \text{is palindromic},\ 0\le j\le n,\ n-j\ \text{even}.$$

Proof. Since $a_1 = 1$ and $a_3 = \frac{1}{6}(k^2+1)$, then using (1.24), equations (8.10)–(8.12) follow from (1.9), (1.11), and (1.12), respectively. Equation

(8.13) follows from (1.13) and (1.24) and the fact that $a_{2n+1} = \dfrac{\hat{P}_{2n}(k)}{2n+1}$, using (8.5). Equation (8.14)–(8.16) are specializations of (1.27), (1.18), and (1.25), respectively. Equations (8.17) and (8.18) come from substituting (8.8) into (3.11) and equating corresponding coefficients. Equation (8.19) follows inductively from (8.15), using the fact that the initial terms have the stated form. Finally, (8.20) follows from (8.9). $\square$

Theorem 8.4. If $2 \le j \le n$, where $n-j$ is even, and $\alpha_r = \dfrac{\hat{P}_{2r}(k)}{2r+1}$, $r \ge 0$, then

$$(8.21) \quad d_j^{(n)}(k) = \frac{n!}{j!} \sum_{s_{j-1}=0}^{\frac{n-j}{2}} \sum_{s_{j-2}=0}^{s_{j-1}} \sum_{s_{j-3}=0}^{s_{j-2}}$$

$$\cdots \sum_{s_2=0}^{s_3} \sum_{s_1=0}^{s_2} \alpha_{s_1} \alpha_{s_2-s_1} \alpha_{s_3-s_2} \cdots \alpha_{s_{j-1}-s_{j-2}} \alpha_{\frac{n-j}{2}-s_{j-1}}.$$

Proof. Using (8.1) and (8.3), we have for $j \ge 2$ that

$$\left(\mathrm{sn}^{-1}(t,k) \right)^j = \left(\sum_{m=0}^{\infty} \alpha_m t^{2m+1} \right)^j = \sum_{n=j}^{\infty} \sigma_n^{(j)} t^n$$

as in (1.2). Multiplying the series by itself j times, we find that

$$\sum_{n=j}^{\infty} \sigma_n^{(j)} t^n = \sum_{n=0}^{\infty} \beta_n t^{2n+j},$$

where by induction we see that

$$\beta_n = \sum_{s_{j-1}=0}^{n} \sum_{s_{j-2}=0}^{s_{j-1}} \sum_{s_{j-3}=0}^{s_{j-2}}$$

$$\cdots \sum_{s_2=0}^{s_3} \sum_{s_1=0}^{s_2} \alpha_{s_1} \alpha_{s_2-s_1} \alpha_{s_3-s_2} \cdots \alpha_{s_{j-1}-s_{j-2}} \alpha_{n-s_{j-1}}.$$

Thus, for $n \ge j$, we have that $\sigma_n^{(j)} = \beta_{\frac{n-j}{2}}$. But then, by (1.8), we obtain $d_j^{(n)}(k) = \dfrac{n!}{j!} \sigma_n^{(j)} = \dfrac{n!}{j!} \beta_{\frac{n-j}{2}}$. $\square$

As an example consider $d_6^{(9)}(k) = 336 \left\{ 10[\hat{P}_2(k)]^2 + 9\hat{P}_4(k) \right\}$.

Corollary 8.5. For $m \ge 1$, we have

$$(8.22) \qquad d_2^{(2m)}(k) = (2m-1)! \sum_{i=0}^{m-1} \frac{\hat{P}_{2i}(k)\hat{P}_{2m-2i-2}(k)}{2i+1}.$$

Proof. From (8.21), we have

$$d_2^{(2m)}(k) = \frac{(2m)!}{2} \sum_{i=0}^{m-1} \frac{\hat{P}_{2i}(k)\,\hat{P}_{2m-2i-2}(k)}{(2i+1)(2m-2i-1)}$$

$$= \frac{(2m-1)!}{2} \sum_{i=0}^{m-1} \left(\frac{1}{2i+1} + \frac{1}{2m-2i+1}\right) \hat{P}_{2i}(k)\,\hat{P}_{2m-2i-2}(k)$$

$$= \frac{(2m-1)!}{2} \left\{ \sum_{i=0}^{m-1} \frac{\hat{P}_{2i}(k)\,\hat{P}_{2m-2i-2}(k)}{2i+1} + \sum_{i=0}^{m-1} \frac{\hat{P}_{2i}(k)\,\hat{P}_{2m-2i-2}(k)}{2m-2i-2} \right\}$$

$$= (2m-1)! \sum_{i=0}^{m-1} \frac{\hat{P}_{2i}(k)\,\hat{P}_{2m-2i-2}(k)}{2i+1},$$

where the two penultimate sums are equal by $i \to m - i$. $\square$

Corollary 8.6. For $3 \le j \le n$ and $n - j$ even, we have

$$(8.23) \qquad d_j^{(n)}(k) = \frac{1}{j} \sum_{i=0}^{\frac{n-j}{2}} \binom{n}{j+2i-1} d_1^{(n-j-2i+1)}(k)\, d_{j-1}^{(j+2i-1)}(k).$$

Proof. With the notation of Theorem 8.4, let $\hat{n} = j + 2s_{j-1} - 1$, where $0 \le s_{j-1} \le \frac{n-j}{2}$, and $\hat{j} = j - 1$, so $\hat{n} \ge \hat{j} \ge 2$. Then by Theorem 8.4,

$$d_{\hat{j}}^{\hat{n}}(k) = d_{j-1}^{(j+2s_{j-1}-1)}(k)$$

$$= \frac{(j+2s_{j-1}-1)!}{(j-1)!} \sum_{s_{j-2}=0}^{s_{j-1}} \sum_{s_{j-3}=0}^{s_{j-2}} \cdots \sum_{s_2=0}^{s_3} \sum_{s_1=0}^{s_2} \alpha_{s_1} \alpha_{s_2-s_1} \cdots \alpha_{s_{j-1}-s_{j-2}}.$$

Again, using Theorem 8.4, we have

$$\frac{j!}{n!} d_j^{(n)}(k) = \sum_{s_{j-1}=0}^{\frac{n-j}{2}} \sum_{s_{j-2}=0}^{s_{j-1}} \cdots \sum_{s_1=0}^{s_2} \alpha_{s_1} \alpha_{s_2-s_1} \cdots \alpha_{s_{j-1}-s_{j-2}} \alpha_{\frac{n-j}{2}-s_{j-1}}$$

$$= \sum_{i=0}^{\frac{n-j}{2}} \frac{(j-1)!}{(j+2i-1)!}\, d_{j-1}^{(j+2i-1)}(k)\, \alpha_{\frac{n-j}{2}},$$

using $i = s_{j-1}$. Thus,

$$d_j^{(n)}(k) = \frac{1}{j} \sum_{i=0}^{\frac{n-j}{2}} \frac{n!}{(j+2i-1)!}\, d_{j-1}^{(j+2i-1)}(k)\, \alpha_{\frac{n-j}{2}}$$

$$= \frac{1}{j} \sum_{i=0}^{\frac{n-j}{2}} \frac{n!}{(j+2i-1)!}\, d_{j-1}^{(j+2i-1)}(k)\, \frac{d_1^{(n-j-2i+1)}(k)}{(n-j-2i+1)!}$$

$$= \frac{1}{j} \sum_{i=0}^{\frac{n-j}{2}} \binom{n}{j+2i-1} d_1^{(n-j-2i+1)}(k)\, d_{j-1}^{(j+2i-1)}(k). \quad \square$$

Observe that setting $j = 2$ and $n = 2m + 2$ in (8.23) gives the second equation in (8.14).

Recall, using (1.37) and (1.38), that for $m \geq 0$,

$$(8.24) \qquad A_m(x^2, k) = \frac{\delta_{2m+1}(x, k)}{x} \quad \text{and} \quad B_m(x^2, k) = \frac{\delta_{2m+2}(x, k)}{x^2}.$$

Table 8.3 $A_m(z, k)$, $0 \leq m \leq 5$

m	$A_m(z, k)$
0	1
1	$z + (k^2 + 1)$
2	$z^2 + 10(k^2 + 1)z + 3(3k^4 + 2k^2 + 3)$
3	$z^3 + 35(k^2 + 1)z^2 + 7(37k^4 + 38k^2 + 37)z$ $\quad + 45(5k^6 + 3k^4 + 3k^2 + 5)$
4	$z^4 + 84(k^2 + 1)z^3 + 42(47k^4 + 58k^2 + 47)z^2$ $\quad + 4(k^2 + 1)(3229k^4 + 86k^2 + 3229)z$ $\quad + 315(35k^8 + 20k^6 + 18k^4 + 20k^2 + 35)$
5	$z^5 + 165(k^2 + 1)z^4 + 462(19k^4 + 26k^2 + 19)z^3$ $\quad + 110(k^2 + 1)(1571k^4 + 514k^2 + 1571)z^2$ $\quad + 99(10679k^8 + 10932k^6 + 10922k^4 + 10932k^2 + 10679)z$ $\quad + 14175(k^2 + 1)(63k^8 - 28k^6 + 58k^4 - 28k^2 + 63)$

Table 8.4 $B_m(z, k)$, $0 \leq m \leq 5$

m	$B_m(z, k)$
0	1
1	$z + 4(k^2 + 1)$
2	$z^2 + 20(k^2 + 1)z + 8(8k^4 + 7k^2 + 8)$
3	$z^3 + 120(k^2 + 1)z^2 + 112(7k^4 + 8k^2 + 7)z$ $\quad + 384(k^2 + 1)(6k^4 - k^2 + 6)$
4	$z^4 + 120(k^2 + 1)z^3 + 336(13k^4 + 17k^2 + 13)z^2$ $\quad + 320(k^2 + 1)(164k^4 + 31k^2 + 164)z$ $\quad + 1152(128k^8 + 104k^6 + 99k^4 + 104k^2 + 128)$
5	$z^5 + 220(k^2 + 1)z^4 + 528(31k^4 + 44k^2 + 31)z^3$ $\quad + 3520(k^2 + 1)(139k^4 + 62k^2 + 139)z^2$ $\quad + 1408(3832k^8 + 4636k^6 + 4821k^4 + 4636k^2 + 3832)z$ $\quad + 23040(k^2 + 1)(640k^8 - 128k^6 + 603x^4 - 128x^2 + 640)$

Note. Formula (7.42) is a special case of (7.44) with $k = 1$. Also, since $c_2 = k^2 > 0$, observe that the limiting case $F(t,0) = \sin^{-1} t \in \mathcal{F}_1 - \mathcal{F}_2$. On the other hand, when $k = 1$, then $\Delta = 0$, which shows that Class III functions are related to these Class I functions.

The polynomial coefficients of the Maclaurin expansion of $\operatorname{sn}(t, k)$ and $\operatorname{sn}^2(t, k)$ have been obtained by various methods, though no simple formula for them is known (see [TaM, v.3, pp. 52–53, 62–63], [Sc2], [Du], [Wr], [Vi1]). Here we write these expansions for $0 \le k \le 1$ as

$$(8.25a) \qquad \operatorname{sn}(t, k) \stackrel{\text{def}}{=} \sum_{n=0}^{\infty} \left\{ -C_{2n+1}(k) \right\} \frac{t^{2n+1}}{(2n+1)!} =$$

$$(8.25b) \quad t - (k^2 + 1)\frac{t^3}{3!} + (k^4 + 14k^2 + 1)\frac{t^5}{5!}$$
$$- (k^2 + 1)(k^4 + 134k^2 + 1)\frac{t^7}{7!} + \cdots$$

and

$$(8.26a) \qquad \operatorname{sn}^2(t, k) \stackrel{\text{def}}{=} \sum_{n=0}^{\infty} D_{2n+2}(k) \frac{t^{2n+2}}{(2n+2)!} =$$

$$(8.26b) \quad t^2 - 8(k^2 + 1)\frac{t^4}{4!} + 16(2k^4 + 13k^2 + 2)\frac{t^6}{6!}$$
$$- 128(k^2 + 1)(k^4 + 29k^2 + 1)\frac{t^8}{8!} + \cdots .$$

In (8.25a), we have expressed the coefficients as $-C_{2n+1}(k)$, since $C_{2n+1}(1) = C_{2n+1}$, in the notation of Nörlund [No, p. 27] (cf. (9.3) and (9.1)).

Remark 8.7. By comparing (5.33) and (5.34), respectively, with (8.25a) and (8.26a), we have for $n \ge 0$ that

$$(8.27) \qquad C_{2n+1}(k) = -P_n(-k^2 - 1, k^4 + 14k^2 + 1)$$

and

$$(8.28) \qquad D_{2n+2}(k) = 2Q_n(-k^2 - 1, k^4 + 14k^2 + 1). \quad \square$$

In what follows, we will give three recursive methods for computing $C_{2n+1}(k)$. From (7.26) and (8.27), we have that

$$(8.29) \qquad \mu_{A(k)}(n) = -C_{2n+1}(k), \ n \ge 0.$$

Thus, specializing the recursion in (7.18) gives us the following recursion formula for $C_{2n+1}(k)$:

$$(8.30) \qquad C_{2n+3}(k) = -\sum_{j=0}^{n} \varphi_{2j+3}^{(2n+1)} C_{2j+1}(k), \quad 0 \le k \le 1, \, n \ge 0,$$

where $C_1(k) = -1$. The φ-coefficients in this formula can be computed recursively using (3.35).

The second method of computing $C_{2n+1}(k)$ and $D_{2n+2}(k)$ is given by the following recursion formulas.

Theorem 8.8. For $n \ge 0$, we have

$$(8.31) \quad C_{2n+5}(k) = -(k^2+1) C_{2n+3}(k)$$
$$+ 12k^2 \sum_{j=0}^{n} \binom{2n+2}{2j+2} C_{2n+1-2j}(k) \sum_{s=0}^{j} \binom{2j+1}{2s} C_{2s+1}(k) C_{2j+1-2s}(k),$$

where $C_1(k) = -1$ and $C_3(k) = k^2+1$. Also,

$$(8.32) \quad D_{2n+6}(k) =$$
$$- 4(k^2+1) D_{2n+4}(k) + 6k^2 \sum_{j=0}^{n} \binom{2n+4}{2j+2} D_{2j+2}(k) D_{2n-2j+2}(k),$$

where $D_2(k) = 2$ and $D_4(k) = -8(k^2+1)$.

Proof. Set $x = -(k^2+1)$ and $y = k^4 + 14k^2 + 1$. Then the recursion in (8.31) comes from (4.4) using (4.2), while that in (8.32) comes from (4.33). $\square$

Remark. Observe that the identities that involve both $C_{2n+2}(k)$ and $D_{2n+2}(k)$ can be obtained from (4.1), (4.4)–(4.6), and (4.33)–(4.42) by the substitution $x = -k^2 - 1$ and $y = k^4 + 14k^2 + 1$ and then using Theorem 8.7.

The third method for computing $C_{2n+1}(k)$ is due to Wrigge [Wr, Theorem 8].

Theorem 8.9. For $n \ge 1$,

$$(8.33) \quad 2(n-2)C_{2n+1}(k) = 2n(k^2+1)C_{2n-1}(k)$$
$$- \sum_{j=2}^{n} \left[\binom{2n+2}{2j-1} - 3\binom{2n}{2j-1}\right] C_{2j-1}(k) C_{2n-2j+3}(k)$$
$$- (k^2+1) \sum_{j=2}^{n} \binom{2n}{2j-1} C_{2j-1}(k) C_{2n-2j+1}(k).$$

Proof. From (4.39), we have that

$$Q_{n+1} - xQ_n = 3(n+1)P_{n+1} + \frac{3}{2} \sum_{j=1}^{n} \binom{2n+2}{2j+1} P_j P_{n-j+1}.$$

Replacing both Q's using (4.6) gives

$$2(n-1)P_{n+1} = -2(n+1)xP_n$$
$$+ \sum_{j=1}^{n}\left[\binom{2n+4}{2j+1} - 3\binom{2n+2}{2j+1}\right]P_jP_{n-j+1} - x\sum_{j=1}^{n}\binom{2n+2}{2j+1}P_jP_{n-j}.$$

Replacing n by $n-1$ and then shifting the j index gives

$$2(n-2)P_n = -2nxP_{n-1}$$
$$+ \sum_{j=2}^{n}\left[\binom{2n+2}{2j-1} - 3\binom{2n}{2j-1}\right]P_{j-1}P_{n-j+1} - x\sum_{j=2}^{n}\binom{2n}{2j-1}P_{j-1}P_{n-j}.$$

Setting $x = -(k^2+1)$, $y = k^4 + 14k^2 + 1$, and using (8.27) gives the result. □

Corollary 8.10. For $n \geq 0$, we have

$$(8.34) \qquad C_{2n+1}(\sqrt{z}) \equiv (-1)^{n+1}(z+1)^n \pmod{12}.$$

Proof. True for $n = 0, 1$, and 2. For $n \geq 3$ the result follows by induction from (8.31). □

Using the recursions in (8.31) and (8.32), we can readily compute the following tables.

An examination of Tables 8.5 and 8.6 shows that $k^2 + 1$ and $k^4 + 14k^2 + 1$ divide certain of the coefficient polynomials in a periodic way. We prove these observations in the following theorem. It is not clear, however, if the quotient obtained by removing these factors is irreducible. (see Chapter 18, Question 2.)

Theorem 8.11. For $n \geq 0$, we have

$$(8.35) \qquad C_{2n+1}(k) \text{ and } D_{2n+2}(k) \text{ are palindromic (cf. [Ha, p. 252]),}$$

$$(8.36) \qquad deg\{C_{2n+1}(k)\} = deg\{D_{2n+2}(k)\} = 2n,$$

$$(8.37) \qquad k^2 + 1 \nmid C_{4n+1}(k) \text{ and } k^2 + 1 \| C_{4n+3}(k),$$

$$(8.38) \qquad k^2 + 1 \nmid D_{4n+2}(k) \text{ and } k^2 + 1 \| D_{4n+4}(k),$$

and

$$(8.39) \quad k^4 + 14k^2 + 1 \nmid C_{6n+1}(k)\, C_{6n+3}(k) \text{ and } k^4 + 14k^2 + 1 \| C_{6n+5}(k).$$

Proof. (8.35) The result follows from (4.85) with $z = k^2$ using (8.27) and (8.28).

(8.36) This follows from (4.86) with $z = k^2$.

(8.37) From (4.13), we have that $P_{2n}(x,y) = x^2 \sum_{s=0}^{n-1} a_s^{(2n)} x^{2n-2s-2} y^s + a_n^{(2n)} y^n$, where $a_n^{(2n)} \neq 0$ by (4.54). Then using (8.27), we have

$$-C_{4n+1}(k) = P_{2n}(-k^2 - 1, k^4 + 14k^2 + 1)$$
$$\equiv a_n^{(2n)}(k^4 + 14k^2 + 1)^n \not\equiv 0 \pmod{k^2 + 1},$$

since $(k^2 + 1, k^4 + 14k^2 + 1) = 1$. Also, from (4.13), we have that $P_{2n+1}(x,y) = x(\cdots + a_n^{(2n+1)} y^n)$, where $a_n^{(2n+1)} \neq 0$ by (4.55).

Table 8.5 $\quad -C_{2n+1}(k)$, $0 \leq n \leq 11$

n	$-C_{2n+1}(k)$
0	1
1	$-(k^2 + 1)$
2	$k^4 + 14k^2 + 1$
3	$-(k^6 + 135k^4 + 135k^2 + 1) = -(k^2 + 1)(k^4 + 134k^2 + 1)$
4	$k^8 + 1228k^6 + 5478k^4 + 1228k^2 + 1$
5	$-(k^{10} + 11069k^8 + 165826k^6 + 165826k^4 + 11069k^2 + 1)$ $= -(k^2 + 1)(k^4 + 14k^2 + 1)(k^4 + 11054k^2 + 1)$
6	$k^{12} + 99642k^{10} + 4494351k^8 + 13180268k^6 + 4494351k^4$ $+99642k^2 + 1$
7	$-(k^2 + 1)(k^{12} + 896802k^{10} + 115397871k^8 + 719289308k^6$ $+115397871k^4 + 896802k^2 + 1)$
8	$(k^4 + 14k^2 + 1)(k^{12} + 8071242k^{10} + 2836967631k^8$ $+7426506188k^6 + 2836967631k^4 + 8071242k^2 + 1)$
9	$-(k^2 + 1)(k^{16} + 72641336k^{14} + 74124438940k^{12}$ $+2429931455624k^{10} + 9536185484614k^8 + 2429931455624k^6$ $+74124438940k^4 + 72641336k^2 + 1)$
10	$k^{20} + 653772070k^{18} + 1859539731885k^{16} + 128453495887560k^{14}$ $+1171517154238290k^{12} + 2347836365864484k^{10}$ $+1171517154238290k^8 + 128453495887560k^6$ $+1859539731885k^4 + 653772070k^2 + 1$
11	$-(k^2 + 1)(k^4 + 14k^2 + 1)(k^{16} + 5883948656k^{14}$ $+46446978770380k^{12} + 5763908464385744k^{10}$ $+20111273799129574k^8 + 5763908464385744k^6$ $+46446978770380k^4 + 5883948656k^2 + 1)$

It then follows that

$$-C_{4n+3}(k) = P_{2n+1}(-k^2 - 1, k^4 + 14k^2 + 1)$$
$$= -(k^2 + 1)\left[\cdots + \alpha_n^{(2n+1)}(k^4 + 14k^2 + 1)^n\right],$$

so $k^2 + 1 \| C_{4n+3}(k)$. This also follows from (8.33) by induction.

Table 8.6 $D_{2n+2}(k)$, $0 \le n \le 11$

n	$D_{2n+2}(k)$
0	2
1	$-8(k^2 + 1)$
2	$16(2k^4 + 13k^2 + 2)$
3	$-128(k^2 + 1)(k^4 + 29k^2 + 1)$
4	$256(2k^8 + 251k^6 + 876k^4 + 251k^2 + 2)$
5	$-2048(k^2 + 1)(k^8 + 508k^6 + 4443k^4 + 508k^2 + 1)$
6	$2048(4k^{12} + 8178k^{10} + 199644k^8 + 513917k^6 + 199644k^4$ $+8178k^2 + 4)$
7	$-32768(k^2 + 1)(k^{12} + 8187k^{10} + 467601k^8 + 2250713k^6$ $+467601k^4 + 8187k^2 + 1)$
8	$65536(2k^{16} + 65527k^{14} + 8818040k^{12} + 102844678k^{10}$ $+220404668k^8 + 102844678k^6 + 8818040k^4 + 65527k^2 + 2)$
9	$-524288(k^2 + 1)(k^{16} + 131066k^{14} + 40139875k^{12} + 871223879k^{10}$ $+2899126879k^8 + 871223879k^6 + 40139875k^4 + 131066k^2 + 1)$
10	$524288(4k^{20} + 2097130k^{18} + 1460501460k^{16} + 62121595755k^{14}$ $+464623310280k^{12} + 880352352396k^{10} + 464623310280k^8$ $+62121595755k^6 + 1460501460k^4 + 2097130k^2 + 4)$
11	$-8388608(k^2 + 1)(k^{20} + 2097145k^{18} + 3296089665k^{16}$ $+255059819220k^{14} + 3089406237645k^{12} + 8022138626799k^{10}$ $+3089406237645k^8 + 255059819220k^6 + 3296089665k^4$ $+2097145k^2 + 1)$

(8.38) From (4.14), we have $Q_{2n}(x,y) = \sum_{s=0}^{n-1} \beta_s^{(2n)} x^{2n-2s} y^s + \beta_n^{(2n)} y^n$, so using (8.28), we find that

$$D_{4n+2}(k) = 2 Q_{2n}(-k^2 - 1, k^4 + 14k^2 + 1) \equiv 2\beta_n^{(2n)}(k^4 + 14k^2 + 1)^n$$
$$\not\equiv 0 \pmod{k^2 + 1},$$

since $\beta_0^{(0)} = 1$ and $\beta_n^{(2n)} \neq 0$ for $n \ge 1$ by (4.56). Similarly we have that

$$D_{4n+4}(k) = 2 Q_{2n+1}(-k^2 - 1, k^4 + 14k^2 + 1)$$
$$\equiv -2(k^2 + 1)\beta_n^{(2n+1)}(k^4 + 14k^2 + 1)^n \not\equiv 0 \pmod{(k^2 + 1)^2},$$

since $\beta_n^{(2n+1)} \neq 0$ by (4.57).

(8.39) From (4.13), it follows that

$$P_{3n}(x,y)P_{3n+1}(x,y) = [\alpha_0^{(3n)}x^{3n} + y(\cdots)][\alpha_0^{(3n+1)}x^{3n+1} + y(\cdots)]$$
$$\equiv \alpha_0^{(3n)}\alpha_0^{(3n+1)}x^{6n+1} \not\equiv 0 \pmod{y},$$

using (4.76), so from (8.27),

$$C_{6n+1}(k)\,C_{6n+3}(k) =$$
$$P_{3n}(-k^2 - 1,\, k^4 + 14k^2 + 1)P_{3n+1}(-k^2 - 1,\, k^4 + 14k^2 + 1) \equiv$$
$$- \alpha_0^{(3n)}\alpha_0^{(3n+1)}(k^2 + 1)^{6n+1} \not\equiv 0 \pmod{k^4 + 14k^2 + 1}.$$

Finally, since $P_{3n+2}(x,y) = y\,[\alpha_1^{(3n+2)}x^{3n} + \cdots]$ by (4.13), then by (8.27)

$$-C_{6n+5}(k) = P_{3n+2}(-k^2 - 1,\, k^4 + 14k^2 + 1)$$
$$\equiv (k^4 + 14k^2 + 1)\,\alpha_1^{(3n+2)}(k^2 + 1)^{3n} \pmod{(k^4 + 14k^2 + 1)^2},$$

from which the result follows. $\quad\square$

The next theorem gives a determinant representation for $A_m(z,k)$ and $B_m(z,k)$ and the evaluation of two related determinants.

Theorem 8.12. For $m \geq 1$ and $0 < k \leq 1$,

$$(8.40) \quad A_m(z,k) = \frac{1}{r_m(k)}\begin{vmatrix} C_1(k) & C_3(k) & \cdots & C_{2m+1}(k) \\ C_3(k) & C_5(k) & \cdots & C_{2m+3}(k) \\ \vdots & \vdots & \ddots & \vdots \\ C_{2m-1}(k) & C_{2m+1}(k) & \cdots & C_{4m-1}(k) \\ 1 & z & \cdots & z^m \end{vmatrix},$$

where

$$(8.41) \quad r_m(k) = \begin{vmatrix} C_1(k) & C_3(k) & \cdots & C_{2m-1}(k) \\ C_3(k) & C_5(k) & \cdots & C_{2m+1}(k) \\ \vdots & \vdots & \ddots & \vdots \\ C_{2m-1}(k) & C_{2m+1}(k) & \cdots & C_{4m-3}(k) \end{vmatrix} = (-1)^m k^{m(m-1)} \prod_{j=1}^{2m-1} j!$$

and

$$(8.42) \quad B_m(z,k) = \frac{1}{s_m(k)}\begin{vmatrix} D_2(k) & D_4(k) & \cdots & D_{2m+2}(k) \\ D_4(k) & D_6(k) & \cdots & D_{2m+4}(k) \\ \vdots & \vdots & \ddots & \vdots \\ D_{2m}(k) & D_{2m+2}(k) & \cdots & D_{4m}(k) \\ 1 & z & \cdots & z^m \end{vmatrix},$$

where

$$(8.43) \quad s_m(k) =$$

$$\begin{vmatrix} D_2(k) & D_4(k) & \cdots & D_{2m}(k) \\ D_4(k) & D_6(k) & \cdots & D_{2m+2}(k) \\ \vdots & \vdots & \ddots & \vdots \\ D_{2m}(k) & D_{2m+2}(k) & \cdots & D_{4m-2}(k) \end{vmatrix} = \frac{k^{m(m-1)}}{2m} \prod_{j=1}^{2m} j!.$$

Proof. These formulas follow from (7.31), (7.32), and Remark 8.7 by setting $x = -k^2 - 1$ and $y = k^4 + 14k^2 + 1$. $\square$

We next remark that the orthogonal sequences $\{A_m(z,k)\}_{m=0}^{\infty}$ and $\{B_m(z,k)\}_{m=0}^{\infty}$ do not have a weight function for $0 < k < 1$, but discrete measures on $(-\infty, 0]$ instead. This was proved by Carlitz [Ca2]. (The Carlitz results are given here with their original formula numbers, which are placed to the right on the line. An asterisk on a formula number indicates that a typographical error in the original formula has been corrected.):

For the sequence $\{A_m(z,k)\}_{m=0}^{\infty}$, $0 < k < 1$, we have the following:

At the points $z_j \in (-\infty, 0)$, where

$$(8.44) \qquad z_j \overset{\text{def}}{=} -\left(\frac{(2j+1)\pi}{2k}\right)^2, \ j \geq 0, \qquad (4.2)^*$$

define the measure

$$(8.45) \qquad \nu(\{z_j\}) = \frac{1}{k}\left(\frac{\pi}{K}\right)^2 (2j+1)\frac{q^{j+\frac{1}{2}}}{1-q^{2j+1}}, \ j \geq 0, \qquad (4.1)^*$$

where $q = e^{-\frac{\pi K'}{K}}$. Note, using the sum after (4.3) in [Ca2], that

$$\nu((-\infty, 0]) = \frac{1}{k}\left(\frac{\pi}{K}\right)^2 \sum_{j=0}^{\infty}(2j+1)\frac{q^{j+\frac{1}{2}}}{1-q^{2j+1}} = 1,$$

as it should be.

This discrete measure provides an inner product and we have the formula

$$(8.46) \qquad (A_m(**,k), A_n(**,k))_A = (2m)!\,(2m+1)!\,k^{2m}\delta_{m,n}. \qquad (4.4)^*$$

Also, by (8.29), we have

$$C_{2m+1}(k) = -\mu_{A(k)}(m) = \sum_{j=0}^{\infty}(z_j)^m \nu(\{z_j\})$$

$$= (-1)^{m+1}\frac{4}{k}\left(\frac{\pi}{2K}\right)^{2m+2}\sum_{j=0}^{\infty}(2j+1)^{2m+1}\frac{q^{j+\frac{1}{2}}}{1-q^{2j+1}}.$$

For the sequence $\{B_m(z)\}_{m=0}^{\infty}$, $0 < k < 1$, we have:

At the points $z_j \in (-\infty, 0)$, where

(8.47) $$z_j \overset{\text{def}}{=} -\left(\frac{\pi j}{k}\right)^2, \ j \ge 1,$$ (5.15)*

define the measure

(8.48) $$\nu(\{z_j\}) = \frac{1}{k^2}\left(\frac{\pi}{K}\right)^4 j^3 \frac{q^j}{1-q^{2j}}, \ j \ge 1.$$ (5.14)*

Here again (using the corrected sum after (5.7) in [Ca2]), we have

$$\nu((-\infty, 0]) = \frac{1}{k^2}\left(\frac{\pi}{K}\right)^4 \sum_{j=1}^{\infty} j^3 \frac{q^j}{1-q^{2j}} = 1,$$

as it should be. Also, we have the inner product

$$(B_m(**, k), B_n(**, k))_B = \frac{1}{2}(2m+1)!\,(2m+2)!\,k^{2m}\delta_{m,n}.$$ (5.17)*

The existence of these discrete measures implies that there are no weight functions for these two sequences of orthogonal polynomials on $(-\infty, 0]$. It also follows that the zeros of $A_m(z, k)$ and $B_m(z, k)$ are distinct and lie in the interval $(-\infty, 0)$. Consequently, they are Hurwitz polynomials [Mard, p. 181], [Ch, p. 27, Theorem 5.2].

8 (b) Class II functions – The polynomials $f^*(t, k)$

In this section we will consider the one-parameter family of Class II functions, determined by $a_1 = 1$, $a_3 = \frac{1}{3}(1 - 2k^2)$, and $a_5 = \frac{1}{5}(1 - 6k^2 + 6k^4)$, where $0 < k < 1$. Here $c_1 = 2(1 - 2k^2)$, $c_2 = 1$, and $\Delta = 2(kk')^2 > 0$.

The integral in (3.24) in this case is, using (3.29),

(8.49) $$f^*(t, k) = \int_0^t \frac{du}{\sqrt{1 - 2(1 - 2k^2)u^2 + u^4}}$$

$$= \int_0^t \frac{du}{\sqrt{(1 - 2k'u + u^2)(1 + 2k'u + u^2)}} = \sum_{m=0}^{\infty} P_m(1 - 2k^2)\frac{t^{2m+1}}{2m+1},$$

which is the function we previously encountered in (5.29). Here $P_m(x)$ is the Legendre polynomial.

The polynomials derived from $f^*(t, k)$ will be denoted by $\delta_n^*(x, k)$, $A_m^*(z, k)$, and $B_m^*(z, k)$. The first few of these polynomials are listed in the next two tables, obtained using the recursion formulas in (3.14) and (3.15), respectively.

We next evaluate $A_m^*(0, k)$ and $B_m^*(0, k)$.

Theorem 8.13. For $m \ge 0$ and $0 < k < 1$, we have that

(8.50) $$A_m^*(0, k) = (2m)!\,P_m(1 - 2k^2)$$

$$= (2m)!\sum_{j=0}^{m}(-1)^j\binom{2j}{j}\binom{m+j}{m-j}k^{2j},$$

and

(8.51) $B_m^*(0, k) =$

$$\frac{(2m+2)!}{2} \sum_{j=0}^{m} \frac{1}{(2j+1)(2m-2j+1)} P_j(1-2k^2) P_{m-j}(1-2k^2).$$

Table 8.7 $A_m^*(z, k)$, $0 \le m \le 5$

m	$A_m^*(z, k)$
0	1
1	$z - 4k^2 + 2$
2	$z^2 + 20(1 - 2k^2)z + 24(6k^4 - 6k^2 + 1)$
3	$z^3 + 70(1 - 2k^2)z^2 + 112(7 - 37k^2 + 37k^4)z$ $+720(1 - 2k^2)(1 - 10k^2 + 10k^4)$
4	$z^4 + 168(1 - 2k^2)z^3 + 336(19 - 94k^2 + 94k^4)z^2$ $+128(1 - 2k^2)(409 - 3229k^2 + 3229k^4)z$ $+40320(1 - 20k^2 + 90k^4 - 140k^6 + 70k^8)$
5	$z^5 + 330(1 - 2k^2)z^4 + 7392(4 - 19k^2 + 19k^4)z^3$ $+1760(1 - 2k^2)(457 - 3142k^2 + 3142k^4)z^2$ $+12672(423 - 6706k^2 + 28064k^4 - 42716k^6 + 21358k^8)z$ $+3628800(1 - 2k^2)(1 - 28k^2 + 154k^4 - 252k^6 + 126k^8)$

Table 8.8 $B_m^*(z, k)$, $0 \le m \le 5$

m	$B_m^*(z, k)$
0	1
1	$z + 8(1 - 2k^2)$
2	$z^2 + 40(1 - 2k^2)z + 8(128k^4 - 128k^2 + 23)$
3	$z^3 + 112(1 - 2k^2)z^2 + 224(11 - 56k^2 + 56k^4)z$ $+768(1 - 2k^2)(11 - 96k^2 + 96k^4)$
4	$z^4 + 240(1 - 2k^2)z^3 + 336(43 - 208k^2 + 208k^4)z^2$ $+640(1 - 2k^2)(359 - 2624k^2 + 2624k^4)z$ $+1152(563 - 9856k^2 + 42624k^4 - 65536k^6 + 32768k^8)$
5	$z^5 + 440(1 - 2k^2)z^4 + 1056(53 - 248k^2 + 248k^4)z^3$ $+28160(1 - 2k^2)(85 - 556k^2 + 556k^4)z^2$ $+1408(21757 - 319424k^2 + 1300416k^4 - 1961984k^6 + 980992k^8)z$ $+46080(1 - 2k^2)$ $\times (1627 - 38912k^2 + 202752k^4 - 327680k^6 + 163840k^8)$

Also, for $n \geq 0$,

$$(8.52) \quad B^*_{2n+1}(0,k) = (4n+4)! \sum_{i=0}^{n} \frac{4i+3}{2n+2i+3} \frac{\binom{2n-2i}{n-i}}{\binom{2n+2i+2}{n+i+1}}$$

$$\times \left\{ \sum_{j=0}^{i} \frac{\binom{2j}{j}\binom{4i-2j+2}{2i-j+1}}{(2n-2i+2j+1)(2n+2i-2j+3)} \right\} P_{2i+1}(1-2k^2),$$

and for $n \geq 1$,

$$(8.53) \quad B^*_{2n}(0,k) = \frac{(4n+2)!}{2} \left[\frac{1}{(2n+1)^3} + \sum_{i=1}^{n} \frac{4i+1}{2n+2i+1} \frac{\binom{2n-2i}{n-i}}{\binom{2n+2i}{n+i}} \right.$$

$$\times \left\{ \frac{\binom{2i}{i}^2}{(2n+1)^2} + 2\sum_{j=0}^{i-1} \frac{\binom{2j}{j}\binom{4i-2j}{2i-j}}{(2n-2i+2j+1)(2n+2i-2j+1)} \right\} P_{2i}(1-2k^2) \right].$$

Proof. (8.50) From (1.1) and (8.49), we obtain the value $a_{2m+1} = \dfrac{P_m(1-2k^2)}{2m+1}$, so that (1.40) implies

$$A^*_m(0,k) = (2m+1)!\, a_{2m+1} = (2m)!\, P_m(1-2k^2),$$

which gives the result using [WhW, p. 312].

(8.51) This follows in a similar way from (1.41).

(8.52), (8.53) In (8.51) let

$$(8.54) \qquad S_m = \frac{2}{(2m+2)!}\, B^*_m(0,k)$$

$$= \sum_{j=0}^{m} \frac{1}{(2j+1)(2m-2j+1)}\, P_j P_{m-j}, \quad m \geq 0.$$

From [As2, p. 39], we have for $m \geq 0$ and $0 \leq j \leq m$ that

$$(8.55) \qquad P_j P_{m-j} = \sum_{i=0}^{min\{j,m-j\}} C_i^{(m,j)} P_{m-2i}.$$

When $m = 2n+1$, $n \geq 0$, the sum in (8.54) is palindromic with no middle term, so using (8.55) the sum becomes

$$S_m = 2\sum_{j=0}^{n} \frac{1}{(2j+1)(4n-2j+3)} \sum_{i=0}^{j} C_i^{(m,j)} P_{2n-2i+1}.$$

Interchanging the order of summation and re-indexing gives

$$S_m = 2\sum_{i=0}^{n} \sum_{j=0}^{i} \frac{C_{n-i}^{(2n+1,n+j-i)}}{(2n-2i+2j+1)(2n+2i-2j+3)} P_{2i+1},$$

from which the result follows.

When $m = 2n$, $n \geq 1$, the sum in (8.54) is palindromic with a middle term, so we have

$$S_m = 2 \sum_{j=0}^{n-1} \frac{1}{(2j+1)(4n-2j+1)} P_j P_{2n-j} + \frac{P_n^2}{(2n+1)^2}.$$

Using (8.55), we obtain

$$S_m = 2 \sum_{j=0}^{n-1} \frac{1}{(2j+1)(4n-2j+1)} \sum_{i=0}^{j} C_i^{(2n,j)} P_{2n-2i}$$

$$+ \frac{1}{(2n+1)^2} \sum_{i=0}^{n} C_{n-i}^{(2n,n)} P_{2i}$$

$$= 2 \sum_{i=1}^{n} \sum_{j=0}^{i-1} \frac{C_{n-i}^{(2n,n+j-i)}}{(2n-2i+2j+1)(2n+2i-2j+1)} P_{2i}$$

$$+ \frac{1}{(2n+1)^2} \sum_{i=0}^{n} C_{n-i}^{(2n,n)} P_{2i},$$

which implies the result. $\square$

Note. If we evaluate $\Delta = 2(kk')^2$ at the endpoints $k = 0$ and $k = 1$, we get $\Delta = 0$ and the functions $f(t) = \tanh^{-1} t$ and $f(t) = \tan^{-1} t$, respectively. These are Class III functions, a class we will discuss in Chapters 10–14. If, on the other hand, we use the familiar singular modulus $k = \frac{1}{\sqrt{2}}$ (cf. [BoB, p. 26]), then $c_1 = 0$, $c_2 = 1$, and we have from (8.49) the following expressions for the inverse hyperbolic sinelemniscate function $\mathrm{slh}^{-1}(t)$ [BeB], [BoB, p. 259], [Ay]:

$$(8.56) \qquad f^*\left(t, \frac{1}{\sqrt{2}}\right) = \int_0^t \frac{du}{\sqrt{1+u^4}} = \sum_{n=0}^{\infty} \frac{(-1)^n}{2^{2n}} \binom{2n}{n} \frac{t^{4n+1}}{4n+1}.$$

Recall that we obtained the Maclaurin expansion of $\mathrm{slh}\,(t)$ and $\mathrm{slh}^2(t)$ in (5.56) and (5.57).

The next two tables contain some of the associated orthogonal polynomials.

Table 8.9 $A_m^*(z, \frac{1}{\sqrt{2}})$, $0 \le m \le 5$

m	$A_m^*(z, \frac{1}{\sqrt{2}})$
0	1
1	z
2	$z^2 - 12$
3	$z(z^2 - 252)$
4	$z^4 - 1512z^2 + 15120$
5	$z(z^4 - 5544z^2 + 1031184)$

Table 8.10 $B_m^*(z, \frac{1}{\sqrt{2}})$, $0 \le m \le 5$

m	$B_m^*(z, \frac{1}{\sqrt{2}})$
0	1
1	z
2	$z^2 - 72$
3	$z(z^2 - 672)$
4	$z^4 - 3024z^2 + 169344$
5	$z(z^4 - 9504z^2 + 4523904)$

We end this chapter with one of the main results of this work, the construction of countably additive measures, defined on $(-\infty, \infty)$, with respect to which the sequences $\{A_m^*(z, k)\}_{m=0}^\infty$ and $\{B_m^*(z, k)\}_{m=0}^\infty$, $0 < k < 1$, are orthogonal. The method used is modeled on Carlitz's development mentioned in the preceding section. (see also Theorems 8.21 and 8.29.)

In what follows, we will use the standard notation: For $0 < k < 1$,

$$K = K(k) = \int_0^1 \frac{du}{\sqrt{(1 - u^2)(1 - k^2 u^2)}}, \quad k' = \sqrt{1 - k^2}, \quad K' = K(k'),$$

$$q = e^{-\pi \frac{K'}{K}}, \quad q' = e^{-\pi \frac{K}{K'}} \quad E = E(k) = \int_0^1 \frac{\sqrt{1 - k^2 u^2}}{\sqrt{1 - u^2}} \, du, \quad E' = E(k').$$

We begin with the $\{A_m^*(z, k)\}_{m=0}^\infty$ sequence.

Theorem 8.14. For $f \in \mathcal{F}_0$, we have

$$(8.57) \quad f^{-1}[xf(t)] = \sum_{n=0}^\infty \left\{ \sum_{r=0}^n (2r + 1)! \, \bar{a}_{2r+1} \varphi_{2r+1}^{(2n+1)} x^{2r+1} \right\} \frac{t^{2n+1}}{(2n + 1)!}.$$

Proof. Since f^{-1} is odd, we can use (2.1) with Corollary 1.13, (1.2), and then (1.8) to obtain

$$f^{-1}[xf(t)] = \sum_{r=0}^{\infty} \bar{a}_{2r+1}x^{2r+1}[f(t)]^{2r+1} = \sum_{r=0}^{\infty} \bar{a}_{2r+1}x^{2r+1} \sum_{n=r}^{\infty} a_{2n+1}^{(2r+1)} t^{2n+1}$$

$$= \sum_{n=0}^{\infty} \left\{ \sum_{r=0}^{n} \bar{a}_{2r+1} a_{2n+1}^{(2r+1)} x^{2r+1} \right\} t^{2n+1}$$

$$= \sum_{n=0}^{\infty} \left\{ \sum_{r=0}^{n} \bar{a}_{2r+1} \left(\frac{(2r+1)!}{(2n+1)!} \varphi_{2r+1}^{(2n+1)} \right) x^{2r+1} \right\} t^{2n+1}$$

$$= \sum_{n=0}^{\infty} \left\{ \sum_{r=0}^{n} (2r+1)! \, \bar{a}_{2r+1} \varphi_{2r+1}^{(2n+1)} x^{2r+1} \right\} \frac{t^{2n+1}}{(2n+1)!}. \quad \square$$

Corollary 8.15. For $f^*(t,k)$ in (8.49), we have

$$(8.58) \quad (f^*)^{-1}[x f^*(t)] = x \left(\frac{\pi}{K'} \right)^2 \sum_{n=0}^{\infty} \left\{ 2 \sum_{m=1}^{\infty} \alpha_m A_n^* \left(-\left(\frac{\pi m x}{K'} \right)^2, k \right) \right.$$

$$\left. + \frac{1}{4} \int_0^{\infty} \frac{u}{\sinh\left(\frac{\pi u}{2} \right)} A_n^* \left(\left(\frac{\pi u x}{2K'} \right)^2, k \right) du \right\} \frac{t^{2n+1}}{(2n+1)!},$$

where $\alpha_m = \dfrac{m\,(q')^m}{1+(-q')^m}$, $m \geq 1$.

Proof. The proof of this formula comes from evaluating the coefficient, Ψ_n, inside the braces in (8.57) in the case when $f(t) = f^*(t)$. We have from (5.31) and [Han, p. 232, (14.17.6)] or [GrR, p. 912, 15] that

$$(8.59) \quad (f^*)^{-1}(t) = -\frac{d}{dt} \log(\mathrm{cn}\,(t,k'))d$$

$$= \frac{2\pi}{K'} \sum_{m=1}^{\infty} \kappa_m \sin\left(\frac{m\pi t}{K'} \right) + \frac{\pi}{2K'} \tan\left(\frac{\pi t}{2K'} \right),$$

where $\kappa_m = \dfrac{(q')^m}{1+(-q')^m}$.

From [GrR, p. 349, 2] or [GrH, p. 161, 3(a)], we have that

$$\tan(z) = \int_0^{\infty} \frac{\sinh(zu)}{\sinh\left(\frac{\pi u}{2} \right)} du, \quad |z| < \frac{\pi}{2},$$

so

$$(f^*)^{-1}(t) = \frac{2\pi}{K'} \sum_{m=1}^{\infty} \kappa_m \sin\left(\frac{m\pi t}{K'} \right) + \frac{\pi}{2K'} \int_0^{\infty} \frac{\sinh\left(\frac{\pi t u}{2K'} \right)}{\sinh\left(\frac{\pi u}{2} \right)} du.$$

Now

$$\sum_{m=1}^{\infty} \kappa_m \sin\left(\frac{m\pi t}{K'}\right) = \sum_{m=1}^{\infty} \kappa_m \sum_{r=0}^{\infty} \frac{(-1)^r}{(2r+1)!}\left(\frac{m\pi t}{K'}\right)^{2r+1}$$

$$= \sum_{r=0}^{\infty} \frac{(-1)^r}{(2r+1)!}\left(\frac{\pi t}{K'}\right)^{2r+1} \sum_{m=1}^{\infty} m^{2r+1} \kappa_m$$

and

$$\int_0^{\infty} \frac{\sinh\left(\frac{\pi t u}{2K'}\right)}{\sinh\left(\frac{\pi u}{2}\right)} du = \sum_{r=0}^{\infty} \frac{1}{(2r+1)!}\left(\frac{\pi t}{2K'}\right)^{2r+1} \int_0^{\infty} \frac{u^{2r+1} \, du}{\sinh\left(\frac{\pi u}{2}\right)}.$$

Thus,

$$(8.60) \quad (f^*)^{-1}(t) = \frac{2\pi}{K'} \sum_{r=0}^{\infty} \frac{1}{(2r+1)!}\left(\frac{\pi t}{K'}\right)^{2r+1}$$

$$\times \left\{ (-1)^r \sum_{m=1}^{\infty} m^{2r+1} \kappa_m + \frac{1}{2^{2r+3}} \int_0^{\infty} \frac{u^{2r+1} \, du}{\sinh\left(\frac{\pi u}{2}\right)} \right\}$$

$$= \sum_{r=0}^{\infty} \left[\frac{2}{(2r+1)!}\left(\frac{\pi}{K'}\right)^{2r+2} \left\{ (-1)^r \sum_{m=1}^{\infty} m^{2r+1} \kappa_m \right.\right.$$

$$\left.\left. + \frac{1}{2^{2r+3}} \int_0^{\infty} \frac{u^{2r+1} \, du}{\sinh\left(\frac{\pi u}{2}\right)} \right\} \right] t^{2r+1},$$

where the coefficient of t^{2r+1} is $\bar{a}_{2r+1}$ by (2.1).

We now evaluate Ψ_n. If we substitute the coefficient in the series in (8.60) into Ψ_n, we obtain

$$\Psi_n = \sum_{r=0}^{n} (2r+1)! \, \bar{a}_{2r+1} \varphi_{2r+1}^{(2n+1)} x^{2r+1}$$

$$= \sum_{r=0}^{n} (2r+1)! \left[\frac{2}{(2r+1)!}\left(\frac{\pi}{K'}\right)^{2r+2} \left\{ (-1)^r \sum_{m=1}^{\infty} m^{2r+1} \kappa_m \right.\right.$$

$$\left.\left. + \frac{1}{2^{2r+3}} \int_0^{\infty} \frac{u^{2r+1} \, du}{\sinh\left(\frac{\pi u}{2}\right)} \right\} \right] \varphi_{2r+1}^{(2n+1)} x^{2r+1}$$

$$= 2 \sum_{r=0}^{n} (-1)^r \left(\frac{\pi}{K'}\right)^{2r+2} \sum_{m=1}^{\infty} m^{2r+1} \kappa_m \, \varphi_{2r+1}^{(2n+1)} x^{2r+1}$$

$$+ \sum_{r=0}^{n} \left(\frac{\pi}{K'}\right)^{2r+2} \varphi_{2r+1}^{(2n+1)} x^{2r+1} \frac{1}{2^{2r+2}} \int_0^{\infty} \frac{u^{2r+1} \, du}{\sinh\left(\frac{\pi u}{2}\right)}$$

$$= 2x \left(\frac{\pi}{K'}\right)^2 \sum_{m=1}^{\infty} m \, \kappa_m \sum_{r=0}^{n} \varphi_{2r+1}^{(2n+1)} \left(-\left(\frac{\pi m x}{K'}\right)^2\right)^r$$

$$+ \frac{x}{4} (\frac{\pi}{K'})^2 \int_0^\infty \left\{ \frac{u}{\sinh(\frac{\pi u}{2})} \sum_{r=0}^n \varphi_{2r+1}^{(2n+1)} \left(\left(\frac{\pi x u}{2K'} \right)^2 \right)^r \right\} du$$

$$= 2x \left(\frac{\pi}{K'} \right)^2 \sum_{m=1}^\infty \alpha_m A_n^* \left(-\left(\frac{\pi m x}{K'} \right)^2, k \right)$$

$$+ \frac{x}{4} \left(\frac{\pi}{K'} \right)^2 \int_0^\infty \frac{u}{\sinh(\frac{\pi u}{2})} A_n^* \left(\left(\frac{\pi x u}{2K'} \right)^2, k \right) du.$$

Substituting this expression for Ψ_n gives the result. $\square$

Corollary 8.16. For $n \geq 0$,

$$(8.61) \quad 2 \left(\frac{\pi}{K'} \right)^2 \sum_{m=1}^\infty \alpha_m A_n^* \left(-(\frac{\pi m}{K'})^2, k \right) + \frac{1}{2} \int_0^\infty \frac{A_n^*(x,k)}{\sinh(K'\sqrt{x})} dx = \delta_{n,0}.$$

Proof. If we put $x = 1$ into (8.58) and equate corresponding coefficients of t^{2n+1}, we obtain the result using the change of variable $u = \frac{2K'}{\pi} \sqrt{x}$ in the integral. $\square$

We next construct a measure for the sequence $\{A_m^*(z,k)\}_{m=0}^\infty$, $0 < k < 1$.

Definition 8.17. Let $z_m = -\left(\frac{\pi m}{K'} \right)^2$, $m \geq 1$. Then let the mapping $\nu : \underline{B}(\mathbb{R}) \longrightarrow [0, \infty)$ be the countably additive measure such that

(i) $\nu(\{z_m\}) = 2 \left(\frac{\pi}{K'} \right)^2 \alpha_m$, $m \geq 1$,

(ii) $\nu \left((-\infty, 0) \setminus \{z_1, z_2, \cdots\} \right) = 0$,

(iii) $\nu(B) = \frac{1}{2} \int_B \frac{dx}{\sinh(K'\sqrt{x})}$, $B \in \underline{B} \left([0, \infty) \right)$.

Note that this measure is discrete on the negative axis, assigning a positive measure to the points $\{z_m\}$. The measure is defined by a weight function on $[0, \infty)$.

Lemma 8.18.

$$(8.62) \qquad \int_{\mathbb{R}} A_n^*(z, k) \, d\nu = \delta_{n,0}, \; n \geq 0$$

and

$$(8.63) \qquad \nu(\mathbb{R}) = 1.$$

Proof. For $n = 0$,

$$\int_{\mathbb{R}} d\nu = \nu(\mathbb{R}) = \sum_{m=1}^\infty \nu(\{z_m\}) + \frac{1}{2} \int_0^\infty \frac{dx}{\sinh(K'\sqrt{x})}$$

$$= 2 \left(\frac{\pi}{K'} \right)^2 \sum_{m=1}^\infty m \kappa_m + \frac{1}{2} \int_0^\infty \frac{dx}{\sinh(K'\sqrt{x})} = 1,$$

by (8.60). This also verifies (8.63) and shows that ν is a probability measure.

If $n \geq 1$, then again using (8.60),

$$\int_{\mathbb{R}} A^*(z,k)\, d\nu(z) = \sum_{m=1}^{\infty} \nu(\{z_m\}) A_n^*(z_m, k) + \frac{1}{2} \int_0^{\infty} \frac{A_n^*(z,k)}{\sinh(K'\sqrt{z})} dz$$

$$= 2\left(\frac{\pi}{K'}\right)^2 \sum_{m=1}^{\infty} \alpha_m A_n^*(z_m, k) + \frac{1}{2} \int_0^{\infty} \frac{A_n^*(z,k)}{\sinh(K'\sqrt{z})} dz = 0. \quad \square$$

Lemma 8.19. For $n \geq 1$ and $0 \leq m < n - 1$,

$$(8.64) \qquad \int_{\mathbb{R}} z^m A_n^*(z,k)\, d\nu(z) = 0.$$

Proof. If $m = 0$, then $n \geq 1$, so (8.64) is just (8.62). Assume (8.64) is true up to some $m \geq 0$. Then rewriting (3.14), we have that

$$(8.65) \qquad z A_n^* = A_{n+1}^* - 2(2n+1)^2(1-2k^2)A_n^* + 4n^2(4n^2-1)A_{n-1}^*.$$

Thus,

$$\int_{\mathbb{R}} z^{m+1} A_n^*\, d\nu = \int_{\mathbb{R}} z^m (z A_n^*)\, d\nu$$

$$= \int_{\mathbb{R}} z^m \left\{ A_{n+1}^* - 2(2n+1)^2(1-2k^2)A_n^* + 4n^2(4n^2-1)A_{n-1}^* \right\} d\nu = 0,$$

since each term is zero by assumption. $\quad \square$

Lemma 8.20. For $n \geq 0$,

$$(8.66) \qquad \int_{\mathbb{R}} z^n A_n^*(z,k)\, d\nu(z) = (2n)!\,(2n+1)!.$$

Proof. Let $I_n = \int_{\mathbb{R}} z^n A_n^*(z,k)\, d\nu(z)$, $n \geq 0$. For $n = 0$, $I_0 = \int_{\mathbb{R}} d\nu = \nu(\mathbb{R}) = 1$ by (8.63). For $n \geq 1$, we find from (8.65), using (8.64), that

$$I_n = \int_{\mathbb{R}} \left\{ z^{n-1} A_{n+1}^* - 2(2n+1)^2(1-2k^2)z^{n-1}A_n^* \right.$$

$$\left. + 4n^2(4n^2-1)z^{n-1}A_{n-1}^* \right\} d\nu(z)$$

$$= 4n^2(4n^2-1)\int_{\mathbb{R}} z^{n-1}A_{n-1}^* d\nu(z) = 4n^2(4n^2-1)I_{n-1} = \prod_{j=1}^{n} 4j^2(4j^2-1)$$

$$= \left\{ \prod_{j=1}^{n} (2j-1)(2j) \right\} \left\{ \prod_{j=1}^{n} (2j)(2j+1) \right\} = (2n)!\,(2n+1)!. \quad \square$$

Remark. This is the expected result according to (7.12). Note that the value of the integral in (8.66) is independent of k because $c_2 = 1$.

Theorem 8.21. For $m, n \geq 0$,

$$(8.67) \qquad \int_{\mathbf{R}} A_m^*(z, k) A_n^*(z, k) \, d\nu(z) = (m+n)! \, (m+n+1)! \, \delta_{m,n}.$$

Proof. Without loss of generality assume that $0 \leq m \leq n$. For $m < n$, equation (8.67) follows linearly from (8.58). For $m = n$, write $A_n^*(z, k) = z^n + T_{n-1}(z)$, where $\deg\{T_{n-1}(z)\} \leq n - 1$. Then using (8.66) and (8.69), we have that

$$\int_{\mathbf{R}} (A_n^*)^2 d\nu(z) = \int_{\mathbf{R}} z^n \, A_n^* \, d\nu(z) + \int_{\mathbf{R}} T_{n-1}(z) A_n^* \, d\nu(z) = (2n)! \, (2n+1)!. \quad \square$$

Next consider the sequence $\{B_m^*(z, k)\}_{m=0}^{\infty}$.

Theorem 8.22. If $f \in \mathcal{F}_0$, then

$$(8.68) \quad \left\{ f^{-1}[xf(t)] \right\}^2 =$$

$$\sum_{n=0}^{\infty} \left\{ \sum_{r=0}^{n} (2r+2)! \, \bar{a}_{2r+2}^{(2)} \, \varphi_{2r+2}^{(2n+2)} x^{2r+2} \right\} \frac{t^{2n+2}}{(2n+2)!}.$$

Proof. Since f^{-1} is odd and $f^{-1}(0) = 0$, we can write the even function $(f^{-1})^2$ as in (2.2), viz.,

$$(8.69) \qquad [f^{-1}(z)]^2 = \sum_{r=0}^{\infty} \bar{a}_{2r+2}^{(2)} z^{2r+2}.$$

Then

$$\left\{ f^{-1}[xf(t)] \right\}^2 = \sum_{r=0}^{\infty} \bar{a}_{2r+2}^{(2)} x^{2r+2} [f(t)]^{2r+2}$$

$$= \sum_{r=0}^{\infty} \bar{a}_{2r+2}^{(2)} x^{2r+2} \sum_{n=r}^{\infty} a_{2n+2}^{(2r+2)} t^{2n+2}$$

$$= \sum_{r=0}^{\infty} \bar{a}_{2r+2}^{(2)} x^{2r+2} \sum_{n=r}^{\infty} \left\{ \frac{(2r+2)!}{(2n+2)!} \varphi_{2r+2}^{(2n+2)} \right\} t^{2n+2}$$

$$= \sum_{n=0}^{\infty} \left\{ \sum_{r=0}^{n} (2r+2)! \, \bar{a}_{2r+2}^{(2)} \, \varphi_{2r+2}^{(2n+2)} x^{2r+2} \right\} \frac{t^{2n+2}}{(2n+2)!},$$

where we have used (1.8). $\quad \square$

The next theorem specializes the above result.

Theorem 8.23. For $f^*(t, k)$ in (8.49), we have

$$(8.70) \quad \left\{ (f^*)^{-1}[xf^*(t)] \right\}^2 =$$

$$2 x^2 \left(\frac{\pi}{K'} \right)^4 \sum_{n=0}^{\infty} \left\{ \sum_{m=1}^{\infty} \beta_m B_n^* \left(- \left(\frac{\pi m x}{K'} \right)^2, k \right) \right.$$

$$\left. + \frac{1}{32} \int_0^{\infty} \frac{u^3}{\sinh\left(\frac{\pi u}{2} \right)} B_n^* \left(\left(\frac{\pi x u}{2K'} \right)^2, k \right) du \right\} \frac{t^{2n+2}}{(2n+2)!},$$

where $\beta_m = \dfrac{m^3 (q')^m}{1 - (-q')^m}$.

Proof. The proof of this formula comes from evaluating the coefficient Ψ_n inside the braces in (8.68) for the function $f(t) = f^*(t, k)$.

We begin by preparing some results [see WhW, pp. 492–493]:

$$\frac{d}{dt}\left(\frac{\operatorname{sn} \operatorname{dn}}{\operatorname{cn}}\right) = \frac{\operatorname{cn} \operatorname{sn} \operatorname{dn}' + \operatorname{cn} \operatorname{dn} \operatorname{sn}' - \operatorname{sn} \operatorname{dn} \operatorname{cn}'}{\operatorname{cn}^2}$$

$$= \frac{\operatorname{dn}^2 - k^2 \operatorname{cn}^2 \operatorname{sn}^2}{\operatorname{cn}^2} = \frac{1 - k^2(1 - \operatorname{cn}^4)}{\operatorname{cn}^2} = \frac{1 - k^2}{\operatorname{cn}^2} + k^2 \operatorname{cn}^2.$$

Also,

$$\left(\frac{\operatorname{sn} \operatorname{dn}}{\operatorname{cn}}\right)^2 = \frac{(1 - \operatorname{cn}^2)(1 - k^2 + k^2\operatorname{cn}^2)}{\operatorname{cn}^2} = \frac{1 - k^2}{\operatorname{cn}^2} + 2k^2 - 1 - k^2\operatorname{cn}^2.$$

Subtracting these two equations gives

$$(8.71) \qquad \left(\frac{\operatorname{sn} \operatorname{dn}}{\operatorname{cn}}\right)^2 = 2k^2 - 1 + \frac{d}{dt}\left(\frac{\operatorname{sn} \operatorname{dn}}{\operatorname{cn}}\right) - 2k^2\operatorname{cn}^2.$$

We next find the Fourier expansions of the last two terms of this equation. If we combine (5.31) and (8.59) and replace k' by k, we get

$$\frac{\operatorname{sn}(t, k)\operatorname{dn}(t, k)}{\operatorname{cn}(t, k)} = \frac{2\pi}{K} \sum_{m=1}^{\infty} \frac{q^m}{1 + (-q)^m} \sin\left(\frac{m\pi t}{K}\right) + \frac{\pi}{2K}\tan\left(\frac{\pi t}{2K}\right).$$

Then

$$(8.72) \qquad \frac{d}{dt}\left(\frac{\operatorname{sn}(t, k)\operatorname{dn}(t, k)}{\operatorname{cn}(t, k)}\right)$$

$$= 2\left(\frac{\pi}{K}\right)^2 \sum_{m=1}^{\infty} \frac{m q^m}{1 + (-q)^m} \cos\left(\frac{m\pi t}{K}\right) + \left(\frac{\pi}{2K}\right)^2 \sec^2\left(\frac{\pi t}{2K}\right).$$

From [Ki, p. 247] we have

$$(8.73) \quad -2k^2\operatorname{cn}^2 = 2\frac{(1 - k^2)K - E}{K} - 4\left(\frac{\pi}{K}\right)^2 \sum_{m=1}^{\infty} \frac{m q^m}{1 - q^{2m}} \cos\left(\frac{m\pi t}{K}\right).$$

Substituting (8.72) and (8.73) into (8.71), we obtain

$$\left(\frac{\operatorname{sn} \operatorname{dn}}{\operatorname{cn}}\right)^2 = 1 - \frac{2E}{K} + 2\left(\frac{\pi}{K}\right)^2 \sum_{m=1}^{\infty} \left\{\frac{m q^m}{1 + (-q)^m} - \frac{2m q^m}{1 - q^{2m}}\right\} \cos\left(\frac{m\pi t}{K}\right)$$

$$+ \left(\frac{\pi}{2K}\right)^2 \sec^2\left(\frac{\pi t}{2K}\right)$$

$$= 1 - \frac{2E}{K} - 2\left(\frac{\pi}{K}\right)^2 \sum_{m=1}^{\infty} \hat{k}_m \cos\left(\frac{m\pi t}{K}\right) + \left(\frac{\pi}{2K}\right)^2 \sec^2\left(\frac{\pi t}{2K}\right),$$

where $\hat{\kappa}_m = \dfrac{m\,q^m}{1-(-q)^m}$, $m \geq 1$. Now, since $\sec^2 z = \displaystyle\int_0^\infty \dfrac{u\,\cosh(zu)}{\sinh\left(\frac{\pi u}{2}\right)}du$,

$|z| < \dfrac{\pi}{2}$ [GrR, p. 350, #8], we have

$$(8.74) \qquad \left(\frac{\text{sn dn}}{\text{cn}}\right)^2 = 1 - \frac{2E}{K} - 2\left(\frac{\pi}{K}\right)^2 \sum_{m=1}^\infty \hat{\kappa}_m \cos\left(\frac{m\pi t}{K}\right)$$

$$+ \left(\frac{\pi}{2K}\right)^2 \int_0^\infty \frac{u\,\cosh\left(\frac{\pi t u}{2K}\right)}{\sinh\left(\frac{\pi u}{2}\right)}du.$$

Setting $t = 0$ in (8.74) gives

$$0 = 1 - \frac{2E}{K} - 2\left(\frac{\pi}{K}\right)^2 \sum_{m=1}^\infty \hat{\kappa}_m + \left(\frac{\pi}{2K}\right)^2 \int_0^\infty \frac{u\,du}{\sinh\left(\frac{\pi u}{2}\right)}.$$

Subtracting this equation from (8.74) gives

$$\left(\frac{\text{sn dn}}{\text{cn}}\right)^2 = -2\left(\frac{\pi}{K}\right)^2 \sum_{m=1}^\infty \hat{\kappa}_m \left[\cos\left(\frac{m\pi t}{K}\right) - 1\right]$$

$$+ \left(\frac{\pi}{2K}\right)^2 \int_0^\infty \frac{u\,[\cosh\left(\frac{\pi t u}{2K}\right) - 1]du}{\sinh\left(\frac{\pi u}{2}\right)}$$

$$= -2\left(\frac{\pi}{K}\right)^2 \sum_{m=1}^\infty \hat{\kappa}_m \sum_{n=1}^\infty \frac{(-1)^n}{(2n)!}\left(\frac{m\pi t}{K}\right)^{2n}$$

$$+ \left(\frac{\pi}{2K}\right)^2 \int_0^\infty \frac{u}{\sinh\left(\frac{\pi u}{2}\right)} \sum_{n=1}^\infty \frac{1}{(2n)!}\left(\frac{\pi t u}{2K}\right)^{2n} du$$

$$= -2\left(\frac{\pi}{K}\right)^2 \sum_{n=1}^\infty \frac{(-1)^n}{(2n)!}\left(\frac{\pi t}{K}\right)^{2n} \sum_{m=1}^\infty m^{2n}\hat{\kappa}_m$$

$$+ \left(\frac{\pi}{2K}\right)^2 \sum_{n=1}^\infty \frac{1}{(2n)!}\left(\frac{\pi t}{2K}\right)^{2n} \int_0^\infty \frac{u^{2n+1}\,du}{\sinh\left(\frac{\pi u}{2}\right)}$$

$$= \sum_{n=0}^\infty \left\{\frac{2}{(2n+2)!}\left(\frac{\pi}{K}\right)^{2n+4}\right.$$

$$\left. \times \left\{(-1)^n \sum_{m=1}^\infty m^{2n+2}\hat{\kappa}_m + \frac{1}{2^{2n+5}}\int_0^\infty \frac{u^{2n+3}\,du}{\sinh\left(\frac{\pi u}{2}\right)}\right\}\right\}t^{2n+2}.$$

Thus, replacing k by k' and using (5.31), we have

$$[(f^*)^{-1}(t,k)]^2 = \sum_{n=0}^\infty \left\{\frac{2}{(2n+2)!}\left(\frac{\pi}{K'}\right)^{2n+4}\left\{(-1)^n \sum_{m=1}^\infty m^{2n+2}\hat{\kappa}'_m\right.\right.$$

$$\left.\left. + \frac{1}{2^{2n+5}}\int_0^\infty \frac{u^{2n+3}\,du}{\sinh\left(\frac{\pi u}{2}\right)}\right\}\right\}t^{2n+2}$$

where $\hat{\kappa}'_m = \dfrac{m(q')^m}{1-(-q')^m}$. Thus,

(8.75) $\bar{a}^{(2)}_{2n+2} =$

$$\frac{2}{(2n+2)!}\left(\frac{\pi}{K'}\right)^{2n+4}\left\{(-1)^n\sum_{m=1}^{\infty}m^{2n+2}\,\hat{\kappa}'_m + \frac{1}{2^{2n+5}}\int_0^{\infty}\frac{u^{2n+3}\,du}{\sinh\left(\frac{\pi u}{2}\right)}\right\}.$$

We now evaluate Ψ_n. Substituting (8.75) into Ψ_n gives

$$\Psi_n = \sum_{r=0}^{\infty}(2r+2)!\,\bar{a}^{(2)}_{2r+2}\,\varphi^{(2n+2)}_{2r+2}\,x^{2r+2}$$

$$= 2x^2\left(\frac{\pi}{K'}\right)^4\sum_{m=1}^{\infty}m^2\hat{\kappa}'_m\sum_{r=0}^{n}\varphi^{(2n+2)}_{2r+2}\left(-\left(\frac{\pi m x}{K'}\right)^2\right)^r$$

$$+\frac{x^2}{16}\left(\frac{\pi}{K'}\right)^4\int_0^{\infty}\frac{u^3\,du}{\sinh\left(\frac{\pi u}{2}\right)}\sum_{r=0}^{n}\varphi^{(2n+2)}_{2r+2}\left(\left(\frac{\pi u x}{2K'}\right)^2\right)^r = 2x^2\left(\frac{\pi}{K'}\right)^4$$

$$\times\left\{\sum_{m=1}^{\infty}\beta_m\,B^{\star}_n\left(-\left(\frac{\pi m x}{K'}\right)^2,k\right) + \frac{1}{32}\int_0^{\infty}\frac{u^3\,B^{\star}_n\left(\left(\frac{\pi x u}{2K'}\right)^2,k\right)\,du}{\sinh\left(\frac{\pi u}{2}\right)}\right\}.$$

Substituting this result into (8.68) proves the formula. $\square$

Corollary 8.24. For $n \geq 0$,

(8.76) $\left(\dfrac{\pi}{K'}\right)^4\displaystyle\sum_{m=1}^{\infty}\beta_m\,B^{\star}_n\left(-\left(\dfrac{\pi m}{K'}\right)^2,k\right) + \dfrac{1}{4}\int_0^{\infty}\dfrac{x\,B^{\star}_n(x,k)}{\sinh(K'\sqrt{x})}dx = \delta_{n,0}.$

Proof. Putting $x = 1$ into (8.70), equating corresponding coefficients of t^{2n+2}, and transforming the integral by $x = \left(\dfrac{\pi}{2K'}\right)^2 u^2$ gives the result. $\square$

We next construct a measure for the sequence $\{B^{\star}_n(z,k)\}^{\infty}_{m=0}$, $0 < k < 1$.

Definition 8.25. Let $z_m = -\left(\dfrac{\pi m}{K'}\right)^2$, $m \geq 1$. Then let the mapping $\nu: \underline{B}(\mathbb{R}) \longrightarrow [0,\infty)$ be the countably additive measure such that

(i) $\nu(\{z_m\}) = \left(\dfrac{\pi}{K'}\right)^4\beta_m$, $m \geq 1$,

(ii) $\nu\left((-\infty,0)\setminus\{z_1,z_2,\cdots\}\right) = 0$,

(iii) $\nu(B) = \dfrac{1}{4}\displaystyle\int_B\dfrac{x\,dx}{\sinh(K'\sqrt{x})}$, $B \in \underline{B}\left([0,\infty)\right)$.

Note that this measure is discrete on the negative axis, assigning a positive measure to the points $\{z_m\}$. The measure is defined by a weight function on $[0,\infty)$.

Lemma 8.26.

(8.77) $\displaystyle\int_{\mathbb{R}}B^{\star}_n(z,k)\,d\nu = \delta_{n,0}$, $n \geq 0$,

and

(8.78) $$\nu(\mathbb{R}) = 1.$$

Proof. Essentially the same proof as for Lemma 8.18. $\square$

Lemma 8.27. For $n \geq 1$ and $0 \leq m < n - 1$,

(8.79) $$\int_{\mathbb{R}} z^m B_n^\star(z, k) \, d\nu(z) = 0.$$

Proof. Essentially the same proof as for Lemma 8.19, but using the recursion in (3.15). $\square$

Lemma 8.28. For $n \geq 0$,

(8.80) $$\int_{\mathbb{R}} z^n B_n^\star(z, k) \, d\nu(z) = \frac{1}{2}(2n + 1)! \, (2n + 2)!.$$

Proof. Essentially the same proof as in (8.66). $\square$

Remark. This is the expected result according to (7.12). Note that the evaluation in (8.80) is independent of k, because $c_2 = 1$.

Theorem 8.29. For $m, n \geq 0$,

(8.81) $$\int_{\mathbb{R}} B_m^\star(z, k) B_n^\star(z, k) \, d\nu(z) = \frac{1}{2}(m + n + 1)! \, (m + n + 2)! \, \delta_{m,n}.$$

Proof. Essentially the same proof as in (8.67). $\square$

Chapter 9

The Tangent Numbers

In this chapter we give a few properties of the tangent numbers which play an important part in this work. We also give a collection of identities that are specializations of earlier identities.

Definition 9.1. The tangent numbers $\{C_{2n+1}\}_{n=0}^{\infty} \subset \mathbb{Z}$ can be defined by

$$(9.1) \qquad \tanh(t) = -\sum_{n=0}^{\infty} C_{2n+1} \frac{t^{2n+1}}{(2n+1)!}.$$

The first few C's are: $C_1 = -1$, $C_3 = 2$, $C_5 = -16$, $C_7 = 272$, and $C_9 = -7936$. By definition, $C_0 = 1$. Note for $n \geq 0$, that $C_{2n+1} \neq 0$.

Let the numbers $\{D_{2n+2}\}_{n=0}^{\infty} \subset \mathbb{Z}$ be defined by

$$(9.2) \qquad \tanh^2(t) = \sum_{n=0}^{\infty} 2D_{2n+2} \frac{t^{2n+2}}{(2n+2)!}.$$

Since $\operatorname{sn}(t, 1) = \tanh(t)$, we find, respectively, from (8.25a) and (9.1) for $n \geq 0$, that

$$(9.3) \qquad C_{2n+1}(1) = C_{2n+1}$$

and from (8.6), that

$$(9.4) \qquad D_{2n+2}(1) = D_{2n+2}.$$

Theorem 9.2. For $n \geq 0$,

$$(9.5) \qquad D_{2n+2} = \frac{1}{2}C_{2n+3}.$$

Proof. We have that

$$(9.6) \quad \sum_{n=0}^{\infty} 2D_{2n+2}\frac{t^{2n+2}}{(2n+1)!} = \tanh^2 t = 1 - \operatorname{sech}^2 t$$

$$= 1 - D_t(\tanh t) = 1 + \sum_{n=0}^{\infty} C_{2n+1}\frac{t^{2n}}{(2n)!}$$

$$= \sum_{n=1}^{\infty} C_{2n+1}\frac{t^{2n}}{(2n)!} = \sum_{n=0}^{\infty} C_{2n+3}\frac{t^{2n+2}}{(2n+2)!},$$

from which the result follows by equating corresponding coefficients. $\square$

Note from (9.5) that D_{2n+2} for $\neq 0$.

Corollary 9.3. We have the expansion

$$(9.7) \qquad \tanh^2 t = \sum_{n=0}^{\infty} C_{2n+3}\frac{t^{2n+2}}{(2n+2)!}.$$

Proof. Equation (9.6). $\square$

Corollary 9.4. For $n \geq 0$,

$$(9.8) \qquad P_n(-2, 16) = -C_{2n+1}$$

and

$$(9.9) \qquad Q_n(-2, 16) = \frac{1}{2}C_{2n+3}.$$

Proof. (9.8) By (8.27), we have that $P_n(-k^2 - 1, k^4 + 14k^2 + 1) = -C_{2n+1}(k)$, so using (9.3) we get $P_n(-2, 16) = -C_{2n+1}(1) = -C_{2n+1}$.
(9.9) By (8.28), we have that $2Q_n(-k^2 - 1, k^4 + 14k^2 + 1) = D_{2n+2}(k)$, so using (9.4) and (9.5) we obtain $2Q_n(-2, 16) = D_{2n+2}(1) = C_{2n+3}$. $\square$

We next deduce some of the many recursions for the tangent numbers.

Theorem 9.5. For $n \geq 0$,

$$(9.10) \qquad C_{2n+5} = -2C_{2n+3} + 6\sum_{j=0}^{n}\binom{2n+2}{2j}C_{2j+1}C_{2n-2j+3},$$

$$(9.11) \qquad C_{2n+5} = -2C_{2n+3} + 6\sum_{j=0}^{n}\binom{2n+2}{2j+2}C_{2j+3}C_{2n-2j+1},$$

(9.12) $\quad C_{2n+3} = 2 \sum_{j=0}^{n} \binom{2n+1}{2j+1} C_{2j+1} C_{2n-2j+1}$

$$= \sum_{j=0}^{n} \binom{2n+2}{2j+1} C_{2j+1} C_{2n-2j+1} \quad \text{(cf. [Jo, p. 299]).}$$

(9.13) $\quad C_{2n+7} = -8C_{2n+5} + 6 \sum_{j=0}^{n} \binom{2n+4}{2j+2} C_{2j+3} C_{2n-2j+3},$

(9.14) $\quad C_{2n+5} = -2C_{2n+3} + 2 \sum_{j=0}^{n} \binom{2n+3}{2j+1} C_{2j+1} C_{2n-2j+3},$

(9.15) $\quad C_{2n+3} = 4C_{2n+1} + 6 \sum_{j=0}^{n} \binom{2n}{2j} C_{2j+1} C_{2n-2j+1},$

(9.16) $\quad C_{2n+5} = -8C_{2n+3} + 6 \sum_{j=0}^{n+1} \binom{2n+2}{2j} C_{2j+1} C_{2n-2j+3}$

$$+ 12 \sum_{j=0}^{n} \binom{2n+2}{2j+1} C_{2j+1} C_{2n-2j+1},$$

(9.17) $\quad C_{2n+7} =$

$$-8\,C_{2n+5} + 12 \sum_{j=0}^{n} \binom{2n+3}{2j+3} (C_{2j+5} + 2C_{2j+3}) C_{2n-2j+1},$$

(9.18) $\quad C_{2n+1} = \dfrac{1}{2n+1} \sum_{j=0}^{n} (n - 3j - 1) \binom{2n+1}{2j+1} C_{2j+1} C_{2n-2j+1},$

(9.19) $\quad C_{2n+5} = -2C_{2n+3} + 3 \sum_{j=0}^{n} \binom{2n+2}{2j+1} C_{2j+1} C_{2n-2j+3},$

(9.20) $\quad C_{2n+5} = -\dfrac{1}{2} \sum_{j=0}^{n+2} \binom{2n+4}{2j} C_{2j+1} C_{2n-2j+5}$

$$+ \dfrac{1}{2} \sum_{j=0}^{n} \binom{2n+4}{2n+2} C_{2j+3} C_{2n-2j+3},$$

(9.21) $C_{2n+5} =$

$$-\frac{1}{(n+1)(2n+5)}\sum_{j=0}^{n}(2n-3j+2)\binom{2n+5}{2j+2}C_{2j+3}C_{2n-2j+3},$$

and

(9.22) $$\sum_{j=0}^{n}(n-3j)\binom{2n+3}{2j+1}C_{2j+1}C_{2n-2j+3} = 0.$$

Proof. To prove (9.10), put $x = -2$ and $y = 16$ into (4.1) and use (9.8) and (9.9). The proofs of the other formulas are similar successively using (4.4)–(4.6) and (4.33)–(4.42). $\square$

We conclude this chapter with a specialization of a previous formula, which gives a formula for the Euler number E_{2n} ($E_0 = 1$, $E_2 = -1$, $E_4 = 5$, $E_6 = -61$).

Corollary 9.6. For $n \geq 1$,

(9.23) $$E_{2n} = -\frac{1}{2^{2n-1}}\sum_{j=0}^{n-1}2^{2j}\binom{2n}{2j+1}C_{2j+1}C_{2n-2j-1}.$$

Proof. From [Han, p. 344, 15.3.2], [AbS, p. 571, 16.6.3], and using (5.37), (4.8), and (9.8), we have that

$$\sum_{n=0}^{\infty}E_{2n}\frac{t^{2n}}{(2n)!} = \text{sech}\,(t) = \text{dn}\,(t,1)$$

$$= 1 - \sum_{n=0}^{\infty}\frac{1}{2^{n+1}}\left\{\sum_{j=0}^{n}2^{j}\binom{2n+2}{2j+1}P_j(-2,16)\,P_{n-j}(-1,4)\right\}\frac{t^{2n+2}}{(2n+2)!}$$

$$= 1 - \sum_{n=0}^{\infty}\frac{1}{2^{2n+1}}\left\{\sum_{j=0}^{n}2^{2j}\binom{2n+2}{2j+1}P_j(-2,16)\,P_{n-j}(-2,16)\right\}\frac{t^{2n+2}}{(2n+2)!}.$$

The result follows by equating corresponding coefficients of $\frac{t^{2n}}{(2n)!}$. $\square$

Chapter 10

Class III Functions –
The $\delta_n(x)$ Polynomials

This chapter is a short treatise on the Mittag–Leffler polynomials, which are the primary sequence of elliptic polynomials of the first kind for $f(t) = \tanh^{-1} t$, the main function in Class III. These polynomials appear in many parts of mathematics and have interesting forms and properties, as is evidenced in what follows.

We have developed some of their operator properties and have expressed these polynomials in terms of rising and falling factorial polynomials, binomial coefficient polynomials, the hypergeometric function, Jacobi polynomials, and several integrals. Also entering here are the Lah, Stirling, and Delannoy numbers. These polynomials, which are monic with positive integer coefficients, also have curious properties, such as those in (10.84). We also give several forms for double-argument and half-argument formulas, a theorem giving their structure constants, a formula for $\delta_{m+n}(x)$, and a Rodrigues–Toscano type formula for $\delta_n(x)$.

The function we use here is in Class III (a), obtained by taking the initial values $a_1 = 1$, $a_3 = \dfrac{1}{3}$, $a_5 = \dfrac{1}{5}$. Since in this case $c_1 = 2$ and $c_2 = 1$, we have from (8.42) that

$$(10.1) \qquad f(t) = \tanh^{-1} t = \frac{1}{2} \log\left(\frac{1+t}{1-t}\right) = \sum_{n=0}^{\infty} \frac{t^{2n+1}}{2n+1}.$$

Note that this is actually a special case of (8.1) with $k = 1$. Here we write $\delta_n(x) = \delta_n(x, 1)$ (cf. Table 8.2).

Definition 10.1 [Ra1, p. 290 (7)] [BoBu, p. 38] [Rom, p. 75] Let the sequence $\{\delta_n(x)\}_{n=0}^{\infty}$ be generated by

$$(10.2) \qquad G = G(x, t) \overset{\text{def}}{=} e^{x \tanh^{-1} t} = \sum_{n=0}^{\infty} \delta_n(x) \frac{t^n}{n!}.$$

(Here $\delta_n(x)$ is written for $f_n(x)$ in (1.3).) Analytically this equality holds for any $x \in \mathbb{R}$ and $|t| < 1$. A proof of this fact will be given in Corollary 10.15.

Alternatively we have that

$$(10.3) \qquad G(x, t) = \left(\frac{1+t}{1-t} \right)^{\frac{x}{2}}.$$

Also observe that $\tanh^{-1} t \in \mathcal{F}_2$ and that $a_{2k+1} = \dfrac{1}{2k+1}$, $k \geq 0$.

Before continuing our study of this interesting primary sequence, we should mention the related function in Class III (b), viz.,

$$(10.4) \qquad f(t) = \tan^{-1} t = \sum_{n=0}^{\infty} \frac{(-1)^n t^{2n+1}}{2n+1},$$

where $a_1 = 1$, $a_3 = -\frac{1}{3}$, $a_5 = \frac{1}{5}$, and $c_1 = -2$.

Actually, this function leads to essentially the same set of polynomials as in (10.2), so it can be ignored. To see this, let

$$e^{x \tan^{-1} t} \overset{\text{def}}{=} \sum_{n=0}^{\infty} \hat{\delta}_n(x) \frac{t^n}{t!},$$

and for $m \geq 0$, as in (1.38) and (1.39), let

$$\delta_{2m+1}(x) = x G_m(x^2), \quad \delta_{2m+2}(x) = x^2 H_m(x^2),$$

and

$$\hat{\delta}_{2m+1}(x) = x \hat{G}_m(x^2), \quad \hat{\delta}_{2m+2}(x) = x^2 \hat{H}_m(x^2).$$

Using the relationship, $\tanh^{-1} t = -i \tan^{-1}(it)$, we have that

$$\sum_{n=0}^{\infty} \delta_n(x) \frac{t^n}{t!} = e^{x \tanh^{-1} t} = e^{-ix \tan^{-1}(it)}$$

$$= \sum_{n=0}^{\infty} \hat{\delta}_n(-ix) \frac{(it)^n}{n!} = \sum_{n=0}^{\infty} i^n \hat{\delta}_n(-ix) \frac{t^n}{n!}.$$

Thus, for $n \geq 0$

$$\delta_n(x) = i^n \hat{\delta}_n(-ix),$$

so

$$xG_m(x^2) = \delta_{2m+1}(x) = i^{2m+1}\hat{\delta}_{2m+1}(-ix) = (-1)^m x \hat{G}_m(-x^2)$$

and

$$x^2 H_m(x^2) = \delta_{2m+2}(x) = i^{2m+2}\hat{\delta}_{2m+2}(-ix) = (-1)^m x^2 \hat{H}_m(-x^2).$$

Finally, putting $x^2 = z$, we find that

$$\hat{G}_m(z) = (-1)^m G_m(-z) \text{ and } \hat{H}_m(z) = (-1)^m H_m(-z),$$

which shows the two sets of polynomials are essentially the same.

Comments.

1. Our decision to use $f(t) = \tanh^{-1} t$ instead of $f(t) = \tan^{-1} t$ was based primarily on the fact that $\tanh^{-1} t = \mathrm{sn}^{-1}(t,1)$ and the simplicity of having the coefficients of $\{\delta_n(x)\}_{n=0}^\infty$ all be positive. That $\tanh^{-1} t$ is essentially the only function in this class is clear from the first equation in (8.42), where we note that the other functions only differ from it by minor changes in multipliers.

2. Note for $f(t) = \tan^{-1}(t)$ that $\Delta = 0$, so by (3.37) the sequence $\{f_n(x)\}_{n=0}^\infty$ will satisfy the recursion

$$f_{n+1}(x) = x f_n(x) - n(n-1)f_{n-1}(x), \; n \geq 0.$$

Carlitz [Ca4, p. 123] remarks that this sequence is orthogonal. Although this recursion has the form of (6.1) in Favard's Theorem, it does not satisfy (6.2) (because $\gamma_1 = 0$), so the sequence is certainly not orthogonal, as we already saw from Theorem 6.10.

Specializing (1.5), (1.23), and (1.34) to this case, we obtain the following properties of $\delta_n(x)$:
For $n \geq 0$,

(10.5)
$$\delta_n(x+y) = \sum_{j=0}^n \binom{n}{j}\delta_j(x)\delta_{n-j}(y),$$

(10.6)
$$\delta_n(-x) = (-1)^n \delta_n(x),$$

and for $n \geq 1$,

(10.7)
$$\delta_n(x) = \begin{vmatrix} x & 1 & 0 & 0 & 0 \cdots \\ 0 & x & 2 & 0 & 0 \cdots \\ x & 0 & x & 3 & 0 \cdots \\ 0 & x & 0 & x & 4 \cdots \\ x & 0 & x & 0 & x \cdots \\ \vdots & \vdots & \vdots & \vdots & \vdots & \vdots \end{vmatrix}_{n \times n}$$

We can also obtain the following table from Table 8.2.

Table 10.1 $\delta_n(x)$, $0 \le n \le 11$

n	$\delta_n(x)$
0	1
1	x
2	x^2
3	$x(x^2 + 2)$
4	$x^2(x^2 + 8)$
5	$x(x^4 + 20x^2 + 24)$
6	$x^2(x^4 + 40x^2 + 184)$
7	$x(x^6 + 70x^4 + 784x^2 + 720)$
8	$x^2(x^6 + 112x^4 + 2464k^2 + 8448) = x^2(x^2 + 24)(x^4 + 88x^2 + 352)$
9	$x(x^8 + 168x^6 + 6384x^4 + 52352x^2 + 40320)$
10	$x^2(x^8 + 240x^6 + 1448x^4 + 229760x^2 + 648576)$
11	$x(x^{10} + 330x^8 + 29568x^6 + 804320x^4 + 5360256x^2 + 3628800)$

The connections that $\delta_n(x)$ has with some related polynomials in the literature are given next. First, there is

$$\delta_n(x) = n!\, g_n\left(\frac{x}{2}\right) = n!\, b_n\left(\frac{x}{2}\right) = M_n\left(\frac{x}{2}\right)$$

where $g_n(x)$ is the original Mittag-Leffler polynomial [ML, pp. 210–215] [Bat, pp. 491–496] [Er, v. 3, p. 248] [MJ, pp. 194–196] [AgJ, pp. 235–242]; where $b_n(x)$ is in Belorizky [Be, pp. 1222–1224]; and where $M_n(x)$ is in [Rom, p. 76]. (Our $\delta_n(x)$ is monic with positive integer coefficients.) Also, for $P_n(x)$ in [Har, p. 5],

$$\delta_n(x) = (-1)^n\, n!\,\left[P_n\left(\frac{x+1}{2}\right) + P_{n-1}\left(\frac{x+1}{2}\right)\right],\ n \ge 0,$$

and, for $\mu_n(x)$ in Pidduck [Pi, pp. 347–356],

$$\delta_n(x) = n!\left(\mu_n\left(\frac{x}{2}\right) - \mu_{n-1}\left(\frac{x}{2}\right)\right).$$

(The μ notation is from [BaB, pp. 273–276]. Also see [Ba].) Finally, $\delta_n(x) = R_n(x,1)$ is in Carlitz [Ca4, p. 122] and $\delta_n(x) = Q_n(x,0)$ is the modified Krawtchouk polynomial in [SW, p. 176].

In what follows, we give a rather extensive development relating to the Mittag-Leffler polynomials, partly because they are of importance in various studies and partly because they have interesting properties in their own right.

From Corollary 3.9 we can obtain the second-order recursion satisfied by $\delta_n(x)$.

Corollary 10.2. Belorizky [Be] For $n \geq 0$,

(10.8) $$\delta_{n+2}(x) = x\delta_{n+1}(x) + n(n + 1)\delta_n(x),$$

where

(10.9) $$\delta_0(x) = 1 \text{ and } \delta_1(x) = x.$$

Proof. Since $\Delta = 0$ in this case, the results follow from (3.37) using $a_1 = 1$ and $c_1 = 2$. $\square$

The next result is a double-step recursion formula for $\delta_n(x)$.

Corollary 10.3. For $n \geq 2$, we have that

(10.10) $$\delta_{n+2}(x) = (x^2 + 2n^2)\,\delta_n(x) - n(n - 1)^2(n - 2)\,\delta_{n-2}(x).$$

Proof. This formula is derived by writing down equation (10.8) and two other equations, obtained from (10.8) by replacing n by $n + 1$ and $n + 2$, and then forming a linear combination so as to eliminate the $\delta_{n+1}(x)$ and $\delta_{n-1}(x)$ terms. $\square$

Theorem 10.4. For $n \geq 0$,

(10.11) $$\delta_{n+1}(x) = n!\,x \sum_{k=0}^{[\frac{n}{2}]} \frac{\delta_{n-2k}(x)}{(n - 2k)!},$$

(10.12) $$\delta_{2n+1}(x) = (2n)!\,x \sum_{j=0}^{n} \frac{\delta_{2j}(x)}{(2j)!},$$

and

(10.13) $$\delta_{2n+2}(x) = (2n + 1)!\,x \sum_{j=0}^{n} \frac{\delta_{2j+1}(x)}{(2j + 1)!}.$$

Proof. If in (1.6) we replace $f_j(x)$ by $\delta_j(x)$ and set $j = n - 2k$, we obtain

$$\delta_{n+1}(x) = n!\,x \sum_{k=0}^{[\frac{n}{2}]} \frac{(2k + 1)\,a_{2k+1}}{(n - 2k)!}\,\delta_{n-2k}(x).$$

Since $f(t) = \tanh^{-1} t$, then $a_{2k+1} = \dfrac{1}{2k + 1}$. The other two identities are specializations of (10.11). $\square$

There is also a different type of formula for $\delta_n(x)$.

Theorem 10.5. [Ba, (35a)] For $n \geq 0$,

(10.14) $x\,\delta_n(x+2) = 4n\,\delta_n(x) + x\,\delta_n(x-2).$

Proof. Observe from (10.2) and (10.3) that

(10.15) $$\sum_{n=0}^{\infty} \delta_n(x+2)\frac{t^n}{n!} = \left(\frac{1+t}{1-t}\right)^{\frac{x+2}{2}} = \left(\frac{1+t}{1-t}\right)G$$

and

$$\sum_{n=0}^{\infty} \delta_n(x-2)\frac{t^n}{n!} = \left(\frac{1-t}{1+t}\right)G,$$

as well as

(10.16) $$\sum_{n=0}^{\infty} \delta_{n+1}(x)\frac{t^n}{n!} = D_t G = \frac{x}{1-t^2}G$$

and using (10.16),

(10.17) $$\sum_{n=0}^{\infty} n\delta_n(x)\frac{t^n}{n!} = \sum_{n=1}^{\infty} \delta_n(x)\frac{t^n}{(n-1)!}$$

$$= t\sum_{n=0}^{\infty} \delta_{n+1}(x)\frac{t^n}{n!} = tD_t G = \frac{xt}{1-t^2}G.$$

Thus, from (10.15) and (10.17), we find that

$$\sum_{n=0}^{\infty} \{x\delta_n(x+2) - 4n\delta_n(x) - x\delta_n(x-2)\}\frac{t^n}{n!}$$

$$= x\left(\frac{1+t}{1-t}\right)G - \frac{4xt}{1-t^2}G - x\left(\frac{1-t}{1+t}\right)G = 0. \quad \square$$

Equation (10.14) is useful in evaluating $\delta_n(x)$, especially at an even integer. Because of (10.6), x can be restricted to non-negative values. (Equations (10.2), (10.8), and (10.5) can also be used.)

The next corollary shows that δ_n is an eigenvector of a certain linear operator.

Corollary 10.6. Let $\mathcal{P}$ be the vector space $\mathbb{R}[x]$ and $\mathcal{H}(\mathcal{P})$ be the vector space of linear operators on $\mathcal{P}$. Also, let Q, D, $e^{\pm 2D}$, and $\sinh(2D)$ be the operators in $\mathcal{H}(\mathcal{P})$ defined by $(Qf)(x) = xf(x)$, $Df = f'$,

$$e^{\pm 2D}f = \sum_{k=0}^{\infty} \frac{(\pm 2)^k}{k!}f^{(k)}, \quad \text{and} \quad \sinh(2D)f = \sum_{k=0}^{\infty} \frac{2^{2k+1}}{(2k+1)!}f^{(2k+1)}.$$

Then for $n \geq 0$,

(10.18) $$Q \sinh (2D) \delta_n = 2n \, \delta_n.$$

Proof. Substituting

$$(e^{2D} \delta_n)(x) = \sum_{k=0}^{\infty} \frac{2^k}{k!} \delta_n^{(k)}(x) = \sum_{k=0}^{\infty} \frac{\delta_n^{(k)}(x)}{k!} 2^k = \delta_n(x+2)$$

and

$$(e^{-2D} \delta_n)(x) = \delta_n(x-2)$$

into identity (10.14) gives

$$Q e^{2D} \delta_n = 4n \delta_n + Q e^{-2D} \delta_n,$$

which implies (10.18). $\square$

Theorem 10.7. We have that

(10.19) the set of eigenvalues of $Q \sinh (2D)$ is $\{2n : n = 0, 1, 2, \cdots\}$

and

(10.20) each eigenvalue of $Q \sinh (2D)$ has geometric multiplicity 1.

Proof. (10.19) Using the notation of Theorem 10.5, let $f \in \mathcal{P}$ be a function such that $Q \sinh (2D) f = \lambda f$, $\deg f = n \geq 0$. Since $\{\delta_j\}_{j=0}^{\infty}$ is a simple sequence of polynomials, we can write $f = \sum_{j=0}^{n} a_j \, \delta_j$, $a_n \neq 0$, $n \geq 0$.

Then

$$\lambda \sum_{j=0}^{n} a_j \, \delta_j = \lambda f = Q \sinh (2D) f = \sum_{j=0}^{n} a_j Q \sinh (2D) \delta_j = \sum_{j=0}^{n} a_j \, 2j \, \delta_j,$$

so equating corresponding coefficients of δ_j, we find that

(10.21) $$\lambda \, a_j = 2j \, a_j, \quad j = 0, \cdots, n.$$

In particular, for $j = n$, we have that $\lambda = 2n$, since $a_n \neq 0$.

(10.20) For $n = 0$, we have $f = a_0 \, \delta_0$. For $n \geq 1$, since $\lambda = 2n$, then equation (10.21) becomes $(n - j) \, a_j = 0$, $j = 0, 1, \cdots, n - 1$. Thus, $a_j = 0$ for $j = 0, 1, \cdots, n - 1$, so $f = a_n \delta_n$. $\square$

We next obtain raising and lowering operators for $\{\delta_n(x)\}_{n=0}^{\infty}$.

Lemma 10.8. If $[A, B] = AB - BA$, then

(10.22) $$[D^n, Q] = nD^{n-1}, \quad n \geq 1,$$

(10.23) $$[\sinh (2D), Q] = 2 \cosh(2D),$$

and

(10.24) $[\cosh(2D), Q] = 2\sinh(2D)$.

Proof. (10.22) For $n = 1$, we have

$$(DQ)f(x) = [xf(x)]' = xf'(x) + f(x) = (QD + I)f(x),$$

so $DQ = QD + I$.

Assume the result is true for some $n \geq 1$, i.e., $QD^n = D^nQ - nD^{n-1}$. Then

$$[D^{n+1}, Q] = D^{n+1}Q - Q D^{n+1} = D^{n+1}Q - (Q D^n)D$$
$$= D^{n+1}Q - (D^nQ - nD^{n-1})D = D^{n+1}Q - D^n(QD) + nD^n$$
$$= D^{n+1}Q - D^n(DQ - I) + nD^n = (n+1)D^n.$$

(10.23) Using (10.22), we have that

$$[\sinh(2D), Q] = \sinh(2D)Q - Q\sinh(2D)$$
$$= \sum_{n=0}^{\infty} \frac{(2D)^{2n+1}}{(2n+1)!}Q - Q\sum_{n=0}^{\infty}\frac{(2D)^{2n+1}}{(2n+1)!}$$
$$= \sum_{n=0}^{\infty} \frac{2^{2n+1}}{(2n+1)!}\left(D^{2n+1}Q - Q D^{2n+1}\right) = 2\sum_{n=0}^{\infty}\frac{(2D)^{2n}}{(2n)!} = 2\cosh(2D).$$

(10.24) We have

$$[\cosh(2D), Q] = \cosh(2D)Q - Q\cosh(2D)$$
$$= \sum_{n=0}^{\infty} \frac{(2D)^{2n}}{(2n)!}Q - Q\sum_{n=0}^{\infty}\frac{(2D)^{2n}}{(2n)!} = \sum_{n=0}^{\infty}\frac{2^{2n}}{(2n)!}\left(D^{2n}Q - Q D^{2n}\right)$$
$$= \sum_{n=1}^{\infty} \frac{2^{2n}}{(2n-1)!}D^{2n-1} = 2\sum_{n=0}^{\infty}\frac{(2D)^{2n+1}}{(2n+1)!} = 2\cosh(2D). \quad \square$$

Definition 10.9.

(10.25) $b_{\pm} = \frac{1}{2}Q[I \pm \cosh(2D)]$.

Lemma 10.10. We have

(10.26) $[Q\sinh(2D), b_+] = 2b_+$

and

(10.27) $[Q\sinh(2D), b_-] = -2b_-$.

Proof. (10.26) We have

$$Q \sinh(2D) b_+ = \frac{1}{2} Q[\sinh(2D) Q][I + \cosh(2D)]$$

$$= \frac{1}{2} Q [Q \sinh(2D) + 2 \cosh(2D)][I + \cosh(2D)]$$

$$= \frac{1}{2} Q \{Q [I + \cosh(2D)]\} \sinh(2D) + Q \cosh(2D)[I + \cosh(2D)]$$

$$= \frac{1}{2} Q \{[I + \cosh(2D)] Q - 2\sinh(2D)\} \sinh(2D)$$
$$+ Q \cosh(2D)[I + \cosh(2D)]$$

$$= b_+ Q \sinh(2D) + Q \left[\cosh^2(2D) - \sinh^2(2D) \right] + Q \cosh(2D).$$

Thus, $Q \sinh(2D) b_+ - b_+ Q \sinh(2D) = Q [I + \cosh(2D)] = 2b_+$. A similar proof establishes (10.27). $\square$

Theorem 10.11. We have

(10.28) $$\qquad\qquad b_+ \delta_n = \delta_{n+1}, \; n \geq 0,$$

and

(10.29) $$\qquad\qquad b_- \delta_n = -n(n-1) \delta_{n-1}, \; n \geq 1.$$

Proof. (10.28) Using (10.26) and (10.18), we have

$$Q \sinh(2D)[b_+ \delta_n] = \left(Q \sinh(2D) b_+ \right) \delta_n = [b_+ Q \sinh(2D) + 2b_+] \delta_n$$
$$= b_+ (Q \sinh(2D)\delta_n) + 2b_+ \delta_n = 2(n+1)[b_+ \delta_n].$$

Thus, by (10.20), we have that there is a $\lambda \in \mathbb{R}$ such that

$$b_+ \delta_n = \lambda \delta_{n+1}.$$

Then, since $\delta_n(x)$ is monic for $n \geq 0$, we have

$$\lambda(x^{n+1} + \cdots) = b_+(x^n + \cdots) = \frac{1}{2} Q [I + \cosh(2D)](x^n + \cdots)$$
$$= Q [I + D^2 + \cdots](x^n + \cdots) = x^{n+1} + \cdots,$$

so $\lambda = 1$.

(10.29) Using (10.27) and (10.18), we have

$$Q \sinh(2D)[b_- \delta_n] = \left(Q \sinh(2D) b_- \right) \delta_n = \left(b_- Q \sinh(2D) - 2b_- \right) \delta_n$$
$$= b_- (Q \sinh(2D)\delta_n) - 2b_- \delta_n = 2(n-1)[b_- \delta_n].$$

Thus, by (10.20), we have that there is a $\lambda \in \mathbb{R}$ such that

$$b_-\delta_n = \lambda\delta_{n-1}.$$

Also, $\lambda(x^{n-1}+\cdots) = b_-\delta_n = -Q(D^2+\cdots)(x^n+\cdots)$, $n \geq 1$. When $n = 1$, then $\lambda = 0$. When $n \geq 2$, then $\lambda(x^{n-1}+\cdots) = -n(n-1)x^{n-1}+\cdots$, so $\lambda = -n(n-1)$. $\square$

Corollary 10.12. We have

(10.30) $$[b_+, b_-] = Q \sinh(2D).$$

Proof. Using (10.28) and (10.29), we have that

$$(b_+ b_- - b_- b_+)\delta_n = b_+(b_-\delta_n) - b_-(b_+\delta_n) = -n(n-1)[b_+\delta_{n-1}] - b_-\delta_{n+1}$$
$$= 2n\delta_n = Q \sinh(2D)\delta_n. \square$$

We next introduce some familiar material from combinatorial analysis (see [Ri1], [Ri2]).

Definition 10.13. The "falling factorial" polynomials of degree $n \geq 0$ and step-size $r \geq 1$ are

(10.31) $$(x)_{0,r} = 1 \text{ and } (x)_{n,r} = \prod_{j=1}^{n}[x - (j-1)r], \ n \geq 1.$$

The "rising factorial" polynomials of degree $n \geq 0$ and step-size $r \geq 1$ are

(10.32) $$[x]_{0,r} = 1 \text{ and } [x]_{n,r} = \prod_{j=1}^{n}[x + (j-1)r], \ n \geq 1.$$

When $r = 1$, we write only $(x)_n$ or $[x]_n$. (The $[x]_n$'s are also known as the Pochhammer polynomials.)

In the first case for $n \geq 0$, we have the expressions:

(10.33) $$(x)_n = \sum_{k=0}^{n} s(n, k) x^k$$

and its inverse

(10.34) $$x^n = \sum_{k=0}^{n} S(n, k) (x)_k,$$

where $s(n, k)$ and $S(n, k)$ are the Stirling numbers of the first and second kind, respectively. (Here the notation $s(n, k)$ should not be confused with that in (8.25a).)

Theorem 10.14. For $n \geq 0$ and $x \geq 0$, we have

(10.35) $$\delta_n(x) \leq [x]_n.$$

Proof. The inequality is true for $n = 0$ and $n = 1$. Assume it is true for n and $n+1$ for some $n \geq 0$. Then using (10.8) and (10.32) we have

$$\delta_{n+2}(x) = x\delta_{n+1}(x) + n(n+1)\delta_n(x) \leq x[x]_{n+1} + n(n+1)[x]_n$$
$$\leq x[x]_{n+1} + n(n+1)[x]_n + (n+1)x[x]_n = [x]_{n+2}. \quad \square$$

Corollary 10.15. Equation (10.2) is valid for any $x \in \mathbb{R}$ and $|t| < 1$.

Proof. Using (10.35), we have for $x \geq 0$ that

$$\sum_{n=0}^{\infty} \frac{\delta_n(x)}{n!} t^n \leq \sum_{n=0}^{\infty} \frac{[x]_n}{n!} t^n.$$

By the ratio test we find that

$$\lim_{n\to\infty} \left| \frac{[x]_{n+1}}{(n+1)!} t^{n+1} \middle/ \frac{[x]_n}{n!} t^n \right| = \lim_{n\to\infty} \frac{|x+n|}{n+1} |t| = |t|.$$

But, since $\delta_n(-x) = (-1)^n \delta_n(x)$ by (10.6), it follows that the series converges absolutely for all $x \in \mathbb{R}$ and $|t| < 1$. $\quad \square$

In what follows, we will also need the Lah numbers.

Definition 10.16. The Lah numbers $L_{n,k}$ are defined by

(10.36) $$L_{n,k} = (-1)^n \frac{n!}{k!} \binom{n-1}{k-1} = (-1)^n \frac{(n-1)!}{(k-1)!} \binom{n}{k}, \quad 1 \leq k \leq n,$$

and $L_{n,0} = \delta_{n,0}$ for $n \geq 0$.

A useful property of the Lah numbers is the following [Ri2, p. 49]:

Given the sequences $\{u_n\}_{n=0}^{\infty}$ and $\{v_n\}_{n=0}^{\infty}$. Then for $n \geq 0$

(10.37) $$u_n = \sum_{k=0}^{n} L_{n,k} v_k \iff v_n = \sum_{k=0}^{n} L_{n,k} u_k.$$

Also [Ri2, p. 48], for $0 \leq k \leq n$

(10.38) $$\sum_{j=k}^{n} L_{n,j} L_{j,k} = \delta_{n,k}.$$

In the next theorems we relate $\delta_n(x)$ and $\delta_n(2x)$ to the falling and rising factorial polynomials. The Lah numbers appear in the next theorem as connection coefficients [As2, Lecture 7].

Theorem 10.17. For $n \geq 0$,

$$(10.39) \qquad \delta_n(x) = (-1)^n \sum_{k=0}^{n} L_{n,k} (x)_{k,2},$$

$$(10.40) \qquad (x)_{n,2} = \sum_{k=0}^{n} (-1)^k L_{n,k} \, \delta_k(x),$$

$$(10.41) \qquad \delta_n(x) = \sum_{k=0}^{n} (-1)^k L_{n,k} [x]_{k,2},$$

and

$$(10.42) \qquad [x]_{n,2} = (-1)^n \sum_{k=0}^{n} L_{n,k} \, \delta_k(x).$$

Proof. (10.39) The equation is true for $n = 0$ and 1. For $n \geq 2$, we show that the RHS of (10.39) satisfies (10.8). Substituting, we obtain after some manipulation that

$$\sum_{k=0}^{n+2} L_{n+2,k} (x)_{k,2} = -\sum_{k=1}^{n+2} L_{n+1,k-1} [(x)_{k,2} + 2(k-1)(x)_{k-1,2}]$$

$$+ n(n+1) \sum_{k=0}^{n} L_{n,k} (x)_{k,2},$$

where we have used the identity

$$x(x)_{k-1,2} = (x)_{k,2} + 2(k-1)(x)_{k-1,2}, \ k \geq 1.$$

Equating the corresponding coefficients of $(x)_{k,2}$ on the two sides for $k = 0$, $1 \leq k \leq n$, $k = n+1$, and $k = n+2$ verifies the result. We have also used the identity

$$L_{n+2,k} = -L_{n+1,k-1} - 2kL_{n+1,k} + n(n+1)L_{n,k}, \ n \geq 0, \ 1 \leq k \leq n+2.$$

(10.40) Use (10.37) with $u_n = (-1)^n \delta_n(x)$ and $v_k = (x)_{k,2}$.
(10.41) Using (10.6) and (10.39), we have

$$\delta_n(x) = (-1)^n \delta_n(-x) = (-1)^n \left[(-1)^n \sum_{k=0}^{n} L_{n,k}(-x)_{k,2} \right]$$

$$= \sum_{k=0}^{n} (-1)^k L_{n,k} [x]_{k,2}.$$

(10.42) Use (10.37) on equation (10.41). $\quad \square$

Corollary 10.18. For $n \geq 1$, we have that

(10.43) $$\delta_n(2x) = n! \sum_{k=1}^{n} 2^k \binom{n-1}{k-1} \binom{x}{k}$$

and

(10.44) $$\binom{x}{n} = \frac{(-1)^n}{2^n} \sum_{k=1}^{n} \frac{(-1)^k}{k!} \binom{n-1}{k-1} \delta_k(2x),$$

Proof. (10.43) From (10.39) and (10.36), we have

$$\delta_n(2x) = n! \sum_{k=1}^{n} \frac{2^k}{k!} \binom{n-1}{k-1} x(x-1)\cdots(x-k+1)$$

$$= n! \sum_{k=1}^{n} 2^k \binom{n-1}{k-1} \frac{(x)_k}{k!} = n! \sum_{k=1}^{n} 2^k \binom{n-1}{k-1} \binom{x}{k}.$$

(10.44) Replacing $L_{n,k}$ in (10.40) by the first product in (10.36) gives

$$x(x-2)\cdots(x-2n+2) = (-1)^n n! \sum_{k=1}^{n} \frac{(-1)^k}{k!} \binom{n-1}{k-1} \delta_k(x),$$

where we have used $L_{n,0} = 0$ for $n \geq 1$. Replacing x by $2x$ and using $\binom{x}{n} = \frac{(x)_n}{n!}$ produces the result. $\square$

In the next results, we give the connection coefficients for the $\delta_n(x)$'s and the continuous Hahn polynomials $\{S_n(x)\}_{n=0}^{\infty}$, which are introduced in [BeMP] by the generating function

(10.45) $$G_B(x,t) = \frac{1}{\sqrt{1+t^2}} e^{x \tan^{-1}(t)} \stackrel{\text{def}}{=} \sum_{n=0}^{\infty} S_n(x) t^n.$$

From this definition, it follows that

$$S_n(x) = (-i)^n \, {}_3F_2\left(-n, n+1, \frac{1-ix}{4}; \frac{1}{2}, 1; 1\right).$$

By (10.2), we have that $G(x,t) = \sqrt{1-t^2} \, G_B(-ix, it)$, from which we obtain the formulas for $n \geq 0$:

(10.46) $$\delta_{2n}(x) = -\frac{(2n)!}{4^n} \sum_{k=0}^{n} \frac{(-4)^k}{2n-2k-1} \binom{2n-2k}{n-k} S_{2k}(-ix),$$

$$(10.47) \quad \delta_{2n+1}(x) = -i\frac{(2n+1)!}{4^n} \sum_{k=0}^{n} \frac{(-4)^k}{2n-2k-1}\binom{2n-2k}{n-k} S_{2k+1}(-ix),$$

$$(10.48) \qquad S_{2n}(x) = \frac{(-1)^n}{4^n} \sum_{k=0}^{n} \frac{4^k}{(2k)!}\binom{2n-2k}{n-k} \delta_{2k}(ix),$$

$$(10.49) \qquad S_{2n+1}(x) = i\frac{(-1)^{n+1}}{4^n} \sum_{k=0}^{n} \frac{4^k}{(2k+1)!}\binom{2n-2k}{n-k} \delta_{2k+1}(ix),$$

Theorem 10.19. [Ca4, (9.11)] For $n \geq 1$,

$$(10.50) \qquad \delta_n(2x) = x \sum_{k=0}^{n} \binom{n}{k}(x+k-1)_{n-1}.$$

Proof. Let $S_{n-1}(x) = \sum_{k=0}^{n} \binom{n}{k}(x+k-1)_{n-1}$, $n \geq 1$. By Vandermonde's identity [Mil, p. 134],

$$(x-1+k)_{n-1} = \sum_{l=0}^{n-1} \binom{n-1}{l}(x-1)_l (k)_{n-1-l}$$

$$= \sum_{l=0}^{n-1}(n-l-1)!\binom{n-1}{l}\binom{k}{n-l-1}(x-1)_l,$$

so

$$S_{n-1}(x) = \sum_{l=0}^{n-1}(n-l-1)!\binom{n-1}{l}(x-1)_l \sum_{k=0}^{n}\binom{n}{k}\binom{k}{n-l-1}.$$

From [Go, p. 36, 3.118], we have

$$\sum_{k=0}^{n}\binom{n}{k}\binom{k}{n-l-1} = \binom{n}{l+1}2^{l+1},$$

so that

$$S_{n-1}(x) = \sum_{l=0}^{n-1} 2^{l+1}(n-l-1)!\binom{n-1}{l}\binom{n}{l+1}(x-1)_l$$

$$= (-1)^n \sum_{l=0}^{n-1} 2^{l+1} L_{n,l+1}(x-1)_l.$$

Therefore, by (10.39) we have

$$
\begin{aligned}
xS_{n-1}(x) &= (-1)^n \sum_{l=0}^{n-1} 2^{l+1} L_{n,l+1}(x)_{l+1} \\
&= (-1)^n \sum_{k=0}^{n} L_{n,k}(2x)_{k,2} = \delta_n(2x). \quad \square
\end{aligned}
$$

Lemma 10.20. For $n \geq 1$ and $0 \leq s \leq n$, we have

(10.51) $\qquad x(x + n - s - 1)_{n-1} = (-1)^{n-s}(x)_s(-x)_{n-s}.$

Proof. For $s = 0$ we have

$$
\begin{aligned}
(-1)^n(-x)_n &= (-1)^n(-x)(-x-1)\cdots(-x-n+1) \\
&= x(x+1)\cdots(x+n-1) = (x+n-1)_n = x(x+n-1)_{n-1}.
\end{aligned}
$$

For $s = n$ the result is immediate. For $1 \leq s \leq n-1$ we have

$$
\begin{aligned}
(x)_s(-x)_{n-s} &= [x(x-1)\cdots(x-s+1)] \\
&\quad \cdot [(-x)(-x-1)\cdots(-x-n+s+1)] \\
&= (-1)^n - sx(x+n-s-1)_{n-1}. \quad \square
\end{aligned}
$$

Theorem 10.21. For $n \geq 0$,

(10.52) $\qquad \delta_n(2x) = (-1)^n \sum_{k=0}^{n} (-1)^k \binom{n}{k} (x)_k(-x)_{n-k},$

(10.53) $\qquad \delta_n(2x) = (-1)^n n! \sum_{k=0}^{n} (-1)^k \binom{x}{k}\binom{-x}{n-k},$

(10.54) $\qquad \delta_n(2x) = (-1)^n n! \sum_{k=0}^{[\frac{n}{2}]} (-1)^k \binom{x}{k}\binom{-2x}{n-2k}$

(10.55) $\delta_n(2x) = \begin{cases} (-1)^{[\frac{n}{2}]+1} n! \sum_{k=0}^{[\frac{n}{2}]} (-1)^k \binom{x}{[\frac{n}{2}]-k}\binom{-2x}{2k+1}, & n \text{ odd}, \\[4mm] (-1)^{\frac{n}{2}} n! \sum_{k=0}^{\frac{n}{2}} (-1)^k \binom{x}{\frac{n}{2}-k}\binom{-2x}{2k}, & n \text{ even}, \end{cases}$

and
(10.56)

$$\delta_n(2x) = \begin{cases} (-1)^{[\frac{n}{2}]} n! \displaystyle\sum_{k=0}^{[\frac{n}{2}]} (-1)^k \binom{x}{[\frac{n}{2}]-k} \binom{2x+2k}{2k+1}, & n \text{ odd}, \\[4mm] (-1)^{\frac{n}{2}} n! \displaystyle\sum_{k=0}^{\frac{n}{2}} (-1)^k \binom{x}{\frac{n}{2}-k} \binom{2x+2k-1}{2k}, & n \text{ even}. \end{cases}$$

Proof. (10.52) True for $n = 0$. For $n \geq 1$, if we replace k by $n - k$ in (10.50) and use (10.51), we obtain the result.

(10.53) Using the fact that $\binom{n}{k} = \dfrac{n!}{k!(n-k)!}$, we can transform (10.52) into (10.53) by the equation

$$\binom{n}{k}(x)_k(-x)_{n-k} = n! \frac{(x)_k}{k!} \frac{(-x)_{n-k}}{(n-k)!} = n! \binom{x}{k}\binom{-x}{n-k}.$$

(10.54) From [Go, p. 25, 3.31], we have for $n \geq 0$ that

$$\sum_{k=0}^{n} (-1)^k \binom{x}{k}\binom{y}{n-k} = \sum_{k=0}^{[\frac{n}{2}]} (-1)^k \binom{x}{k}\binom{y-x}{n-2k}.$$

Setting $y = -x$ in this identity and substituting in (10.53) gives (10.54).

(10.55) Replace k by $\left[\frac{n}{2}\right] - k$ in (10.54) and consider the cases where n is even and odd.

(10.56) Let $k \geq 0$. Since

$$(-2x)_{2k+1} = (-2x)(-2x-1)\cdots(-2x-2k)$$
$$= -(2x+2k)(2x+2k-1)\cdots(2x) = -(2x+2k)_{2k+1},$$

then

$$\binom{-2x}{2k+1} = \frac{(-2x)_{2k+1}}{(2k+1)!} = -\frac{(2x+2k)_{2k+1}}{(2k+1)!} = -\binom{2x+2k}{2k+1}.$$

Also, since

$$(-2x)_{2k} = (-2x)(-2x-1)\cdots(-2x-2k+1)$$
$$= (2x)(2x+1)\cdots(2x+2k-1) = (2x+2k-1)_{2k},$$

then similarly $\binom{-2x}{2k} = \binom{2x+2k-1}{2k}$. Substituting these results into (10.55) gives (10.56). □

Corollary 10.22. For $n \geq 0$,

$$(10.57) \quad \sum_{k=0}^{n} (-1)^k \binom{x}{n-k} \binom{2x+2k}{2k+1} = (-1)^n \sum_{k=0}^{2n} 2^{k+1} \binom{2n}{k} \binom{x}{k+1}.$$

Proof. Replacing n by $2n+1$ in (10.56) and using (10.39) and the first equation in (10.36), we find that

$$\sum_{k=0}^{n} (-1)^k \binom{x}{n-k} \binom{2x+2k}{2k+1} = \frac{(-1)^n}{(2n+1)!} \delta_{2n+1}(2x)$$

$$= \frac{(-1)^{n+1}}{(2n+1)!} \sum_{k=1}^{2n+1} L_{2n+1,k} (2x)_{k,2} = \frac{(-1)^{n+1}}{(2n+1)!} \sum_{k=0}^{2n} L_{2n+1,k+1} (2x)_{k+1,2}$$

$$= (-1)^n \sum_{k=0}^{2n} 2^{k+1} \binom{2n}{k} \frac{(x)_{k+1}}{(k+1)!}. \quad \square$$

Theorem 10.23. For $n \geq 0$,

$$(10.58) \quad \delta_{2n+1}(2x) = - \sum_{j=0}^{n} 2^{2j+1} \frac{(2n+1)!}{(n+j+1)!} L_{n+j+1,2j+1}\, \delta_{2j+1}(x)$$

and

$$(10.59) \quad \delta_{2n+2}(2x) = \sum_{j=0}^{n} 2^{2j+2} \frac{(2n+2)!}{(n+j+2)!} L_{n+j+2,2j+2}\, \delta_{2j+2}(x).$$

Proof. The two formulas are readily verified for $n = 0$. Assume they are true for some $n \geq 0$.

(10.58) We have from (10.8) that

$$\delta_{2n+3}(2x) = 2x\delta_{2n+2}(2x) + (2n+1)(2n+2)\delta_{2n+1}(2x),$$

so, using (10.59) and (10.58), we obtain

$$\delta_{2n+3}(2x) = 2x \sum_{j=0}^{n} 2^{2j+2} \frac{(2n+2)!}{(n+j+2)!} L_{n+j+2,2j+2}\, \delta_{2j+2}(x)$$

$$- (2n+1)(2n+2) \sum_{j=0}^{n} 2^{2j+1} \frac{(2n+1)!}{(n+j+1)!} L_{n+j+1,2j+1}\, \delta_{2j+1}(x).$$

But, from (10.8) we have

$$x\delta_{2j+2}(x) = \delta_{2j+3}(x) - (2j+1)(2j+2)\delta_{2j+1}(x), \quad j \geq 0,$$

so we obtain

$$\delta_{2n+3}(2x) =$$

$$\sum_{j=0}^{n} 2^{2j+3} \frac{(2n+2)!}{(n+j+2)!} L_{n+j+2,2j+2} \left(\delta_{2j+3}(x) - (2j+1)(2j+2)\delta_{2j+1}(x) \right)$$

$$- (2n+1)(2n+2) \sum_{j=0}^{n} 2^{2j+1} \frac{(2n+1)!}{(n+j+1)!} L_{n+j+1,2j+1}\, \delta_{2j+1}(x)$$

$$= \sum_{j=0}^{n} 2^{2j+3} \frac{(2n+2)!}{(n+j+2)!} L_{n+j+2,2j+2} \delta_{2j+3}(x)$$

$$- \sum_{j=0}^{n} \frac{2^{2j+1}(2n+2)!}{(n+j+2)!} \Big(4(2j+1)(2j+2)L_{n+j+2,2j+2}$$

$$+ (2n+1)(n+j+2)L_{n+j+1,2j+1} \Big) \delta_{2j+1}(x)$$

$$= \sum_{j=0}^{n+1} 2^{2j+1} \frac{(2n+2)!}{(n+j+1)!} L_{n+j+1,2j}\, \delta_{2j+1}(x)$$

$$- \sum_{j=0}^{n} \frac{2^{2j+1}(2n+2)!}{(n+j+2)!} \Big(4(2j+1)(2j+2)L_{n+j+2,2j+2}$$

$$+ (2n+1)(n+j+2)L_{n+j+1,2j+1} \Big) \delta_{2j+1}(x)$$

$$= 2^{2n+3}\delta_{2n+3}(x) + \sum_{j=0}^{n} \frac{2^{2j+1}(2n+2)!}{(n+j+2)!} \Big((n+j+2)L_{n+j+1,2j}$$

$$- 4(2j+1)(2j+2)L_{n+j+2,2j+2}$$

$$- (2n+1)(n+j+2)L_{n+j+1,2j+1} \Big) \delta_{2j+1}(x).$$

After simplifying the expression in the brackets, the RHS becomes

$$2^{2n+3}\delta_{2n+3}(x) - \sum_{j=0}^{n} 2^{2j+1} \frac{(2n+3)!}{(n+j+2)!} L_{n+j+2,2j+1}\, \delta_{2j+1}(x)$$

$$= - \sum_{j=0}^{n+1} 2^{2j+1} \frac{(2n+3)!}{(n+j+2)!} L_{n+j+2,2j+1}\, \delta_{2j+1}(x).$$

The proof of (10.59) is similar. $\square$

Note that the coefficients in (10.58) and (10.59) are integers.

Theorem 10.24. For $n \geq 0$,

$$(10.60) \qquad \delta_{2n+1}\left(\frac{x}{2}\right) = \frac{1}{2^{2n+1}} \sum_{j=0}^{n} \frac{(2n)!}{(2j)!} \binom{2n+1}{n-j} \delta_{2j+1}(x)$$

and

$$(10.61) \qquad \delta_{2n+2}\left(\frac{x}{2}\right) = \frac{1}{2^{2n+2}} \sum_{j=0}^{n} \frac{(2n+1)!}{(2j+1)!}\binom{2n+2}{n-j}\delta_{2j+2}(x).$$

Proof. The method of proof for these results is essentially the same as in the proof of Theorem 10.23. $\square$

From the above two theorems we can derive a collection of summation formulas.

Corollary 10.25. For $0 \le k \le n$,

$$(10.62) \qquad \sum_{j=k}^{n} \frac{(2j)!}{(n+j+1)!}\binom{2j+1}{j-k}L_{n+j+1,2j+1} = -\frac{1}{2n+1}\delta_{n,k},$$

$$(10.63) \qquad \sum_{j=k}^{n} \frac{2j+1}{(j+k+1)!}\binom{2n+1}{n-j}L_{j+k+1,2j+1} = -\frac{1}{(2n)!}\delta_{n,k},$$

$$(10.64) \qquad \sum_{j=k}^{n} \frac{(-1)^j}{2j+1}\binom{n+j}{2j}\binom{2j+1}{j-k} = \frac{(-1)^n}{2n+1}\delta_{n,k},$$

and

$$(10.65) \qquad \sum_{j=k}^{n}(-1)^j(2j+1)\binom{2n+1}{n-j}\binom{j+k}{2k} = (-1)^n(2n+1)\delta_{n,k}.$$

Proof. (10.62) For $n \ge 0$, let

$$\alpha_j^{(n)} = -2^{2j+1}\frac{(2n+1)!}{(n+j+1)!}L_{n+j+1,2j+1}$$

and

$$\beta_k^{(j)} = \frac{1}{2^{2j+1}}\frac{(2j)!}{(2k)!}\binom{2j+1}{j-k}, \quad j \ge 0.$$

Then we see that (10.58) and (10.60) (with x replaced by $2x$) can also be written, respectively, as $\delta_{2n+1}(2x) = \sum_{j=0}^{n}\alpha_j^{(n)}\delta_{2j+1}(x)$ and $\delta_{2j+1}(x) =$

$\sum_{k=0}^{j}\beta_k^{(j)}\delta_{2k+1}(2x)$. Substituting the second summation into the first gives

$$\delta_{2n+1}(2x) = \sum_{j=0}^{n}\alpha_j^{(n)}\sum_{k=0}^{j}\beta_k^{(j)}\delta_{2k+1}(2x) = \sum_{k=0}^{n}\left\{\sum_{j=k}^{n}\alpha_j^{(n)}\beta_k^{(j)}\right\}\delta_{2k+1}(2x).$$

Thus, $\displaystyle\sum_{j=k}^{n}\alpha_j^{(n)}\beta_k^{(j)} = \delta_{n,k}$, from which (10.62) follows.

(10.63) For $n \geq 0$, let

$$\alpha_j^{(n)} = \frac{1}{2^{2n+1}}\frac{(2n)!}{(2j)!}\binom{2n+1}{n-j}$$

and

$$\beta_k^{(j)} = -2^{2k+1}\frac{(2j+1)!}{(j+k+1)!}L_{j+k+1,2k+1}, \ j \geq 0.$$

Then (10.60) (with x replaced by $2x$) and (10.58) can be written, respectively, as $\displaystyle\delta_{2n+1}(x) = \sum_{j=0}^{n}\alpha_j^{(n)}\delta_{2j+1}(2x)$ and $\displaystyle\delta_{2j+1}(2x) = \sum_{k=0}^{j}\beta_k^{(j)}\delta_{2k+1}(x)$. Substituting the second summation into the first gives

$$\delta_{2n+1}(x) = \sum_{j=0}^{n}\alpha_j^{(n)}\sum_{k=0}^{j}\beta_k^{(j)}\delta_{2k+1}(x) = \sum_{k=0}^{n}\left\{\sum_{j=k}^{n}\alpha_j^{(n)}\beta_k^{(j)}\right\}\delta_{2k+1}(x).$$

Thus, $\displaystyle\sum_{j=k}^{n}\alpha_j^{(n)}\beta_k^{(j)} = \delta_{n,k}$, from which (10.63) follows.

(10.64) Use the definition of the Lah number in (10.36).

(10.65) Same as in (10.64). □

Corollary 10.26. For $0 \leq k \leq n$,

(10.66) $$\sum_{j=k}^{n}\frac{(2j+1)!}{(n+j+2)!}\binom{2j+2}{j-k}L_{n+j+2,2j+2} = \frac{1}{2n+2}\delta_{n,k},$$

(10.67) $$\sum_{j=k}^{n}\frac{2j+2}{(j+k+2)!}\binom{2n+2}{n-j}L_{j+k+2,2k+2} = \frac{1}{(2n+1)!}\delta_{n,k},$$

(10.68) $$\sum_{j=k}^{n}\frac{(-1)^j}{2j+1}\binom{n+j+1}{n-j}\binom{2j+2}{j-k} = \frac{(-1)^n}{2n+2}\delta_{n,k},$$

and

(10.69) $$\sum_{j=k}^{n}(-1)^j(j+1)\binom{2n+2}{n-j}\binom{j+k+1}{j-k} = (-1)^n(n+1)\delta_{n,k}.$$

Proof. This proof is similar to the proof of Corollary 10.25. ◻

Theorem 10.27. For $n \geq 0$ and $x > 0$, we have that

$$(10.70) \qquad \delta_n(x) = x^{1-\frac{z}{2}} \left\{ \frac{\partial^n}{\partial u^n} \left[u^{n-1} (2u - x)^{\frac{z}{2}} \right] \right\}_{u=x}.$$

Proof. It is clear that

$$(10.71) \qquad \frac{d^j u^{n-1}}{du^j} \bigg]_{u=x} = \frac{(n-1)!}{(n-j-1)!} x^{n-j-1}, \ 0 \leq j \leq n-1,$$

and

$$(10.72) \qquad \left[\frac{\partial^j}{\partial u^j} (2u - x)^{\frac{z}{2}} \right]_{u=x} = (x)_{j,2} \, x^{\frac{z}{2}-j}, \ j \geq 0.$$

Then, using Leibniz's formula, (10.71), (10.72), (10.36), and (10.39), we find that

$$\left\{ \frac{\partial^n}{\partial u^n} \left[u^{n-1}(2u - x)^{\frac{z}{2}} \right] \right\}_{u=x}$$

$$= \sum_{j=1}^{n} \binom{n}{j} \left\{ \frac{d^{n-j}}{du^{n-j}} u^{n-1} \right\}_{u=x} \left\{ \frac{\partial^j}{\partial u^j} (2u - x)^{\frac{z}{2}} \right\}_{u=x}$$

$$= x^{\frac{z}{2}-1} (-1)^n \sum_{j=0}^{n} L_{n,j} (x)_{j,2} = x^{\frac{z}{2}-1} \delta_n(x). \quad ◻$$

Theorem 10.28. We have for $n \geq 0$ that

$$(10.73) \qquad \delta'_{2n+1}(x) = (2n+1)! \sum_{k=0}^{n} \frac{\delta_{2k}(x)}{(2n-2k+1)(2k)!}$$

and

$$(10.74) \qquad \delta'_{2n+2}(x) = (2n+2)! \sum_{k=0}^{n} \frac{\delta_{2k+1}(x)}{(2n-2k+1)(2k+1)!}.$$

Proof. From (2.19), we have that $\delta'_n(x) = n! \sum_{k=0}^{n-1} \frac{a_k \, \delta_k(x)}{k!}$, where $f(t) = \tanh^{-1} t$ in (2.3), so $a_k = \frac{1}{k}$ when $k \geq 1$ is odd and $a_k = 0$ when $k \geq 0$ is even. The two results follow. ◻

Note. The above theorem is a special case of Theorem 1.11, using (11.5). Theorem 1.11 also specializes to give a formula for the rth derivative of $\delta_n(x)$.

Corollary 10.29. We have that

$$(10.75) \qquad \delta''_{2n+1}(x) = (2n+1)! \sum_{k=1}^{n} \frac{\xi_1(n-k)\,\delta_{2k-1}(x)}{(n-k+1)(2k-1)!}, \quad n \geq 1,$$

and

$$(10.76) \qquad \delta''_{2n+2}(x) = (2n+2)! \sum_{k=0}^{n} \frac{\xi_1(n-k)\,\delta_{2k}(x)}{(n-k+1)(2k)!}, \quad n \geq 0,$$

where $\xi_1(m) = \displaystyle\sum_{k=0}^{m} \frac{1}{2k+1}$, $m \geq 0$.

Proof. (10.75) Differentiating (10.73) and substituting from (10.74) gives

$$\delta''_{2n+1}(x) = (2n+1)! \sum_{j=0}^{n} \frac{\delta'_{2j}(x)}{(2n-2j+1)(2j)!}$$

$$= (2n+1)! \sum_{j=0}^{n} \frac{1}{(2n-2j+1)(2j)!} \cdot (2j)! \sum_{k=1}^{j} \frac{\delta_{2k-1}(x)}{(2j-2k+1)(2k-1)!}$$

$$= (2n+1)! \sum_{k=1}^{n} \frac{\delta_{2k-1}(x)}{(2k-1)!} \sum_{j=k}^{n} \frac{1}{(2n-2j+1)(2j-2k+1)}$$

$$= (2n+1)! \sum_{k=1}^{n} \frac{\delta_{2k-1}(x)}{2(n-k+1)(2k-1)!} \sum_{j=k}^{n} \left(\frac{1}{2n-2j+1} + \frac{1}{2j-2k+1}\right).$$

But

$$(10.77) \qquad \sum_{j=k}^{n} \frac{1}{2n-2j+1} = \sum_{j=k}^{n} \frac{1}{2j-2k+1} = \sum_{j=0}^{n-k} \frac{1}{2j+1} = \xi_1(n-k),$$

which gives the result.

(10.76) Differentiating (10.74) and substituting from (10.73) gives

$$\delta''_{2n+2}(x) = (2n+2)! \sum_{j=0}^{n} \frac{\delta'_{2j+1}(x)}{(2n-2j+1)(2j+1)!}$$

$$= (2n+2)! \sum_{j=0}^{n} \frac{1}{(2n-2j+1)(2j+1)!} \cdot (2j+1)! \sum_{k=0}^{j} \frac{\delta_{2k}(x)}{(2j-2k+1)(2k)!}$$

$$= (2n+2)! \sum_{k=0}^{n} \frac{\delta_{2k}(x)}{(2k)!} \sum_{j=k}^{n} \frac{1}{(2n-2j+1)(2j-2k+1)}$$

$$= (2n+2)! \sum_{k=0}^{n} \frac{\delta_{2k}(x)}{2(n-k+1)(2k-1)!} \sum_{j=k}^{n} \left(\frac{1}{2n-2j+1} + \frac{1}{2j-2k+1}\right).$$

As before, we obtain the result using (10.77). □

We next relate $\delta_n(x)$ to the hypergeometric function $F(a, b; c; x)$.

Theorem 10.30. [Bat, pp. 491–496] For $n \geq 1$,

$$(10.78) \qquad \delta_n(x) = n!\, x F(1 - n, 1 - \tfrac{1}{2}x; 2; 2).$$

Proof. For $n = 1$, $\delta_1(x) = x$, since $F(0, 1 - \tfrac{1}{2}x; 2, 2) = 1$. For $n \geq 2$, if we put $a = 1 - n$, $b = 1 - \tfrac{1}{2}x$, and $c = x = 2$ into

$$(10.79) \quad F(a, b; c; x) =$$
$$1 + \sum_{k=1}^{\infty} \frac{a(a + 1) \cdots (a + k - 1) b(b + 1) \cdots (b + k - 1)}{c(c + 1) \cdots (c + k - 1)\, k!}\, x^k,$$

we obtain the finite sum

$$x F(1 - n, 1 - \tfrac{1}{2}x; 2; 2)$$
$$= x + x \sum_{k=1}^{n-1} \frac{(1 - n)(2 - n) \cdots (k - n)(1 - \tfrac{x}{2})(2 - \tfrac{x}{2}) \cdots (k - \tfrac{x}{2})}{k!\,(k + 1)!}\, 2^k$$
$$= x + \sum_{k=1}^{n-1} \frac{(n - 1)!}{k!\,(k + 1)!\,(n - k - 1)!}\, (x)_{k+1,2}$$
$$= x + \sum_{l=2}^{n} \frac{(n - 1)!}{(l - 1)!\,l!\,(n - l)!}\, (x)_{l,2} = \sum_{l=1}^{n} \frac{1}{l!}\binom{n - 1}{l - 1} (x)_{l,2}.$$

It follows then that

$$n!\, x\, F(1 - n, 1 - \tfrac{1}{2}x; 2; 2) = (-1)^n \sum_{l=0}^{n} L_{n,l}\, (x)_{l,2} = \delta_n(x),$$

using (6.24) and $L_{n,0} = 0$, $n \geq 1$. □

Corollary 10.31. [GrR, p. 1043, 9.131, 1] [AbS, p. 559, 15.3, 4] We have for $n \geq 1$ that

$$(10.80) \qquad \delta_n(x) = (-1)^{n-1} n!\, x F(1 - n, 1 + \tfrac{1}{2}x; 2; 2).$$

Proof. Replace x by $-x$ in (6.45) and use (10.6). □

Corollary 10.32. For $n \geq 0$,

$$(10.81) \qquad \delta_n(x) = (x)_{n,2}\, F\left(1 - n, -n; 1 - n + \frac{x}{2}; \frac{1}{2}\right).$$

Proof. Trivial for $n = 0$. For $n \geq 1$, we have from [Ca, p. 179, 6.5-3] that

$$F(-n, \beta; c; x) = \frac{[\beta]_n}{[c]_n}(-x)^n F\left(-n, 1 - c - n; 1 - \beta - n; \frac{1}{x}\right).$$

Replacing n by $n-1$, β by $1 - \dfrac{x}{2}$, and setting $c = x = 2$ in this formula and combining with (10.78) gives

$$\delta_n(x) = n!\, x \frac{[1 - \frac{x}{2}]_{n-1}}{[2]_{n-1}} (-2)^{n-1} F\left(1 - n, -n;\, 1 - n + \frac{x}{2};\, \frac{1}{2}\right).$$

Substituting $[2]_{n-1} = n!$ and $x[1 - \frac{x}{2}]_{n-1} = (-\frac{1}{2})^{n-1}(x)_{n,2}$ into this equation gives the result. $\square$

Theorem 10.33. For $n \geq 1$, we have

(10.82) $\qquad \delta_n(2x) = (-1)^n n!\, (F(-n, x;\, 1, 2) - F(-n+1, x;\, 1, 2)).$

Proof. From [AbS, p. 561, 15.4.1], we have for $n \geq 1$ that

$$F(-n, x;\, 1, 2) = \sum_{k=0}^{n} \frac{[-n]_k [x]_k}{(k!)^2} 2^k = \sum_{k=0}^{n} \frac{(-1)^k}{k!} \binom{n}{k} [2x]_{k,2},$$

where the formulas $2^k [x]_k = [2x]_{k,2}$ and $[-n]_k = (-1)^k \dfrac{n!}{(n-k)!}$ were used. We then have that

$$F(-n, x;\, 1; 2) - F(-n+1, x;\, 1, 2)$$

$$= \sum_{k=0}^{n} \frac{(-1)^k}{k!} \binom{n}{k} [2x]_{k,2} - \sum_{k=0}^{n-1} \frac{(-1)^k}{k!} [2x]_{k,2}$$

$$= \frac{(-1)^n}{n!} [2x]_{n,2} + \sum_{k=1}^{n-1} \frac{(-1)^k}{k!} \left[\binom{n}{k} - \binom{n-1}{k} \right] [2x]_{k,2}$$

$$= \sum_{k=1}^{n} \frac{(-1)^k}{k!} \binom{n-1}{k-1} [2x]_{k,2}$$

$$= \frac{(-1)^n}{n!} \sum_{k=0}^{n} (-1)^k L_{n,k} [2x]_{k,2} = \frac{(-1)^n}{n!} \delta_n(2x),$$

where we have used (10.36) and (10.41). $\square$

In the next corollary we obtain an integral representation for $\delta_n(x)$.

Corollary 10.34. For $n \geq 1$, we have

(10.83) $\qquad \delta_n(x) = \dfrac{n!}{\pi} \sin\left(\dfrac{\pi x}{2}\right) \displaystyle\int_{-1}^{1} t^{n-1} G(x, t)\, dt,$

where $|\,\mathrm{Re}\,(x)\,| < 2$.

Proof. Using (10.78) and [AbS, p. 558, 15.3.1], we have

$$\delta_n(x) = n!\, x F(1 - n, 1 - \tfrac{1}{2}x;\, 2;\, 2)$$

$$= n!\, x \frac{1}{\Gamma(1 - \frac{1}{2}x)\Gamma(1 + \frac{1}{2}x)} \int_{0}^{1} u^{-\frac{1}{2}x} (1 - u)^{\frac{1}{2}x} (1 - 2u)^{n-1}\, du.$$

Setting $t = 1 - 2u$, we obtain $du = -\frac{1}{2}dt$, $u = \frac{1}{2}(1 - t)$, $1 - u = \frac{1}{2}(1 + t)$.
Then using the familiar formula

$$\Gamma\left(1 - \frac{x}{2}\right)\Gamma\left(1 + \frac{x}{2}\right) = \frac{\pi x}{2}\csc\left(\frac{\pi x}{2}\right),$$

we have

$$\delta_n(x) = \frac{n!x}{\pi x \csc(\frac{\pi x}{2})}\int_{-1}^{1}\left(\frac{1 + t}{1 - t}\right)^{\frac{x}{2}}t^{n-1}dt$$

$$= \frac{n!}{\pi}\sin\left(\frac{\pi x}{2}\right)\int_{-1}^{1}t^{n-1}G(x, t)dt. \quad \Box$$

The curious property in the next corollary follows from the obvious fact that $F(a, b; c; z) = F(b, a; c; z)$.

Corollary 10.35. [BaB, (10)] [Ba, p. 2071] For $m, n \geq 1$, we have

(10.84) $$(m - 1)!\,\delta_n(2m) = (n - 1)!\,\delta_m(2n).$$

Proof. Using the formula (10.78), we find that

$$(m - 1)!\,\delta_n(2m) = (m - 1)!\,n!\,2m\,F(1 - n, 1 - m; 2; 2)$$
$$= (n - 1)!\,m!\,2n\,F(1 - m, 1 - n; 2; 2) = (n - 1)!\,\delta_m(2n). \quad \Box$$

Lemma 10.36. Let $m, n \geq 1$ with $m \leq n$. Then

(10.85) $$\delta_m(2n) = (m - 1)!\sum_{k=1}^{m}k\,2^k\binom{m}{k}\binom{n}{k}.$$

Proof. By (10.39), we have

$$\delta_m(2n) = (-1)^m\sum_{k=1}^{m}L_{m,k}\,(2n)_{k,2}.$$

Since $m \leq n$, it follows that $(2n)_{k,2} = 2^k\dfrac{n!}{(n - k)!}$. We obtain the result using (10.36). $\quad \Box$

Theorem 10.37. If $m, n \geq 1$, then

(10.86) $$\delta_m(2n) = (m - 1)!\sum_{k=1}^{min\{m,n\}}k\,2^k\binom{m}{k}\binom{n}{k}.$$

Proof. When $m \leq n$, the result is (10.85). When $m \geq n$, then by (10.84) and (10.85) we have that

$$\delta_m(2n) = \frac{(m - 1)!}{(n - 1)!}\,\delta_n(2m) = (m - 1)!\sum_{k=1}^{n}k\,2^k\binom{m}{k}\binom{n}{k}.$$

Since in the two cases the upper index is the minimum of m and n, the result is proved. $\square$

Definition 10.38 [Co, p. 81] Delannoy numbers For $m, n \geq 0$, let

$$(10.87) \qquad \mu_{m,n} = \sum_{j=0}^{\min\{m,n\}} \binom{m}{j}\binom{n}{j} 2^j.$$

It is clear that $\mu_{m,n} \in \mathbb{Z}^+$, $\mu_{m,n} = \mu_{n,m}$, and $\mu_{m,0} = 1$.

Theorem 10.39. For $m \geq 0$ and $n \geq 1$,

$$(10.88) \qquad \delta_n(2m) = n!\,(\mu_{m,n} - \mu_{m,n-1}).$$

Proof. From [StC, p. 278], we have for $m \geq 0$ that

$$\frac{(1+t)^m}{(1-t)^{m+1}} = \sum_{n=0}^{\infty} \mu_{m,n}\, t^n. \quad (\text{cf. [Pi, p. 348]}).$$

Then from (10.2) and (10.3), we have that

$$1 + \sum_{n=1}^{\infty} \frac{\delta_n(2m)}{n!}\, t^n = G(2m,t) = \left(\frac{1+t}{1-t}\right)^m = (1-t)\sum_{n=0}^{\infty}\mu_{m,n} t^n$$

$$= 1 + \sum_{n=1}^{\infty}(\mu_{m,n} - \mu_{m,n-1})\, t^n,$$

from which the result is obtained by equating the corresponding coefficients of t^n. $\square$

We note that (10.88) can also be obtained from (10.86).

Corollary 10.40. If $m \geq 0$ and $n \geq 1$, then

$$(10.89) \qquad \mu_{m,n} = 1 + \sum_{j=1}^{n} \frac{\delta_j(2m)}{j!}.$$

Proof. From (10.88) we have

$$\frac{\delta_j(2m)}{j!} = \mu_{m,j} - \mu_{m,j-1},$$

so summing we find

$$\sum_{j=1}^{n} \frac{\delta_j(2m)}{j!} = \mu_{m,n} - \mu_{m,0} = \mu_{m,n} - 1. \quad \square$$

We next express $\delta_n(x)$ in terms of the Jacobi polynomial $P_n^{(\alpha,\beta)}(z)$.

From [GrR, 8.962, 1], we have the formula

$$(10.90) \quad P_n^{(\alpha,\beta)}(z) = \frac{\Gamma(n+\alpha+1)}{n!\,\Gamma(\alpha+1)} F\left(-n,\; n+\alpha+\beta+1;\; \alpha+1;\; \frac{1-z}{2}\right).$$

Corollary 10.41. [Ba, p. 2072] For $n \geq 1$,

$$(10.91) \quad \delta_n(x)$$
$$= (n-1)!\,x P_{n-1}^{(1,\,-n-\frac{5}{2})}(-3) = (-1)^{n-1}(n-1)!\,x\,P_{n-1}^{(-n-\frac{5}{2},1)}(3).$$

Proof. Combining (10.78) with (10.90) gives the result. The second form in equation (10.91) comes from the formula $P_n^{(\alpha,\beta)}(-x) = (-1)^n P_n^{(\beta,\alpha)}(x)$ [GrR, p. 10.35, 8.961, 1]. □

We can now use (10.91) to transform identities involving the Jacobi polynomial into identities for $\delta_n(x)$. For example, from [GrR, p. 1036, 9] we have the identity

$$(10.92) \quad (2n+\alpha+\beta)P_n^{(\alpha,\,\beta-1)}(x)$$
$$= (n+\alpha+\beta)P_n^{(\alpha,\,\beta)}(x) + (n+\alpha)P_{n-1}^{(\alpha,\,\beta)}(x).$$

Replacing n by $n+1$ in (10.91) and shifting x gives

$$(10.93) \qquad\qquad P_n^{(1,\,-n-\frac{5}{2})}(-3) = \frac{\delta_{n+1}(x-2)}{(x-2)n!}.$$

Thus, if we put $x = -3, \alpha = 1$, and $\beta = -n - \frac{5}{2}$, into (10.92), we obtain

$$(10.94) \quad (x-2n-2)P_n^{(1,\,-(n+1)-\frac{5}{2})}(-3)$$
$$= (x-2)P_n^{(1,\,-n-\frac{5}{2})}(-3) - 2(n+1)P_{n-1}^{(1,\,-n-\frac{5}{2})}(-3).$$

This equation produces (10.105) below when (10.91) and (10.93) are used. Although the other three identities in Theorem 10.46 can be derived in a similar way, we will prove these four identities using (10.2) or (10.3) instead.

In Chapter 15 we will derive further identities for $\delta_n(x)$ using Jacobi polynomial identities (cf. Theorems 15.7–15.10).

Theorem 10.42. [Ba, (38a)] For $n \geq 1$ and $-2 < x < \infty$,

$$(10.95) \qquad \delta_n(x) = \frac{(-1)^{n-1}(n-1)!\,x}{2^{1+\frac{5}{2}}\Gamma(1+\frac{5}{2})} \int_0^\infty e^{-\frac{1}{2}t}\, t^{\frac{5}{2}}\, L_{n-1}^1(t)dt$$

and

$$(10.96) \qquad \delta_n(2x) = \frac{2(-1)^{n-1}(n-1)!}{\Gamma(x)} \int_0^\infty e^{-t}\, t^x\, L_{n-1}^1(2t)dt,$$

where $L_n^1(t)$ is the generalized Laguerre polynomial [Ra1, p. 201].

Also, for $x > 0$, we have

$$(10.97) \qquad \delta_n(x) = \frac{(-1)^{n-1}(n-1)!\,x}{2^{1+\frac{x}{2}}\Gamma(1+\frac{x}{2})} \int_0^\infty (x-t)\,t^{\frac{x}{2}-1}e^{-\frac{1}{2}t}L_n(t)\,dt,$$

where $L_n(t)$ is the Laguerre polynomial.

Proof. (10.95) From [ObB, p. 110, 11.44, 2nd form] we have

$$\int_0^\infty e^{-\frac{1}{2}t}t^{\frac{x}{2}}\,L_{n-1}^1(t)\,dt = n\,2^{1+\frac{x}{2}}\Gamma(1+\frac{x}{2})F(1-n,\,1+\frac{x}{2};\,2\,;2).$$

The result follows using (10.80).

(10.96) This result follows from (10.95) by the change of variable $x \to 2x$ and $t \to 2t$.

(10.97) From [GrR, p. 1037, 8.971, 2] we have for $n \geq 1$ that

$$\frac{d}{dt}L_n^\alpha(t) = -L_{n-1}^{\alpha+1}(t),$$

so in particular,

$$L_{n-1}^1(t) = -\frac{d}{dt}L_n(t), \quad \text{with } L_n^0 = L_n.$$

If we substitute this result in (10.95) we obtain

$$
\begin{aligned}
\delta_n(x) &= \frac{(-1)^n(n-1)!\,x}{2^{1+\frac{x}{2}}\Gamma(1+\frac{x}{2})} \int_0^\infty e^{-\frac{1}{2}t}t^{\frac{x}{2}}\frac{d}{dt}L_n(t)\,dt \\
&= \frac{(-1)^n(n-1)!\,x}{2^{1+\frac{x}{2}}\Gamma(1+\frac{x}{2})} \left\{ \left[e^{-\frac{1}{2}t}t^{\frac{x}{2}}L_n(t)\right]_0^\infty - \int_0^\infty L_n(t)\frac{d}{dt}[e^{-\frac{1}{2}t}t^{\frac{x}{2}}]\,dt\right\} \\
&= \frac{(-1)^{n-1}(n-1)!\,x}{2^{1+\frac{x}{2}}\Gamma(1+\frac{x}{2})} \int_0^\infty L_n(t)\frac{d}{dt}[e^{-\frac{1}{2}t}t^{\frac{x}{2}}]\,dt \\
&= \frac{(-1)^{n-1}(n-1)!\,x}{2^{1+\frac{x}{2}}\Gamma(1+\frac{x}{2})} \int_0^\infty (x-t)\,t^{\frac{x}{2}-1}e^{-\frac{1}{2}t}L_n(t)\,dt. \quad \square
\end{aligned}
$$

Theorem 10.43. For $n \geq 0$ and $x > -1$, we have

$$(10.98) \qquad \delta_n(2x) = \frac{(-1)^n 2\,n!}{\Gamma(x+1)} \int_0^\infty e^{-2t}t^x L_n(2t)\,{}_1F_1(x;\,x+1;\,t)\,dt,$$

where ${}_1F_1$ is the Kummer confluent hypergeometric function.

Proof. From [ObB, p. 200, 21.2], we have for $x > -1$ and $p > 1$ that

$$\int_0^\infty e^{-pt}t^x\,{}_1F_1(x;\,x+1;\,t)\,dt = \frac{\Gamma(x+1)}{p(p-1)^x}.$$

Next, using this result, we have that

$$\int_0^\infty e^{-pt} t^x L_n(2t) \, _1F_1(x; \, x+1; \, t) \, dt$$

$$= \int_0^\infty \left\{ L_n\left(-2\frac{\partial}{\partial p}\right) e^{-pt} \right\} t^x \, _1F_1(x; \, x+1; \, t) dt$$

$$= L_n\left(-2\frac{\partial}{\partial p}\right) \int_0^\infty e^{-pt} t^x \, _1F_1(x; \, x+1; \, t) \, dt$$

$$= \Gamma(x+1) L_n\left(-2\frac{\partial}{\partial p}\right) \left\{ \frac{1}{p(p-1)^x} \right\}.$$

To compute the latter partial derivative, we derive a formula using Leibniz's formula:

$$\frac{\partial^k}{\partial p^k}\left\{ (p-1)^{-x} p^{-1} \right\} = \sum_{j=0}^k \binom{k}{j} \frac{\partial^j}{\partial p^j} \{(p-1)^{-x}\} \frac{\partial^{k-j}}{\partial p^{k-j}} \{p^{-1}\}$$

$$= \sum_{j=0}^k \binom{k}{j} \frac{(-1)^j [x]_j}{(p-1)^{x+j}} \frac{(-1)^{k-j}(k-j)!}{p^{k-j+1}}$$

$$= \frac{(-1)^k k!}{p^{k+1}(p-1)^x} \sum_{j=0}^k \frac{1}{j!} \left(\frac{p}{p-1}\right)^j [x]_j.$$

Thus, since the $L_n(x) = \sum_{k=0}^n \binom{n}{k} \frac{(-x)^k}{k!}$, $n \geq 0$, we can use the previous result to obtain

$$L_n\left(-2\frac{\partial}{\partial p}\right)\left\{ \frac{1}{p(p-1)^x} \right\} = \sum_{k=0}^m \frac{2^k}{k!} \binom{n}{k} \frac{\partial^k}{\partial p^k}\left\{ \frac{1}{p(p-1)^x} \right\}$$

$$= \frac{1}{p(p-1)^x} \sum_{k=0}^n \binom{n}{k} \left(\frac{-2}{p}\right)^k \sum_{j=0}^k \frac{1}{j!} \left(\frac{p}{p-1}\right)^j [x]_j.$$

If we now set $p = 2$ in the preceding formula, we obtain

$$\int_0^\infty e^{-2t} \, t^x L_n(2t) \, _1F_1(x; \, x+1; \, t) \, dt = \frac{1}{2}\Gamma(x+1) \sum_{k=0}^n (-1)^k \binom{n}{k} \sum_{j=0}^k \frac{2^j}{j!} [x]_j$$

$$= \frac{1}{2}\Gamma(x+1) \sum_{k=0}^n (-1)^k \binom{n}{k} \sum_{j=0}^k \frac{1}{j!} [2x]_{j,2}$$

$$= \frac{1}{2}\Gamma(x+1) \sum_{j=0}^n \left\{ \frac{1}{j!} \sum_{k=j}^n (-1)^k \binom{n}{k} \right\} [2x]_{j,2}$$

But, from [Go, p. 1, 1.4], we have that $\sum_{k=j}^{n}(-1)^k \binom{n}{k} = (-1)^j \binom{n-1}{j-1}$, so the RHS becomes

$$\frac{1}{2}\Gamma(x+1) \sum_{j=0}^{n} \frac{(-1)^j}{j!} \binom{n-1}{j-1} [2x]_{j,2}.$$

Finally,

$$\frac{(-1)^n 2n!}{\Gamma(x+1)} \int_0^\infty e^{-2t} t^x L_n(2t)\,_1F_1(x; x+1; t)\, dt$$

$$= (-1)^n \sum_{j=0}^{n} (-1)^j \frac{n!}{j!} \binom{n-1}{j-1} [2x]_{j,2}$$

$$= (-1)^n \sum_{j=0}^{n} (-1)^j L_{n,j}\, [2x]_{j,2} = \delta_n(2x),$$

using (10.41). $\square$

Corollary 10.44. For $n \geq 0$ and $x \in (-\infty, 2)$, we have

$$(10.99) \qquad \delta_n(x) = \frac{2n!}{\Gamma(1-\frac{x}{2})} \int_0^\infty e^{-t} t^{-\frac{x}{2}} L_n(2t)\,_1F_1(1; 1 - \tfrac{x}{2}; -t)\, dt.$$

Proof. Replace x by $-\frac{x}{2}$ in equation (10.98). Then, using (10.6) and the formula $_1F_1(a; b; z) = e^z\,_1F_1(b-a; b; -z)$ from [AbS, p. 505, 13.1.27], the result follows. $\square$

Theorem 10.45. Suppose that $x \in (-1, 1)$ and $t > 0$. Then

$$(10.100) \qquad \frac{e^t t^{-\frac{x}{2}}}{\Gamma(1-\frac{x}{2})}\,_1F_1\left(1; 1 - \tfrac{x}{2}; -t\right) = \sum_{n=0}^{\infty} \frac{L_n(2t)}{n!} \delta_n(x).$$

Proof. Let $x \in (-\infty, 2)$ and define $f : (0, \infty) \to \mathbb{R}$ by

$$(10.101) \qquad f(t) = \frac{2^{\frac{x}{2}} e^{\frac{t}{2}}}{\Gamma(1-\frac{x}{2}) t^{\frac{x}{2}}}\,_1F_1\left(1; 1 - \tfrac{x}{2}; -\tfrac{t}{2}\right).$$

Then, with $c(x) = \frac{2^x x^2}{[\Gamma(1-\frac{x}{2})]^2}$, we have $e^{-t}[f(t)]^2 \sim \frac{c(x)}{t^{x+2}}$ as $t \to \infty$, since [Ol, p. 257, (10.08)] gives that $_1F_1\left(1; 1 - \tfrac{x}{2}; -\tfrac{t}{2}\right) \sim -\tfrac{x}{t}$ as $t \to \infty$. For $x \in (-1,1)$, it follows that $e^{-t}[f(t)]^2 \in L_1([0,\infty))$, since $\frac{1}{t^x}$ is integrable near 0 and $\frac{1}{t^{x+2}}$ is integrable for large t.

Now, from [Le, p. 88, Theorem 3] with $\alpha = 0$, we have the expansion $f(t) = \sum_{n=0}^{\infty} b_n L_n(t)$, where

$$b_n = \int_0^\infty f(t) e^{-t} L_n(t)\, dt$$

$$= \frac{2^{\frac{s}{2}}}{\Gamma(1 - \frac{s}{2})} \int_0^\infty e^{-\frac{1}{2}t} t^{-\frac{s}{2}} {}_1F_1\left(1; 1 - \frac{x}{2}; -\frac{t}{2}\right) L_n(t)\, dt,$$

using (10.101). Thus, replacing t by $2t$, we have

$$b_n = \frac{2}{\Gamma(1 - \frac{x}{2})} \int_0^\infty e^{-t} t^{-\frac{s}{2}} {}_1F_1\left(1; 1 - \frac{x}{2}; -t\right) L_n(2t)\, dt = \frac{1}{n!}\, \delta_n(x),$$

using (10.99), so $f(t) = \sum_{n=0}^{\infty} \frac{L_n(t)}{n!}\, \delta_n(x)$. Equating this result with that in (10.101) and replacing t by $2t$ gives the result. $\square$

Theorem 10.46. For $n \geq 0$,

(10.102) $\delta_{n+1}(x + 2) - \delta_{n+1}(x) = (n + 1)\, [\delta_n(x + 2) + \delta_n(x)],$

(10.103) $2\delta_{n+1}(x) = (x + 2n)\delta_n(x) + x\, \delta_n(x - 2),$

(10.104) $2\delta_{n+1}(x) = x\, \delta_n(x + 2) + (x - 2n)\, \delta_n(x),$

(10.105) $(x - 2 - 2n)\, \delta_{n+1}(x) = x\, \delta_{n+1}(x - 2) - 2n(n + 1)\, \delta_n(x),$

and

(10.106) $x(x + 2n)\, \delta_{n+2}(x + 2)$
$$= (x + 2n + 2)[x^2 + 2(n + 1)x + 4n(n + 2)]\delta_{n+1}(x)$$
$$- n(n + 1)x(x + 2n + 4)\delta_n(x - 2).$$

Proof. (10.102)

$$\sum_{n=0}^{\infty} \left[\frac{\delta_{n+1}(x + 2) - \delta_{n+1}(x)}{n + 1}\right] \frac{t^n}{n!} = \sum_{n=1}^{\infty} \left[\frac{\delta_n(x + 2) - \delta_n(x)}{n}\right] \frac{t^{n-1}}{(n - 1)!}$$

$$= \frac{1}{t} \sum_{n=0}^{\infty} [\delta_n(x + 2) - \delta_n(x)] \frac{t^n}{n!} = \frac{1}{t}\, [G(x + 2, t) - G(x, t)]$$

$$= G(x + 2, t) + G(x, t) = \sum_{n=0}^{\infty} [\delta_n(x + 2) + \delta_n(x)] \frac{t^n}{n!},$$

using (10.3). The result follows by equating corresponding coefficients.

(10.103) Using (10.16), (10.2), and (10.17), we have that

$$\sum_{n=0}^{\infty}[2\delta_{n+1}(x) - (x+2n)\delta_n(x) - x\delta_n(x-2)]\frac{t^n}{n!}$$

$$= 2\sum_{n=0}^{\infty}\delta_{n+1}(x)\frac{t^n}{n!} - x\sum_{n=0}^{\infty}\delta_n(x)\frac{t^n}{n!}$$

$$- 2\sum_{n=0}^{\infty}n\delta_n(x)\frac{t^n}{n!} - x\sum_{n=0}^{\infty}\delta_n(x-2)\frac{t^n}{n!}$$

$$= \frac{2x}{1-t^2}G - xG - \frac{2xt}{1-t^2}G - x\left(\frac{1-t}{1+t}\right)G = 0.$$

(10.104) From (10.14) we find that

$$x\delta_n(x-2) = -4n\delta_n(x) + x\delta_n(x+2).$$

Substituting this in (10.103) gives the result.

(10.105) We begin by computing some results. From (10.16) we obtain

$$\sum_{n=0}^{\infty}n\delta_{n+1}(x)\frac{t^n}{n!} = tD_t^2G = \frac{tx(x+2t)}{(1-t^2)^2}G$$

and

$$\sum_{n=0}^{\infty}\delta_{n+1}(x-2)\frac{t^n}{n!} = \frac{x-2}{1-t^2}G(x-2,t) = \frac{x-2}{(1+t)^2}G,$$

using (10.3). Also, from (10.2),

$$\sum_{n=0}^{\infty}(n+1)\delta_n(x)\frac{t^n}{n!} = D_t(tG),$$

so

$$\sum_{n=0}^{\infty}n(n+1)\delta_n(x)\frac{t^n}{n!} = tD_t^2(tG) = \frac{tx(tx+2)}{(1-t^2)^2}G.$$

Then

$$\sum_{n=0}^{\infty}[(x-2-2n)\delta_{n+1}(x) - x\delta_{n+1}(x-2) + 2n(n+1)\delta_n(x)]\frac{t^n}{n!}$$

$$= (x-2)\sum_{n=0}^{\infty}\delta_{n+1}(x)\frac{t^n}{n!} - 2\sum_{n=0}^{\infty}n\delta_{n+1}(x)\frac{t^n}{n!}$$

$$- x\sum_{n=0}^{\infty}\delta_{n+1}(x-2)\frac{t^n}{n!} + 2\sum_{n=0}^{\infty}n(n+1)\delta_n(x)\frac{t^n}{n!}$$

$$= \frac{x(x-2)}{1-t^2}G - \frac{2tx(x+2t)}{(1-t^2)^2}G - \frac{x(x-2)}{(1+t)^2}G + \frac{2tx(tx+2)}{(1-t^2)^2}G = 0.$$

(10.106) To verify this complicated identity, we first write it in the form

$$2x[n\delta_{n+2}(x+2)] + x^2[\delta_{n+2}(x+2)] - 8[n^3\delta_{n+1}(x)] - (8x+24)[n^2\delta_{n+1}(x)]$$
$$- (4x^2 + 16x + 16)[n\delta_{n+1}(x)] - (x^3 + 4x^2 + 4x)[\delta_{n+1}(x)]$$
$$+ 2x[n^3\delta_n(x-2)] + (x^2 + 6x)[n^2\delta_n(x-2)] + (x^2 + 4x)[n\delta_n(x-2)] = 0.$$

As in the proof of (10.105), for each bracketed term in the above equation we can express the sum $\sum_{n=0}^{\infty}[\cdots]\dfrac{t^n}{n!}$ as some operator acting on $G(x,t)$. The operators for the nine terms in the brackets are listed below after the arrow:

$$n\,\delta_{n+2}(x+2) \to tD_t^3\, G(x+2,t) \qquad \delta_{n+2}(x+2) \to D_t^2\, G(x+2,t)$$
$$n^3\delta_{n+1}(x) \to (tD_t)^3 D_t\, G(x,t) \qquad n^2\delta_{n+1}(x) \to (tD_t)^2 D_t\, G(x,t)$$
$$n\,\delta_{n+1}(x) \to tD_t^2\, G(x,t) \qquad \delta_{n+1}(x) \to D_t\, G(x,t)$$
$$n^3\delta_n(x-2) \to (tD_t)^3 G(x-2,t) \qquad n^2\delta_n(x-2) \to (tD_t)^2 G(x-2,t)$$
$$n\,\delta_n(x-2) \to tD_t\, G(x-2,t).$$

If these derivatives are computed and substituted into the above equation, we obtain zero. This is probably best done using a symbol-manipulating program such as MACSYMA. □

In the next theorem we give another generating function for $\delta_n(x)$.

Theorem 10.47.

$$(10.107) \qquad xe^{-t}\,_1F_1\left(1 + \frac{x}{2}; 2; 2t\right) = \sum_{n=0}^{\infty} \frac{\delta_{n+1}(x)}{(n+1)!}\frac{t^n}{n!}.$$

Proof. From [Han, p. 295, 45.2.2], we have that

$$\sum_{n=0}^{\infty} \frac{1}{[a+1]_n} P_n^{(a,\,b-n)}(x)\, t^n = e^{\frac{(1+x)t}{2}}\,_1F_1\left(-b; a+1; \frac{(1-x)t}{2}\right).$$

Setting $x = -3, a = 1$, and $b = -1 - \dfrac{x}{2}$, with $[2]_n = (n+1)!$, we obtain the result from (10.31) and (10.91) (see also [Fe, p. 120, (12)]). □

Further identities for the $\delta_n(x)$ polynomials can be derived by using other Jacobi polynomial identities.

We next give some evaluations of $\delta_n(x)$.

Theorem 10.48. For $m \geq 0$ and $n \geq 1$, where $c_m = \left(\dfrac{(2m)!}{2^m m!}\right)^2$, we have

$$(10.108) \qquad \delta_n(0) = 0,$$

$$(10.109) \qquad \delta_{2m}(1) = c_m, \quad \delta_{2m+1}(1) = (2m+1)c_m,$$

(10.110) $\delta_n(2) = 2n!,$

(10.111) $\delta_{2m}(3) = (8m+1)c_m, \quad \delta_{2m+1}(3) = (2m+1)(8m+3)c_m,$

(10.112) $\delta_n(4) = 4n \cdot n!,$

(10.113) $\delta_n(6) = 2(2n^2+1)n!,$

and

(10.114) $\delta_n(8) = \dfrac{8}{3}n(n^2+2)n!.$

Proof. (10.108) Put $x = 0$ into (10.14).

(10.109) True for $m = 0$ and 1. The equations are verified inductively by showing that the pair of formulas in (10.109) satisfy (10.8) (at $x = 1$) for $n = 2m$ and for $n = 2m + 1$.

(10.110) Put $x = 2$ in (10.2) and (10.3), which gives

$$1 + \sum_{n=1}^{\infty} \frac{\delta_n(2)}{n!} t^n = \frac{1+t}{1-t} = 1 + \sum_{n=1}^{\infty} 2t^n.$$

Equation (10.110) then follows by equating the corresponding coefficients of t^n.

(10.111) Put $x = 1$ in (10.14) and use (10.6). This gives the equation $\delta_n(3) = [4n + (-1)^n]\delta_n(1)$, from which the results follow using (10.109).

(10.112)–(10.114) These follow successively from (10.14) by putting $x = 2, 4,$ and 6, and using preceding results. $\square$

Remark. In the proof of (10.110), we evaluated $\delta_n(2)$ from the expansion of $\dfrac{1+t}{1-t}$. Conversely, in general we can use the values of $\delta(x)$ to find the series expansion for $\left(\dfrac{1+t}{1-t}\right)^{\frac{x}{2}}$. For example, setting $x = 1$ in (10.2) and (10.3) and using (10.109), we find the interesting Maclaurin expansion

(10.115) $e^{\tanh^{-1}t} = \left(\dfrac{1+t}{1-t}\right)^{\frac{1}{2}} = \sum_{n=0}^{\infty} \delta_n(1)\dfrac{t^n}{n!}$

$$= \sum_{m=0}^{\infty}\left\{\delta_{2m}(1)\frac{t^{2m}}{(2m)!} + \delta_{2m+1}(1)\frac{t^{2m+1}}{(2m+1)!}\right\}$$

$$= \sum_{m=0}^{\infty} \frac{(2m)!}{2^{2m}(m!)^2}\left(t^{2m} + t^{2m+1}\right).$$

In the following theorem we determine the structure constants for the sequence $\{\delta_n(x)\}_{n=0}^{\infty}$.

Theorem 10.49. Let $m, n \geq 0$, where $m + n \neq 0$, and let $\rho(m, n) = min\{m, n\} - \delta_{m,n}$. Then

$$(10.116) \qquad \delta_m(x)\,\delta_n(x) = \sum_{j=0}^{\rho(m,n)} \Delta_j^{(m,n)}\, \delta_{m+n-2j}(x),$$

where

$$(10.117) \qquad \Delta_j^{(m,n)} = (-1)^j \left(j!\right)^2 \binom{m}{j}\binom{n}{j}\binom{m+n-1-j}{j}.$$

Proof. Using (10.2) and the sum formula for $\tanh^{-1} t$ obtained from (10.1), we find that

$$\sum_{m,n=0}^{\infty} \frac{\delta_m(x)}{m!}\frac{\delta_n(x)}{n!}\, t^m u^n = \left(\sum_{m=0}^{\infty} \delta_m(x)\frac{t^m}{m!}\right)\left(\sum_{n=0}^{\infty} \delta_n(x)\frac{u^n}{n!}\right)$$

$$= e^x \tanh^{-1} t\, e^x \tanh^{-1} u = e^x (\tanh^{-1} t + \tanh^{-1} u)$$

$$= e^{x \tanh^{-1}\left(\frac{t+u}{1+tu}\right)} = 1 + \sum_{i=0}^{\infty} \frac{\delta_{i+1}(x)}{(i+1)!}\frac{(t+u)^{i+1}}{(1+tu)^{i+1}}.$$

By [AbS, p. 8.22], we have that

$$\frac{1}{(1+tu)^{i+1}} = \sum_{j=0}^{\infty}(-1)^j \binom{i+j}{i}(tu)^j, \quad i \geq 0.$$

Thus, using the binomial expansion we find that

$$\sum_{m,n=0}^{\infty} \frac{\delta_m(x)}{m!}\frac{\delta_n(x)}{n!}\, t^m u^n$$

$$= 1 + \sum_{i=0}^{\infty}\sum_{j=0}^{\infty}\sum_{k=0}^{i+1} \frac{(-1)^j}{(i+1)!}\binom{i+1}{k}\binom{i+j}{i}\delta_{i+1}(x)\, t^{j+k} u^{i+j-k+1}.$$

For $i, j \geq 0$ and $0 \leq k \leq i+1$, let

$$(10.118) \quad a(i,j,k) = \frac{(-1)^j}{(i+1)!}\binom{i+1}{k}\binom{i+j}{i}\delta_{i+1}(x)$$

$$= \frac{(-1)^j\,(i+j)!}{i!\,j!\,k!\,(i-k+1)!}\,\delta_{i+1}(x).$$

Then

(10.119) $\displaystyle\sum_{m,n=0}^{\infty} \frac{\delta_m(x)}{m!} \frac{\delta_n(x)}{n!} t^m u^n$

$$= 1 + \sum_{i=0}^{\infty}\sum_{j=0}^{\infty}\sum_{k=0}^{i+1} a(i,j,k)\, t^{j+k} u^{i+j-k+1}.$$

Now consider the case $m \geq n$, where $m + n \neq 0$. Set $j + k = m$ and $i + j - k + 1 = n$. Combining the conditions $i, j, k \geq 0$ with $j + k = m$ and $i = m + n - 1 - 2j$ gives that $0 \leq j, k \leq m$ and $0 \leq i \leq m + n - 1$, where $i \equiv m + n - 1 \pmod 2$. However, for any i we conclude from $0 \leq k \leq i + 1$ and $j = n - (i + 1 - k)$ that $0 \leq j \leq n$. In the special case when $m = n = j$, the relation $i = 2n - 2j - 1$ gives $i = -1$, which is below the range $i \geq 0$. We thus write the interval for j as $0 \leq j \leq n - \delta_{m,n}$. The complete set of triples (i, j, k) for which the term $a(i,j,k)\, t^m u^n$ occurs on the right side of (10.119) is then the following: For each $j \in [0, n - \delta_{m,n}]$, we have $i = m + n - 1 - 2j$ and $k = m - j$. Consequently equation (10.119) can be written as

(10.120) $\displaystyle\sum_{m,n=0}^{\infty} \frac{\delta_m(x)}{m!} \frac{\delta_n(x)}{n!} t^m u^n$

$$= 1 + \left\{ \sum_{j=0}^{n-\delta_{m,n}} a(m+n-2j-1, j, m-j) \right\} t^m u^n.$$

Equating coefficients on the two sides and using (10.118), we obtain

(10.121) $\displaystyle \delta_m(x)\,\delta_n(x) = m!\,n! \sum_{j=0}^{n-\delta_{m,n}} a(m+n-2j-1, j, m-j)$

$$= \sum_{j=0}^{n-\delta_{m,n}} \Delta_j^{(m,n)} \delta_{m+n-2j}(x).$$

In the case $m < n$, we can obtain the result by using formula (10.121). Since the final summand in this formula is symmetric in m and n, we can interchange those letters in (10.121) to obtain

(10.122) $\displaystyle \delta_n(x)\,\delta_m(x) = \sum_{j=0}^{m-\delta_{m,n}} \Delta_j^{(m,n)} \delta_{m+n-2j}(x).$

Combining (10.121) and (10.122) gives (10.116). □

The next corollary gives a collection of binomial coefficient identities.

Corollary 10.50. We have for $m, n \geq 1$ that

(10.123) $\displaystyle \sum_{j=0}^{\rho(m,n)} (-1)^j \binom{m+n-1-j}{j}\binom{m+n-2j}{m-j} = 2.$

For $m, n \geq 0$, we have that

$$
(10.124) \quad \sum_{j=0}^{\rho(2m,2n)} (-4)^j \binom{2m+2n-1-j}{j} \binom{2m+2n-2j}{2m-j}
$$
$$
\times \binom{2m+2n-2j}{m+n-j} = \binom{2m}{m}\binom{2n}{n}, \quad m+n \neq 0,
$$

$$
(10.125) \quad \sum_{j=0}^{\rho(2m+1,2n+1)} (-4)^j \binom{2m+2n+1-j}{j} \binom{2m+2n+2-2j}{2m+1-j}
$$
$$
\times \binom{2m+2n+2-2j}{m+n+1-j} = 4\binom{2m}{m}\binom{2n}{n},
$$

and

$$
(10.126) \quad \sum_{j=0}^{\min\{2m,2n+1\}} (-4)^j \binom{2m+2n-j}{j} \binom{2m+2n+1-2j}{2m-j}
$$
$$
\times \binom{2m+2n-2j}{m+n-j} = \binom{2m}{m}\binom{2n}{n}.
$$

Proof. (10.123) Put $x = 2$ into (10.116) and use (10.110).

(10.124) Put $x = 1$, $2m$ for m, $2n$ for n in (10.116) and use (10.109).

(10.125) Put $x = 1$, $2m+1$ for m, $2n+1$ for n in (10.116) and use the results in (10.109).

(10.126) Put $x = 1$, $2m$ for m, $2n+1$ for n in (10.116) and use the results in (10.109). $\square$

Note. Other more complicated identities of a similar kind can be gotten from (10.116) by using other values of $\delta_n(x)$, as in (10.111)–(10.114).

We next develop an addition formula for $\delta_n(x)$.

Lemma 10.51. For $n \geq 0$,

$$
(10.127) \quad \sum_{k=0}^{n} (-1)^k (x-2k) \binom{x}{k} \binom{x-n-k-1}{n-k} = x\,\delta_{n,0}.
$$

Proof. True for $n = 0$. Assume $n \geq 1$. Let

$$
F(n,k) = (-1)^k (x-2k) \binom{x}{n} \binom{x-n-k-1}{n-k}, \quad 0 \leq k \leq n.
$$

Then using Gosper's algorithm as implemented in EKHAD [PWZ, Ch. 5], we find for $0 \leq k \leq n-1$ that

$$
F(n,k) = \frac{1}{n} \Big[G(n,k+1) - G(n,k) \Big],
$$

where

$$G(n,r) = -\frac{r(x-n-r)}{x-2r}F(n,r), \quad 0 \le r \le n,$$

as is readily checked. In general then, for $0 \le N \le n-1$, we have that

$$\sum_{k=0}^{N} F(n,k) = \frac{1}{n}\sum_{k=0}^{N}[G(n,k+1) - G(n,k)]$$

$$= \frac{1}{n}[G(n,N+1) - G(n,0)] = \frac{1}{n}G(n,N+1).$$

Thus, for $N = n-1$, we obtain

$$\sum_{k=0}^{n} F(n,k) = F(n,n) + \sum_{k=0}^{n-1} F(n,k) = F(n,n) + \frac{1}{n}G(n,n) = 0. \quad \Box$$

The following identity is analogous to an identity for Hermite polynomials given by Burchnall [Bur].

Theorem 10.52. Let $m,n \ge 0$, $m+n \ne 0$, and $\rho(m,n) = \min\{m,n\} - \delta_{m,n}$. Then

$$(10.128) \quad \delta_{m+n}(x) = \frac{1}{m+n}\sum_{j=0}^{\rho(m,n)}(j!)^2(m+n-2j)\binom{m}{j}\binom{n}{j}\binom{m+n}{j}$$

$$\times \delta_{m-j}(x)\delta_{n-j}(x).$$

Proof.
Case 1. $0 \le m < n$
If we use (10.116) on the product of deltas in equation (10.128), then we obtain

$$\sum_{j=0}^{m}(j!)^2(m+n-2j)\binom{m}{j}\binom{n}{j}\binom{m+n}{j}\sum_{k=0}^{m-j}\Delta_k^{(m-j,n-j)}\delta_{m+n-2j-2k}(x)$$

$$= \sum_{s=0}^{m}\left\{\sum_{i=0}^{s}(i!)^2(m+n-2i)\binom{m}{i}\binom{n}{i}\binom{m+n}{i}\Delta_{s-i}^{(m-i,n-i)}\right\}\delta_{m+n-2s}(x),$$

where we have used the formula $\sum_{j=0}^{m}\sum_{k=0}^{m-j}f(j,k) = \sum_{s=0}^{m}\sum_{i=0}^{s}f(i,s-i)$.

Now,

$$\Delta_{s-i}^{(m-i,n-i)} = (-1)^{s-i}[(s-i)!]^2\binom{m-i}{s-i}\binom{n-i}{s-i}\binom{m+n-s-i-1}{s-i},$$

which gives

$$\frac{1}{m+n}\sum_{s=0}^{m}\left\{\sum_{i=0}^{s}(-1)^i(i\,!)^2[(s-i)\,!]^2(m+n-2i)\binom{m}{i}\binom{n}{i}\binom{m-i}{s-i}\right.$$
$$\left.\times\binom{n-i}{s-i}\binom{m+n}{i}\binom{m+n-s-i-1}{s-i}\right\}(-1)^s\delta_{m+n-2s}(x)$$

$$=\frac{m\,!\,n\,!}{m+n}\sum_{s=0}^{m}\frac{(-1)^s}{(m-s)\,!\,(n-s)\,!}\,\delta_{m+n-2s}(x)$$
$$\times\sum_{i=0}^{s}(-1)^i(m+n-2i)\binom{m+n}{i}\binom{m+n-s-i-1}{s-i}.$$

Applying Lemma 10.51 to the inner sum gives

$$\frac{m\,!\,n\,!}{m+n}\sum_{s=0}^{m}\frac{(-1)^s}{(m-s)\,!\,(n-s)\,!}\,\delta_{m+n-2s}(x)\,(m+n)\,\delta_{s,0}.$$

Thus, the right side of (10.128) equals $\delta_{m+n}(x)$.

Case 2. $m=n\geq 1$

In this case, again using (10.116), we have on the right:

$$\frac{1}{n}\sum_{j=0}^{n-1}(j\,!)^2(n-j)\binom{n}{j}^2\binom{2n}{j}[\delta_{n-j}(x)]^2$$

$$=\frac{1}{n}\sum_{j=0}^{n-1}(j\,!)^2(n-j)\binom{n}{j}^2\binom{2n}{j}\sum_{k=0}^{n-j-1}\Delta_k^{(n-j,n-j)}\delta_{2n-2j-2k}(x)$$

$$=\frac{1}{n}\sum_{j=0}^{n-1}\left\{\sum_{i=0}^{s}(i\,!)^2(n-i)\binom{n}{i}^2\binom{2n}{i}\Delta_{s-i}^{(n-i,n-i)}\right\}\delta_{2n-2s}(x).$$

Now,

$$\Delta_{s-i}^{(n-i,n-i)}=(-1)^{n-i}[(s-i)\,!]^2\binom{n-i}{s-i}^2\binom{2m-s-i-1}{s-i},$$

so

$$\frac{1}{n}\sum_{s=0}^{n-1}\left\{\sum_{i=0}^{s}(-1)^i(i\,!)^2[(s-i)\,!]^2(n-i)\binom{n}{i}^2\binom{n-i}{s-i}^2\right.$$
$$\left.\times\binom{2n}{i}\binom{2n-2-i-1}{s-i}\right\}(-1)^s\delta_{2n-2s}(x)$$

$$=\frac{(n\,!)^2}{n}\sum_{s=0}^{n-1}\frac{(-1)^s}{[(n-s)\,!]^2}\,\delta_{2n-2s}(x)\sum_{i=0}^{s}(-1)^i(n-i)\binom{2n}{i}\binom{2n-s-i-1}{s-i}.$$

Again, applying Lemma 10.51 to the inner sum gives

$$\frac{(n\,!)^2}{n}\sum_{s=0}^{n-1}\frac{(-1)^s}{[(n-s)\,!]^2}\,\delta_{2n-2s}(x)\,n\,\delta_{s,0},$$

which equals $\delta_{2n}(x)$. □

Corollary 10.53. We have that

$$(10.129)\qquad \sum_{j=0}^{m-1}(m+n-2j)\binom{m+n}{j}=m\binom{m+n}{m},\quad 1\le m\le n.$$

Proof. If $m=n$, then put $x=2$ in (10.128) and use (10.110), which gives (10.129). If $m<n$, then $\rho(m,n)=m$, so we write (10.128) as

$$\delta_{m+n}(x)=\frac{1}{m+n}\left\{(m!)^2(n-m)\binom{n}{m}\binom{m+n}{m}\delta_{n-m}(x)\right.$$
$$\left.+\sum_{j=0}^{m-1}(j!)^2(m+n-2j)\binom{m}{j}\binom{n}{j}\binom{m+n}{j}\delta_{m-j}(x)\,\delta_{n-j}(x)\right\}.$$

Replacing x by 2 and using (10.110) gives (10.129). □

It was observed by C. Krattenthaler (private communication) that the $\delta_n(x)$'s and the Meixner polynomials $\{M_n(x;b,c)\}_{n=0}^\infty$, as defined in [Ch, p. 176], are simply related. From their generating functions we find the relationship is $\delta_0(x)=M_0\left(\frac{x}{2};1,-1\right)$ and $\delta_n(x)=M_n\left(\frac{x}{2};1,-1\right)-nM_{n-1}\left(\frac{x}{2};1,-1\right)$, $n\ge1$, so

$$M_n\left(\frac{x}{2};1,-1\right)=n!\sum_{j=0}^{n}\frac{\delta_j(x)}{j!},\quad n\ge0.$$

To develop a Rodrigues-type formula for $\delta_n(x)$, we observe that $\delta_n(x)$ is a limiting case of the Pollaczek polynomial $P_n^\lambda(x;\varphi)$ on $\mathbb{R}$ as $\lambda\to0$, given by the formula

$$\delta_n(x)=n!\,(-i)^n P_n^0\left(\frac{ix}{2};\frac{\pi}{2}\right).$$

(See [Er, v. 2, p. 218–221].) (We wish to thank Mourad Ismail for pointing out this fact to us.) Using this connection, we can modify Toscano's formula for $P_n^\lambda(x;\varphi)$ to apply to $\delta_n(x)$ (See [Er, v. 2, p. 221]). The following simple polynomials occur in this formula.

Definition 10.54. Using (10.28), let

$$(10.130)\qquad K_n(x)=(-1)^{n+1}\left(\frac{x+n}{2}-1\right)_{n-1},\quad n\ge1.$$

The initial members of this polynomial sequence are: $K_1(x) = 1$, $K_2(x) = -\dfrac{x}{2}$, $K_3(x) = \dfrac{1}{4}(x^2 - 1)$, $K_4(x) = -\dfrac{x}{8}(x^2 - 4)$, $K_5(x) = \dfrac{1}{16}(x^2 - 1)(x^2 - 9)$, and $K_6(x) = -\dfrac{x}{32}(x^2 - 4)(x^2 - 16)$.

Lemma 10.55. For $n \geq 1$,

(10.131) $$K_{n+2}(x) = \frac{1}{4}(x^2 - n^2)K_n(x).$$

Proof. True for $n = 1$. Assume $n \geq 2$. Then using (10.130), we have

$$K_{n+2}(x) = (-1)^{n+3}\left(\frac{x+n}{2}\right)_{n+1}$$

$$= (-1)^{n+3}\left(\frac{x+n}{2}\right)\left(\frac{x+n}{2} - 1\right)\cdots\left(\frac{x+n}{2} - n + 1\right)\left(\frac{x+n}{2} - n\right)$$

$$= (-1)^{n+3}\left(\frac{x+n}{2}\right)(-1)^{n+1}K_n(x)\left(\frac{x-n}{2}\right) = \frac{1}{4}(x^2 - n^2)K_n(x). \quad \square$$

Corollary 10.56. We have that

(10.132) $$K_{2n+1}(x) = \frac{1}{2^{2n}}\prod_{j=1}^{n}\left(x^2 - (2j-1)^2\right), \quad n \geq 1,$$

(10.133) $$K_{2n}(x) = -\frac{x}{2^{2n-1}}\prod_{j=1}^{n-1}\left(x^2 - (2j)^2\right), \quad n \geq 2,$$

and

(10.134) $$K_n(-x) = (-1)^{n+1}K_n(x), \quad n \geq 1.$$

We next define a difference operator η.

Definition 10.57. Let $(\eta f)(x) = f(x+1) - f(x-1)$.

As usual, let $(Ef)(x) = f(x+1)$, then we have that $(E^{-1}f)(x) = f(x-1)$ and $\eta = E - E^{-1}$.

Also, let $e(x) = e^{\frac{i\pi x}{2}}$, so that $e(x \pm 1) = \pm i\, e(x)$.

Lemma 10.58. For $n \geq 1$,

(10.135) $$[\eta^n(fe)](x) = i^n e(x)\sum_{k=0}^{n}\binom{n}{k}f(x+n-2k).$$

Proof. For $n = 1$, we have the operator equation

$$\eta(fe) = i[[(E + E^{-1})f]\,e,$$

since

$$[\eta(fe)](x) = f(x+1)e(x+1) - f(x-1)e(x-1)$$
$$= i[f(x+1) + f(x-1)]e(x) = i\{[(E+E^{-1})f\}](x)e(x).$$

Thus,

$$\eta^n(fe) = i^n[(E+E^{-1})^n f]e = i^n\left[\sum_{k=0}^{n}\binom{n}{k}E^{n-2k}f\right]e.$$

It follows that

$$[\eta^n(fe)](x) = i^n e(x)\sum_{k=0}^{n}\binom{n}{k}f(x+n-2k). \quad \square$$

We now give the $\delta_n(x)$ analogue of the Rodrigues–Toscano formula for $P_n^\lambda(x;\varphi)$.

Theorem 10.59. For $n \geq 1$,

$$(10.136) \qquad \delta_n(x) = -\frac{i^n x}{2}e^{-\frac{i\pi x}{2}}[\eta^n(K_n e)](x).$$

Proof. Using (10.135), (10.130), $k \to n-k$, and (10.50), we find that

$$-\frac{i^n x}{2}e^{-\frac{i\pi x}{2}}[\eta^n(K_n e)](x) = -\frac{i^n x}{2}e^{-\frac{i\pi x}{2}}\left\{i^n e^{\frac{i\pi x}{2}}\sum_{k=0}^{n}\binom{n}{k}K_n(x+n-2k)\right\}$$

$$= \frac{x}{2}\sum_{k=0}^{n}\binom{n}{k}(-1)^{n+1}K_n(x+n-2k) = \frac{x}{2}\sum_{k=0}^{n}\binom{n}{k}\left(\frac{x}{2}+n-k-1\right)_{n-1}$$

$$= \frac{x}{2}\sum_{k=0}^{n}\binom{n}{k}\left(\frac{x}{2}+k-1\right)_{n-1} = \delta_n(x). \quad \square$$

We next prove a useful inversion formula.

Lemma 10.60. Let $\{a_n\}_{n=1}^{\infty} \subset \mathbf{C}$ and define $\{b_n\}_{n=0}^{\infty}$ by $b_0 = a_1$ and $b_n = a_{n+1} - n(n+1)a_n$, $n \geq 1$. Then for $n \geq 1$,

$$(10.137) \qquad a_n = (n-1)!\,n!\sum_{j=1}^{n}\frac{b_{j-1}}{(j-1)!j!}.$$

Proof. Let $\hat{a}_n = (n-1)!\,n!\sum_{j=1}^{n}\frac{b_{j-1}}{(j-1)!j!}$, $n \geq 1$. Then

$$\hat{a}_{n+1} - n(n+1)\hat{a}_n = n!\,(n+1)!\sum_{j=1}^{n+1}\frac{b_{j-1}}{(j-1)!j!}$$

$$- n!\,(n+1)!\sum_{j=1}^{n}\frac{b_{j-1}}{(j-1)!j!} = b_n.$$

Since $\hat{a}_1 - a_1 = 0$ and $\hat{a}_{n+1} - a_{n+1} = n(n+1)(\hat{a}_n - a_n)$, it follows that $\hat{a}_n - a_n = 0$. $\square$

The following is a Christoffel-Darboux type identity for $\delta_n(x)$ (see [Ch, p. 23, 4.9]).

Theorem 10.61. For $n \geq 1$, we have that

$$(10.138) \qquad \left| \begin{matrix} \delta_n(x) & \delta_n(y) \\ \delta_{n+1}(x) & \delta_{n+1}(y) \end{matrix} \right|_+ = (n-1)!\, n!\, (x+y) \sum_{j=1}^{n} \frac{\delta_j(x)\, \delta_j(y)}{(j-1)!\, j!},$$

where the subscript "+" indicates the permanent function.

Proof. If we let

$$a_n = a_n(x,y) = \left| \begin{matrix} \delta_n(x) & \delta_n(y) \\ \delta_{n+1}(x) & \delta_{n+1}(y) \end{matrix} \right|_+, \quad n \geq 1,$$

then using (10.8), we can write

$$a_{n+1} = \left| \begin{matrix} \delta_{n+2}(x) & \delta_{n+2}(y) \\ \delta_{n+1}(x) & \delta_{n+1}(y) \end{matrix} \right|_+$$

$$= \left| \begin{matrix} x\,\delta_{n+1}(x) + n(n+1)\,\delta_n(x) & y\,\delta_{n+1}(y) + n(n+1)\,\delta_n(y) \\ \delta_{n+1}(x) & \delta_{n+1}(y) \end{matrix} \right|_+$$

$$= (x+y)\,\delta_{n+1}(x)\,\delta_{n+1}(y) + n(n+1)\,a_n,$$

so

$$a_{n+1} - n(n+1)\,a_n = (x+y)\,\delta_{n+1}(x)\,\delta_{n+1}(y).$$

If we then let $b_n = b_n(x,y) = (x+y)\,\delta_{n+1}(x)\,\delta_{n+1}(y)$, $n \geq 0$, we find that

$$b_0 = (x+y)\,\delta_1(x)\,\delta_1(y) = \left| \begin{matrix} \delta_1(x) & \delta_1(y) \\ \delta_2(x) & \delta_2(y) \end{matrix} \right|_+ = a_1.$$

The theorem follows from Lemma 10.60. $\square$

We conclude this chapter with a remarkable binomial coefficient identity that was obtained by substituting (10.116) in (10.128). The proof we give was obtained from EKHAD (see the proof of Lemma 10.51).

Theorem 10.62. Suppose that $1 \leq m \leq n$, where $n \geq 2$, and that $1 \leq s \leq min\{m, n-1\}$. Then

$$(10.139) \qquad \sum_{j=0}^{s} (-1)^j (m+n-2j) \binom{m}{j} \binom{m-j}{m-s} \binom{n}{j} \binom{n-j}{n-s}$$

$$\times \binom{m+n}{j} \binom{m+n-s-j-1}{s-j} \Big/ \binom{s}{j}^2 = 0.$$

Proof. Writing the summand as $f(j)$, it is straightforward to check that $-sf(j) = g(j+1) - g(j)$ for $0 \leq j \leq s - 1$, where

$$g(j) = \frac{j(m+n-s-j)}{m+n-2j} f(j).$$

From this we obtain that

$$-s \sum_{j=0}^{s-1} f(j) = g(s) - g(0) = sf(s),$$

so $\displaystyle\sum_{j=0}^{s} f(j) = 0$, since $s \geq 1$. $\square$

Chapter 11

Coefficients of the $\delta_n(x)$ Polynomials

This chapter is an extension of the previous chapter. It is a study of the positive integer coefficients of $\delta_n(x)$.

The basic properties of the coefficients $d_j^{(n)}$ of $\delta_n(x)$ are given in the next theorem. The $d_j^{(n)}$ were written more generally as $\varphi_j^{(n)}$ in (1.4).

Theorem 11.1. Let

$$(11.1) \qquad \delta_n(x) \stackrel{\text{def}}{=} \sum_{j=0}^{n} d_j^{(n)} x^j, \ n \geq 0.$$

Then the numbers $d_j^{(n)}$ are integers. Also, for $n \geq 0$,

$$(11.2) \qquad d_n^{(n)} = 1,$$

$$(11.3) \qquad d_n^{(n+2)} = \frac{1}{3} n(n+1)(n+2),$$

$$(11.4) \qquad d_0^{(n)} = \delta_{n,0},$$

$$(11.5) \qquad d_1^{(2n+1)} = (2n)!,$$

and

$$(11.6) \qquad d_2^{(2n+2)} = (2n+1)! \sum_{j=0}^{n} \frac{1}{2j+1}.$$

Also,

$$(11.7) \qquad d_{r+1}^{(n+1)} = n! \sum_{j=r}^{n} \frac{d_r^{(j)}}{j!}, \quad 0 \le r \le n,$$

$$(11.8) \qquad d_j^{(n)} = 0, \quad 1 \le j < n, \ n-j \ \text{odd},$$

$$(11.9) \qquad d_j^{(n)} \in Z^+, \quad 0 \le j \le n, \ n-j \ \text{even},$$

and

$$(11.10) \qquad d_j^{(n+2)} = d_{j-1}^{(n+1)} + n(n+1) d_j^{(n)}, \quad 1 \le j \le n.$$

Proof. Equations (11.2)–(11.4), (11.8), and (11.9) follow, respectively, from (8.10)–(8.12), (8.16), and (8.19) by setting $k = 1$.

(11.5) If we put $k = 1$ into (8.5) with (8.3), we find

$$\frac{1}{2^{2n}} \sum_{j=0}^{n} \binom{2j}{j} \binom{2n-2j}{n-j} = \hat{P}_{2n}(1) = P_n(1) = 1.$$

(See also [Go, 3.90].) Thus, putting $k = 1$ in (8.13) gives

$$d_1^{(2n+1)} = \frac{(2n)!}{2^{2n}} \sum_{j=0}^{n} \binom{2j}{j} \binom{2n-2j}{n-j} = (2n)!.$$

(11.6) The result follows by setting $k = 1$ in the first equation of (8.14) and using (11.5).

(11.7) Set $k = 1$ in (8.15) and use (11.5).

(11.10) Substituting the sum in (11.1) into (10.8) and equating corresponding coefficients of x for $1 \le j \le n$ gives the result. $\square$

Before proceeding, we give four more formulas like (11.3) without proof. Here $n \ge 1$.

$$(11.3.1) \qquad d_n^{(n+4)} = \frac{4}{3}(5n+13)\binom{n+4}{5}$$

$$(11.3.2) \qquad d_n^{(n+6)} = \frac{8}{9}(35n^2 + 273n + 502)\binom{n+6}{7}$$

$$(11.3.3) \qquad d_n^{(n+8)} = \frac{16}{15}(175n^3 + 2730n^2 + 13589n + 21306)\binom{n+8}{9}$$

(11.3.4) $\quad d_n^{(n+10)} = \dfrac{32}{9}(385n^4 + 10010n^3 + 94259n^2 + 377938n$

$$+538008)\binom{n+10}{11}.$$

Corollary 11.2. For $r \geq 1$, we have

(11.11) $$[\tanh^{-1} t]^r = r! \sum_{n=r}^{\infty} d_r^{(n)} \frac{t^n}{n!}.$$

Proof. If equation (10.2) is differentiated r times with respect to x and $x = 0$ is put in the result, we find that

$$[\tanh^{-1} t]^r = \sum_{n=r}^{\infty} \delta_n^{(r)}(0) \frac{t^n}{n!}.$$

But by (11.1), we have that $\delta_n^{(r)}(0) = r! \, d_r^{(n)}$. □

Table 11.1 $d_j^{(n)}$, $0 \leq n \leq 8$, $0 \leq j \leq n$

n\j	0	1	2	3	4	5	6	7	8
0	1								
1	0	1							
2	0	0	1						
3	0	2	0	1					
4	0	0	8	0	1				
5	0	24	0	20	0	1			
6	0	0	184	0	40	0	1		
7	0	720	0	784	0	70	0	1	
8	0	0	8448	0	2464	0	112	0	1

In the next theorem we express $d_j^{(n)}$ in terms of the Stirling numbers and vice versa.

Theorem 11.3. For $n \geq 0$ and $0 \leq j \leq n$,

(11.12) $$d_j^{(n)} = \frac{(-1)^n}{2^j} \sum_{k=j}^{n} 2^k L_{n,k}\, s(k,j)$$

and

(11.13) $$s(n,j) = \frac{1}{2^{n-j}} \sum_{k=j}^{n} (-1)^k L_{n,k}\, d_k^{(n)}.$$

Proof. We begin by noting that

$$(11.14) \qquad (x)_{k,2} = \prod_{j=1}^{k} (x + 2 - 2j) = 2^k \prod_{j=1}^{k} \left(\frac{x}{2} + 1 - j \right) = 2^k \left(\frac{x}{2} \right)_k.$$

Thus, using (10.33), we get

$$(x)_{k,2} = 2^k \sum_{j=1}^{k} s(k,j) \left(\frac{x}{2} \right)^j = \sum_{j=1}^{k} 2^{k-j} s(k,j)\, x^j.$$

Substituting this result into (10.39), using (10.36), and equating the corresponding coefficients gives the result. Equation (11.13) follows from (10.39) using (10.37). □

Formulas for the non-zero coefficients $d_j^{(n)}$ can be obtained in terms of the following functions (see Theorem 11.7).

Definition 11.4. Let $\xi_0(m) = 1$ and

$$(11.15) \qquad \xi_{2r+1}(m) \stackrel{\text{def}}{=} \sum_{k=r}^{m} \frac{\xi_{2r}(k)}{2k+1},\ 0 \le r \le m,\ m \ge 0,$$

$$(11.16) \qquad \xi_{2r+2}(m) \stackrel{\text{def}}{=} \sum_{k=r}^{m-1} \frac{\xi_{2r+1}(k)}{k+1},\ 0 \le r \le m,\ m \ge 1.$$

Note the special case

$$(11.17) \qquad \xi_1(m) = \sum_{k=0}^{m} \frac{1}{2k+1},\ m \ge 0.$$

Table 11.2 $\xi_r(m)$, $1 \le r \le 6$, $\left[\frac{r}{2} \right] \le m \le 6$

r \ m	0	1	2	3	4	5	6
1	1	$\frac{4}{3}$	$\frac{23}{15}$	$\frac{176}{105}$	$\frac{563}{315}$	$\frac{6508}{3465}$	$\frac{88069}{45045}$
2		1	$\frac{5}{3}$	$\frac{98}{45}$	$\frac{818}{315}$	$\frac{517}{175}$	$\frac{169819}{51975}$
3		$\frac{1}{3}$	$\frac{2}{3}$	$\frac{44}{45}$	$\frac{718}{567}$	$\frac{21757}{14175}$	$\frac{39788}{22275}$
4			$\frac{1}{6}$	$\frac{7}{18}$	$\frac{19}{30}$	$\frac{5027}{5670}$	$\frac{48581}{42525}$
5			$\frac{1}{30}$	$\frac{4}{45}$	$\frac{43}{270}$	$\frac{136}{567}$	$\frac{1991}{6075}$
6				$\frac{1}{90}$	$\frac{1}{30}$	$\frac{44}{675}$	$\frac{4472}{42525}$

Theorem 11.5. Choose $r \geq 0$. Then for $m \geq r$,

(11.18)
$$d_{2r+1}^{(2m+1)} = \frac{(2m)!}{2^r} \xi_{2r}(m),$$

and

(11.19)
$$d_{2r+2}^{(2m+2)} = \frac{(2m+1)!}{2^r} \xi_{2r+1}(m).$$

Proof. Both are true for $r = 0$, by Definition 11.4, (11.5), (11.6), and (11.17). For m odd, replacing m by $2m-1$ and r by $2r+1$ in (11.5) implies that

$$d_{2r+1}^{(2m+1)} - 2m(2m-1) d_{2r+1}^{(2m-1)} = \frac{(2m)!}{2^r} [\xi_{2r}(m) - \xi_{2r}(m-1)]$$

$$= \frac{(2m)!\, \xi_{2r-1}(m-1)}{2^r \quad m} = \frac{(2m-1)!}{2^{r-1}} \xi_{2r-1}(m-1) = d_{2r}^{(2m)},$$

using (11.16) and (11.19).

For m even, replacing m by $2m$ and r by $2r+2$ in (11.5) implies that

$$d_{2r+2}^{(2m+2)} - 2m(2m+1) d_{2r+2}^{(2m)}$$

$$= \frac{(2m+1)!}{2^r} \xi_{2r+1}(m) - 2m(2m+1)\frac{(2m-1)!}{2^r} \xi_{2r+1}(m-1)$$

$$= \frac{(2m+1)!}{2^r}[\xi_{2r+1}(m) - \xi_{2r+1}(m-1)] = \frac{(2m+1)!}{2^r}\frac{\xi_{2r}(m)}{2m+1}$$

$$= \frac{(2m)!}{2^r} \xi_{2r}(m) = d_{2r+1}^{(2m+1)},$$

using (11.15) and (11.18). $\square$

Theorem 11.6. For $0 \leq j \leq m$, we have

(11.20)
$$d_{2j+1}^{(2m+1)} = (2m)! \sum_{k=j}^{m} \frac{d_{2j}^{(2k)}}{(2k)!}$$

and

(11.21)
$$d_{2j+2}^{(2m+2)} = (2m+1)! \sum_{k=j}^{m} \frac{d_{2j+1}^{(2k+1)}}{(2k+1)!}.$$

Proof. If we set $n = 2m$ and re-index the sum in (10.11), we obtain

$$\delta_{2m+1}(x) = (2m)!\, x \sum_{k=0}^{m} \frac{\delta_{2k}(x)}{(2k)!}.$$

From (11.1) and (11.8), this equation becomes

$$\sum_{j=0}^{m} d_{2j+1}^{(2m+1)} x^{2j+1} = (2m)! \sum_{k=0}^{m} \frac{1}{(2k)!} \sum_{j=0}^{k} d_{2j}^{(2k)} x^{2j+1}$$

$$= (2m)! \sum_{j=0}^{m} \left\{ \sum_{k=j}^{m} \frac{d_{2j}^{(2k)}}{(2k)!} \right\} x^{2j+1},$$

from which the result follows by equating corresponding coefficients. The proof of (11.21) is similar. □

Theorem 11.7. For $0 \le j \le m$, we have

(11.22) $$d_{2j+1}^{(2m+1)} = \frac{(2m+1)!}{2j+1} \sum_{k=j}^{m} \frac{d_{2j}^{(2k)}}{(2m-2k+1)(2k)!}$$

and

(11.23) $$d_{2j+2}^{(2m+2)} = \frac{(2m+2)!}{2j+2} \sum_{k=j}^{m} \frac{d_{2j+1}^{(2k+1)}}{(2m-2k+1)(2k+1)!}.$$

Proof. If we set $n = 2m + 1$ and re-index the sum in (10.73), we obtain

$$\delta_{2m+1}'(x) = (2m+1)! \sum_{k-0}^{m} \frac{\delta_{2k}(x)}{(2m-2k+1)(2k)!}.$$

Then using (11.1) and (11.8), this equation becomes

$$\frac{1}{(2m+1)!} \sum_{j=0}^{m} (2j+1) d_{2j+1}^{(2m+1)} x^{2j} = \sum_{k=0}^{m} \frac{1}{(2m-2k+1)(2k)!} \sum_{j=0}^{k} d_{2j}^{(2k)} x^{2j}$$

$$= \sum_{j=0}^{m} \left\{ \sum_{k=j}^{m} \frac{d_{2j}^{(2k)}}{(2m-2k+1)(2k)!} \right\} x^{2j},$$

from which the result follows by equating corresponding coefficients. The proof of (11.23) is similar. □

Theorem 11.8. For $n \ge 0$ and $0 \le j \le n$, we have that

(11.24) $$d_j^{(n)} \le d_{j+1}^{(n+1)},$$

(11.25) $$(2n+1) d_{2j}^{(2n)} \le (2j+1) d_{2j+1}^{(2n+1)},$$

and

(11.26) $$(n+1) d_{2j+1}^{(2n+1)} \le (j+1) d_{2j+2}^{(2n+2)}.$$

Proof. (11.24) From (11.7) we have that

$$d_{j+1}^{(n+1)} = n! \sum_{n=j}^{n} \frac{d_j^{(k)}}{k!} \geq n! \left(\frac{d_j^{(n)}}{n!} \right) = d_j^{(n)}.$$

(11.25) From (11.22) we have that

$$d_{2j+1}^{(2n+1)} = \frac{(2n+1)!}{2j+1} \sum_{k=j}^{n} \frac{d_{2j}^{(2k)}}{(2n-2k+1)(2k)!}$$

$$\geq \frac{(2n+1)! \, d_{2j}^{(2n)}}{(2j+1)(2n)!} = \frac{2n+1}{2j+1} d_{2j}^{(2n)}.$$

(11.26): The proof is the same as in (12.25) using (12.23). □

Chapter 12

The $\lambda_n(x)$ Polynomials

As we saw in Chapter 2, the sequence of polynomials generated by the inverse function f^{-1} is closely related to the sequence generated by f. In this chapter we discuss the sequence $\{\lambda_n(x)\}_{n=0}^{\infty}$, the primary sequence of elliptic polynomials of the second kind, related to $\{\delta_n(x)\}_{n=0}^{\infty}$ in this way. (Here we use the notation $\lambda_n(x)$ instead of $\bar{\delta}_n(x)$.) Some of the results we obtain here are used in Chapter 14. The final topic of the chapter is an investigation into the zeros of $\lambda_n(x)$, where it is shown that they are real and distinct, except for the double zero $x = 0$ of $\lambda_{2n}(x)$, $n \geq 1$.

Definition 12.1. Let $\{\lambda_n(x)\}_{n=0}^{\infty}$ be generated by the function

(12.1)
$$\bar{G} = e^{x \tanh t} \overset{\text{def}}{=} \sum_{n=0}^{\infty} \lambda_n(x)\frac{t^n}{n!}.$$

Specializing (1.5) and (1.23) to this case gives the following properties of $\lambda_n(x)$:

For $n \geq 0$,

(12.2)
$$\lambda_n(x+y) = \sum_{j=0}^{n} \binom{n}{j} \lambda_j(x)\lambda_{n-j}(y)$$

and

(12.3) $$\lambda_n(-x) = (-1)^n \lambda_n(x).$$

From (12.1) we can derive a recursion formula for $\lambda_n(x)$. Unlike the recursion for $\delta_n(x)$, this recursion involves differentiation.   **Theorem 12.2.** For $n \geq 0$,

(12.4) $$\lambda_{n+1}(x) = x\,(I - D^2)\,\lambda_n(x), \quad \lambda_0(x) = 1.$$

Proof. Since $\tanh^{-1} t \in \mathcal{F}_1$ and $\Delta = 0$, the result follows from Corollary 5.6 with $a_1 = 1$ and $a_3 = \frac{1}{3}$. □

Next consider the operators $(Qf)(x) = x\,f(x)$, $\hat{b}_+ = Q(I - D^2)$, $\hat{b}_- = \tanh^{-1} D = \sum_{n=0}^{\infty} \dfrac{D^{2n+1}}{2n+1}$, and $N = \hat{b}_+\hat{b}_-$. Equation (12.4) can be written as the raising operator $\hat{b}_+\lambda_n = \lambda_{n+1}$, $n \geq 0$. In what follows, we will show that $\hat{b}_-$ is a lowering operator for $\{\lambda_n\}$ and that N and $\hat{b}_-\hat{b}_+$ both have λ_n as an eigenvector. The bracket is the commutator.

Lemma 12.3. We have that

(12.5) $$[\hat{b}_-, \hat{b}_+] = I$$

and

(12.6) $$[N, \hat{b}_+] = \hat{b}_+.$$

Proof. (12.5) We have

$$\hat{b}_+\hat{b}_- = Q(I - D^2)\tanh^{-1} D = Q(\tanh^{-1} D)(I - D^2)$$

$$= \left(\sum_{n=0}^{\infty} \frac{1}{2n+1} QD^{2n+1} \right)(I - D^2)$$

$$= \left\{ \sum_{n=0}^{\infty} \frac{1}{2n+1} \left(D^{2n+1}Q - (2n+1)D^{2n} \right) \right\}(I - D^2)$$

$$= \left(\sum_{n=0}^{\infty} \frac{1}{2n+1} D^{2n+1} \right)Q(I - D^2) - \left(\sum_{n=0}^{\infty} D^{2n} \right)(I - D^2) = \hat{b}_-\hat{b}_+ - I,$$

where (10.22) was used.

(12.6) We have

$$N\hat{b}_+ = \hat{b}_+(\hat{b}_-\hat{b}_+) = \hat{b}_+(\hat{b}_+\hat{b}_- + I) = \hat{b}_+(N + I) = \hat{b}_+N + \hat{b}_+. □$$

Theorem 12.4. For $n \geq 1$,

(12.7) $$\hat{b}_-\lambda_n = n\lambda_{n-1}.$$

Proof. For $n = 1$, we have

$$\hat{b}_-\lambda_1 = \left(\sum_{n=0}^{\infty} \frac{D^{2n+1}}{2n+1} \right) x = 1 = \lambda_0.$$

Assume next that the formula is true for some $n \geq 1$. Then we have

$$\hat{b}_-\lambda_{n+1} = \hat{b}_-(\hat{b}_+\lambda_n) = (\hat{b}_-\hat{b}_+)\lambda_n = (\hat{b}_+\hat{b}_- + I)\lambda_n = \hat{b}_+(\hat{b}_-\lambda_n) + \lambda_n$$
$$= \hat{b}_+(n\lambda_{n-1}) + \lambda_n = n(\hat{b}_+\lambda_{n-1}) + \lambda_n = (n+1)\lambda_n,$$

where we used (12.4) and (12.5). $\square$

Theorem 12.5. For $n \geq 0$,

(12.8) $$\hat{b}_+\hat{b}_-\lambda_n = n\lambda_n$$

and

(12.9) $$\hat{b}_-\hat{b}_+\lambda_n = (n+1)\lambda_n.$$

Proof. True for $n = 0$. For $n \geq 1$, we have $\hat{b}_+\hat{b}_-\lambda_n = \hat{b}_+ n\lambda_{n-1} = n\lambda_n$. The proof of the second equation is similar. $\square$

Table 12.1 $\lambda_n(x)$, $0 \leq n \leq 11$

n	$\lambda_n(x)$
0	1
1	x
2	x^2
3	$x(x^2 - 2)$
4	$x^2(x^2 - 8)$
5	$x(x^4 - 20x^2 + 16)$
6	$x^2(x^4 - 40x^2 + 136)$
7	$x(x^6 - 70x^4 + 616x^2 - 272)$
8	$x^2(x^6 - 112x^4 + 2016k^2 - 3968)$
9	$x(x^8 - 168x^6 + 5376x^4 - 28190x^2 + 7936)$
10	$x^2(x^8 - 240x^6 + 12432x^4 - 135680x^2 + 176896)$
11	$x(x^{10} - 330x^8 + 25872x^6 - 508640x^4 + 1805056x^2 - 353792)$

Definition 12.6. For $n \geq 0$, let

(12.10) $$\lambda_n(x) \stackrel{\text{def}}{=} \sum_{j=0}^{n} \lambda_j^{(n)} x^j.$$

Theorem 12.7. For $n \geq 0$

(12.11) $$x^n = \sum_{k=0}^{n} d_k^{(n)} \lambda_k(x) = \sum_{k=0}^{n} \lambda_k^{(n)} \delta_k(x).$$

Proof. This is a specialization of (2.8). $\quad\square$

The coefficients of these polynomials play an important part in studying the orthogonal polynomials that come from $\delta_n(x)$ (cf. Chapter 14), so we will examine them next. We obtain some properties of the $\lambda_j^{(n)}$ by specializing (2.10)–(2.12).

For $0 \leq j \leq n$,

(12.12) $$\sum_{k=j}^{n} d_k^{(n)} \lambda_j^{(k)} = \sum_{k=j}^{n} \lambda_k^{(n)} d_j^{(k)} = \delta_{n,j},$$

(12.13) $$\sum_{k=j}^{n} d_{2k+1}^{(2n+1)} \lambda_{2j+1}^{(2k+1)} = \sum_{k=j}^{n} \lambda_{2k+1}^{(2n+1)} d_{2j+1}^{(2k+1)} = \delta_{n,j},$$

and

(12.14) $$\sum_{k=j}^{n} \lambda_{2k+2}^{(2n+2)} d_{2j+2}^{(2k+2)} = \sum_{k=j}^{n} d_{2k+2}^{(2n+2)} \lambda_{2j+2}^{(2k+2)} = \delta_{n,j}.$$

Substituting $j = 0$ into the second and first sums in (12.13) and (12.14), respectively, and using (11.8) and (11.6), we also have for $n \geq 0$ that

(12.15) $$\sum_{k=0}^{n} (2k)! \, \lambda_{2k+1}^{(2n+1)} = \sum_{k=0}^{n} (2k+2)! \, \xi_1(k) \lambda_{2k+2}^{(2n+2)} = \delta_{n,0}.$$

Finally, for $n \geq 1$ we let

$$\Delta_n = \begin{bmatrix} d_1^{(1)} & 0 & 0 & \cdots & 0 \\ d_1^{(2)} & d_2^{(2)} & 0 & \cdots & 0 \\ \vdots & \vdots & \vdots & & \vdots \\ d_1^{(n)} & d_2^{(n)} & d_3^{(n)} & \cdots & d_n^{(n)} \end{bmatrix}$$

and

$$\Lambda_n = \begin{bmatrix} \lambda_1^{(1)} & 0 & 0 & \cdots & 0 \\ \lambda_1^{(2)} & \lambda_2^{(2)} & 0 & \cdots & 0 \\ \vdots & \vdots & \vdots & & \vdots \\ \lambda_1^{(n)} & \lambda_2^{(n)} & \lambda_3^{(n)} & \cdots & \lambda_n^{(n)} \end{bmatrix},$$

then

(12.16) $$\Lambda_n = \Delta_n^{-1},$$

by specializing (2.13). Further properties of the $\lambda_j^{(n)}$ are contained in the following theorem.

Theorem 12.8. For $0 \le j \le n$, we have

(12.17) $$\lambda_j^{(n)} \text{ is an integer,}$$

(12.18) $$\lambda_0^{(n)} = \delta_{n,0},$$

(12.19) $$\lambda_n^{(n)} = 1,$$

(12.20) $$\lambda_j^{(n)} = 0, \; n - j \text{ odd,}$$

For $n \ge 2$ and $1 \le j \le n - 1$, we have

(12.21) $$\lambda_j^{(n+1)} = \lambda_{j-1}^{(n)} - j(j + 1)\lambda_{j+1}^{(n)}.$$

For $n \ge 0$ and $0 \le j \le n$, we have

(12.22) $$(-1)^{\frac{n-j}{2}}\lambda_j^{(n)} \ge 1, \; n - j \text{ even.}$$

Proof. (12.17) Clear from (12.4).
(12.18) Specialization of (1.12).
(12.19) Specialization of (1.9).
(12.20) Specialization of (1.25).
(12.21) We have that

$$\sum_{j=1}^{n+1} \lambda_j^{(n+1)}\delta_j(x) = \sum_{j=0}^{n+1} \lambda_j^{(n+1)}\delta_j(x) = x^{n+1} = x \cdot x^n = x\sum_{j=0}^{n} \lambda_j^{(n)}\delta_j(x)$$

$$= \sum_{j=0}^{n-1} \lambda_{j+1}^{(n)} x\delta_{j+1}(x) = \sum_{j=0}^{n-1} \lambda_{j+1}^{(n)}[\delta_{j+2}(x) - j(j+1)\delta_j(x)]$$

$$= \sum_{j=2}^{n+1} \lambda_{j-1}^{(n)}\delta_j(x) - \sum_{j=1}^{n-1} j(j+1)\lambda_{j+1}^{(n)}\delta_j(x),$$

using (12.18), (12.11), and (10.8). The result follows from equating in turn the corresponding coefficients of $\delta_j(x)$ for $j = n+1$ and n, for $2 \le j \le n-1$, and then for $j = 1$.

(12.22) Since $\lambda_j^{(n)}$ is an integer, it is sufficient to show that $\sigma(n,j) \overset{\text{def}}{=} (-1)^{\frac{n-j}{2}}\lambda_j^{(n)} > 0$. Multiplying (12.21) by $(-1)^{\frac{n+1-j}{2}}$ gives

$$\sigma(n+1,j) = \sigma(n,j-1) + j(j+1)\sigma(n,j+1),$$

which is true for $n = 0$ and implies the positivity of $\sigma(n,j)$ by induction. $\square$

Table 12.2 $\quad \lambda_j^{(n)}, \; 0 \le n \le 8, \; 0 \le j \le n$

n \ j	0	1	2	3	4	5	6	7	8
0	1								
1	0	1							
2	0	0	1						
3	0	-2	0	1					
4	0	0	-8	0	1				
5	0	16	0	-20	0	1			
6	0	0	136	0	-40	0	1		
7	0	-272	0	616	0	-70	0	1	
8	0	0	-3968	0	2016	0	-112	0	1

The next theorem gives addition formulas for $d_j^{(n)}$ and $\lambda_j^{(n)}$.

Theorem 12.9. For $0 \le j \le n$ and $0 \le k \le n-j$, we have

$$(12.23) \qquad d_{j+k}^{(n)} = \frac{1}{\binom{j+k}{j}} \sum_{s=0}^{n-j-k} \binom{n}{j+s} d_j^{(j+s)} d_k^{(n-j-s)}$$

and

$$(12.24) \qquad \lambda_{j+k}^{(n)} = \frac{1}{\binom{j+k}{j}} \sum_{s=0}^{n-j-k} \binom{n}{j+s} \lambda_j^{(j+s)} \lambda_k^{(n-j-s)}.$$

Proof. We will prove (12.23) by expanding the two sides of (10.5) and equating the corresponding coefficients. To begin,

$$\delta_n(x+y) = \sum_{s=0}^{n} d_s^{(n)}(x+y)^s = \sum_{s=0}^{n} d_s^{(n)} \sum_{j=0}^{s} \binom{s}{j} x^j y^{s-j}$$

$$= \sum_{j=0}^{n} x^j \sum_{s=j}^{n} \binom{s}{j} d_s^{(n)} y^{s-j},$$

so putting $k = s - j$, we get

$$\delta_n(x+y) = \sum_{j=0}^{n} x^j \sum_{k=0}^{n-j} \left\{ \binom{j+k}{j} d_{j+k}^{(n)} \right\} y^k.$$

On the other hand,

$$\sum_{r=0}^{n} \binom{n}{r} \delta_r(x) \delta_{n-r}(y) = \sum_{r=0}^{n} \binom{n}{r} \left(\sum_{j=0}^{r} d_j^{(r)} x^j \right) \left(\sum_{k=0}^{n-r} d_k^{(n-r)} y^k \right)$$

$$= \sum_{r=0}^{n} \sum_{j=0}^{r} \left(\sum_{k=0}^{n-r} \binom{n}{r} d_j^{(r)} d_k^{(n-r)} x^j y^k \right)$$

$$= \sum_{j=0}^{n} x^j \left(\sum_{r=j}^{n} \sum_{k=0}^{n-r} \binom{n}{r} d_j^{(r)} d_k^{(n-r)} y^k \right).$$

Putting $s = r - j$ gives the LHS as

$$\sum_{j=0}^{n} x^j \left(\sum_{s=0}^{n-j} \sum_{k=0}^{n-j-s} \binom{n}{j+s} d_j^{(j+s)} d_k^{(n-j-s)} y^k \right)$$

$$= \sum_{j=0}^{n} x^j \sum_{k=0}^{n-j} \left\{ \sum_{s=0}^{n-j-k} \binom{n}{j+s} d_j^{(j+s)} d_k^{(n-j-s)} \right\} y^k.$$

Equating the expressions in the braces on the two sides gives the result.

The proof of (12.24) can be obtained from the proof of (12.23) by replacing each $d_i^{(m)}$ by $\lambda_i^{(m)}$ and each power z^i by $\delta_i(z)$. $\square$

Note that (12.23) is a special case of (1.21).

The next theorem relates $\lambda_j^{(n)}$ and the Stirling numbers.

Theorem 12.10. For $0 \le j \le n$,

(12.25) $$\lambda_j^{(n)} = (-1)^j 2^n \sum_{k=j}^{n} \frac{1}{2^k} L_{k,j} S(n,k)$$

and

(12.26) $$S(n,j) = 2^{j-n} \sum_{k=j}^{n} (-1)^k L_{k,j} \lambda_k^{(n)}.$$

Proof. From (6.6) and (11.14) note that

$$x^n = 2^n \left(\frac{x}{2}\right)^n = 2^n \sum_{k=0}^{n} S(n,k) \left(\frac{x}{2}\right)_k = \sum_{k=0}^{n} 2^{n-k} S(n,k)(x)_{k,2}.$$

Thus, from (12.11) and (6.20) it follows that

$$\sum_{j=0}^{n} \lambda_j^{(n)} \delta_j(x) = x^n = \sum_{k=0}^{n} 2^{n-k} S(n,k)(x)_{k,2}$$

$$= \sum_{k=0}^{n} 2^{n-k} S(n,k) \sum_{j=0}^{k} (-1)^j L_{k,j} \delta_j(x)$$

$$= \sum_{j=0}^{n} \left\{ (-1)^j 2^n \sum_{k=j}^{n} \frac{1}{2^k} L_{k,j} S(n,k) \right\} \delta_j(x),$$

which gives (12.25). Equation (12.26) follows from (12.25) and (10.37). □

Theorem 12.11. For $n \geq 0$,

(12.27) $\lambda_1^{(2n+1)} = -C_{2n+1}$

and

(12.28) $\lambda_2^{(2n+2)} = \frac{1}{2} C_{2n+3}.$

Also, $\lambda_1^{(2n+1)} \lambda_2^{(2n+2)} \neq 0$.

Proof. (12.27) From the second equation in (7.14), we have

$$\lambda_1^{(2n+1)} = (2n+1)! \, \bar{a}_{2n+1} = (2n+1)! \left(-\frac{C_{2n+1}}{(2n+1)!} \right) = -C_{2n+1},$$

using (9.1).

(12.28) In (12.21), replace n by $2n+2$ and set $j = 1$. Then use (12.18) and (12.27), which gives $\lambda_2^{(2n+2)} = -\frac{1}{2} \lambda_1^{(2n+3)} = \frac{1}{2} C_{2n+3}$. That the product of λ's does not vanish follows from the comment after (9.1). □

Note that the other values of $\lambda_k^{(n)}$ can be computed as above using (12.21).

The results in Theorem 12.11 will be generalized later in Theorem 15.4.

Corollary 12.12. For $n \geq 0$,

(12.29) $\lambda_n^{(n+2)} = -\frac{1}{3} n(n+1)(n+2) = - d_n^{(n+2)}.$

Proof. From (12.25), we have that

(12.30) $\lambda_{n-2}^{(n)} = (-2)^n \sum_{k=n-2}^{n} \frac{1}{2^k} L_{k,n-2} S(n,k).$

From (10.36), we obtain the values $L_{n-2,n-2} = (-1)^n$, $L_{n-1,n-2} = (-1)^{n-1}(n-1)(n-2)$, and $L_{n,n-2} = \frac{1}{2}(-1)^n n(n-1)^2(n-2)$. From

[Ri2, p. 231], we find that $S(n,n) = 1$, $S(n, n-1) = \frac{1}{2}n(n-1)$, and $S(n, n-2) = \frac{1}{24}n(n-1)(n-2)(3n-5)$. Substituting these values in (12.30) gives the result. The second equality follows from (11.3). $\square$

Theorem 12.13. For $n \geq 0$,

$$(12.31) \qquad \lambda_{2n+1}(x) = -x \sum_{j=0}^{n} \binom{2n}{2j} C_{2n-2j+1} \lambda_{2j}(x)$$

and

$$(12.32) \qquad \lambda_{2n+2}(x) = -x \sum_{j=0}^{n} \binom{2n+1}{2j+1} C_{2n-2j+1} \lambda_{2j+1}(x).$$

Proof. Replacing n by $2n$ in (1.6), we obtain

$$f_{2n+1}(x) = (2n)!\, x \sum_{j=0}^{2n} \frac{2n-j+1}{j!} a_{2n-j+1}\, f_j(x).$$

For $f \in \mathcal{F}_0$, this becomes by (1.24)

$$(12.33) \qquad f_{2n+1}(x) = (2n)!\, x \sum_{j=0}^{n} \frac{2n-2j+1}{(2j)!} a_{2n-2j+1}\, f_{2j}(x).$$

If we use (7.22) in (12.1) and substitute the resulting $\{f_n(x)\}$ into (12.33), i.e., we have the equation $a_{2n+1} = -\frac{C_{2n+1}}{(2n+1)!}$ and $f_n(x) = \lambda_n(x)$, the result follows. Equation (12.32) is proved in a similar way. $\square$

Corollary 12.14. For $0 \leq k \leq n$,

$$(12.34) \qquad \lambda_{2k+1}^{(2n+1)} = -\sum_{j=k}^{n} \binom{2n}{2j} C_{2n-2j+1} \lambda_{2k}^{(2j)}$$

and

$$(12.35) \qquad \lambda_{2k+2}^{(2n+2)} = -\sum_{j=k}^{n} \binom{2n+1}{2j+1} C_{2n-2j+1} \lambda_{2k+1}^{(2j+1)}.$$

Proof. (12.34) Using (12.10) in (12.31) gives

$$\sum_{j=0}^{2n+1} \lambda_j^{(2n+1)} x^j = -x \sum_{j=0}^{n} \binom{2n}{2j} C_{2n-2j+1} \sum_{k=0}^{2j} \lambda_k^{(2j)} x^k$$

$$= -\sum_{k=0}^{2n} \left\{ \sum_{j=[\frac{k+1}{2}]}^{n} \binom{2n}{2j} C_{2n-2j+1} \lambda_k^{(2j)} \right\} x^{k+1}.$$

Equating the coefficients of x^{k+1}, we find that

$$\lambda_{k+1}^{(2n+1)} = -\sum_{j=[\frac{k+1}{2}]}^{n} \binom{2n}{2j} C_{2n-2j+1} \lambda_{k}^{(2j)}.$$

Using (12.20) and replacing k by $2k$ gives

$$\lambda_{2k+1}^{(2n+1)} = -\sum_{j=k}^{n} \binom{2n}{2j} C_{2n-2j+1} \lambda_{2k}^{(2j)}.$$

Equation (12.35) is proved in a similar way. $\square$

We next discuss the zeros of $\lambda_n(x)$. From (12.18) and (12.20) we can write (12.10) for $n \geq 0$ as

(12.36) $$\lambda_{2n+1}(x) = x \sum_{j=0}^{n} \lambda_{2j+1}^{(2n+1)} x^{2j}$$

and

(12.37) $$\lambda_{2n+2}(x) = x^2 \sum_{j=0}^{n} \lambda_{2j+2}^{(2n+2)} x^{2j}.$$

Note that x and x^2 are the exact powers of x, respectively, dividing these polynomials, because the constant terms in these sums are not zero by Theorem 12.11.

We next prove two useful theorems.

Theorem 12.15. [FaS, p. 147] If $P(x)$ is a monic, real polynomial of degree $n \geq 2$ with distinct, real zeros, then $P(x) - P''(x)$ is also monic with distinct, real zeros.

Proof. If the zeros of $P(x)$ are $\alpha_1, \alpha_2, \ldots \alpha_n$, then the α's are also the zeros of the function $e^{-x} P(x)$. By Rolle's theorem, the function $\frac{d}{dx}[e^{-x} P(x)]$ has at least $n-1$ distinct, real zeros $\beta_1, \beta_2, \ldots, \beta_{n-1}$, as does the nth degree polynomial $P(x) - P'(x) = -e^x \frac{d}{dx}[e^{-x} P(x)]$. The remaining zero, β_n, will therefore be real as well.

Now suppose that $\beta_n = \beta_i$ for some i, $1 \leq i \leq n-1$. Then β_n will be a double zero of $P(x) - P'(x)$, so $P(\beta_n) = P'(\beta_n) = P''(\beta_n)$. It also follows that $P(\beta_n), \neq 0$, since otherwise $P(\beta_n) = P'(\beta_n) = 0$, which says that β_n is a multiple zero of $P(x)$, contrary to assumption.

From $P(x) = \prod_{j=1}^{n}(x - \alpha_j)$, we find that $P'(x) = P(x)\sum_{j=1}^{n}\frac{1}{x - \alpha_j}$ and

$$P''(x) = P'(x)\sum_{j=1}^{n}\frac{1}{x - \alpha_j} - P(x)\sum_{j=1}^{n}\frac{1}{(x - \alpha_j)^2}.$$ Substituting $x = \beta_n$ and

canceling P and its derivatives evaluated at β_n, we obtain $\displaystyle\sum_{j=1}^{n} \frac{1}{\beta_n - \alpha_j} = 1$

and

$$\sum_{j=1}^{n} \frac{1}{\beta_n - \alpha_j} - \sum_{j=1}^{n} \frac{1}{(\beta_n - \alpha_j)^2} = 1.$$

Combining these results, we obtain the contradiction $\displaystyle\sum_{j=1}^{n} \frac{1}{(\beta_n - \alpha_j)^2} = 0$.

Thus, the zeros of $P(x) - P'(x)$ are distinct.

Next set $Q(x) = P(x) - P'(x)$, so $Q(x) + Q'(x) = P(x) - P''(x)$. Then $\beta_1, \beta_2, , \ldots, \beta_n$ are also zeros of the function $e^x Q(x)$. By Rolle's theorem, the function $\dfrac{d}{dx}[e^x Q(x)]$ has at least $n - 1$ distinct, real zeros $\gamma_1, \gamma_2, \ldots,$ γ_{n-1}, as does the nth degree polynomial $Q(x) + Q'(x) = e^{-x}\dfrac{d}{dx}[e^x Q(x)]$. The remaining zero, γ_n, will be real as well.

Now suppose that $\gamma_n = \gamma_i$ for some i, $1 \le i \le n - 1$. Then γ_n will be a zero of $Q(x) + Q'(x)$, so $Q(\gamma_n) = -Q'(\gamma_n) = Q''(\gamma_n)$. It also follows that $Q(\gamma_n) \ne 0$, for otherwise $Q(\gamma_n) = Q'(\gamma_n) = 0$, which says that γ_n is at least a double zero of $Q(x)$, a contradiction.

Next, from $Q(x) = \displaystyle\prod_{j=1}^{n}(x - \beta_j)$, we find that $Q'(x) = Q(x)\displaystyle\sum_{j=1}^{n}\frac{1}{x - \beta_j}$ and

$$Q''(x) = Q'(x)\sum_{j=1}^{n}\frac{1}{x - \beta_j} - Q(x)\sum_{j=1}^{n}\frac{1}{(x - \beta_j)^2}.$$

Substituting $x = \gamma_n$ and canceling Q and its derivatives evaluated at γ_n, we obtain

$$\sum_{j=1}^{n}\frac{1}{\gamma_n - \beta_j} = -1 \text{ and } \sum_{j=1}^{n}\frac{1}{\gamma_n - \beta_j} + \sum_{j=1}^{n}\frac{1}{(\gamma_n - \beta_j)^2} = -1.$$

Combining these results gives the contradiction $\displaystyle\sum_{j=1}^{n}\frac{1}{(\gamma_n - \beta_j)^2} = 0$. Thus, the zeros of $P(x) - P''(x)$ are distinct. $\square$

Theorem 12.16. If $P(x) = x^2 \displaystyle\prod_{j=1}^{n-2}(x - \alpha_j)$, where the α_j are real, non-zero, and distinct, then $P(x) - P''(x)$ has real, non-zero, distinct zeros.

Proof. Since $P(x)$ has the $n-1$ distinct zeros $0, \alpha_1, \ldots, \alpha_{n-2}$, then so does the function $e^{-x}P(x)$. By Rolle's theorem, the function $\dfrac{d}{dx}[e^{-x}P(x)]$ has at least $n - 2$ non-zero distinct zeros $\beta_1, \beta_2, \ldots, \beta_{n-2}$, as does the nth degree polynomial $P(x) - P'(x) = -e^x\dfrac{d}{dx}[e^{-x}P(x)]$. Since 0 is a double zero of

$P(x)$, it is also a zero of $P(x) - P'(x)$. However, it is a simple zero of the latter, because if $P(x) = x^2 P_1(x)$, where $P_1(0) \neq 0$, and we assume that 0 is a multiple zero of $P(x) - P'(x) = x^2[P_1(x) - P_1'(x)] - 2x P_1(x)$, then it will also be a zero of $P'(x) - P''(x) = x^2[P_1'(x) - P_1''(x)] + 2x[P_1(x) - P_1'(x)] - 2P_1(x) - 2x P_1'(x)$, i.e., $0 = -2P_1(0)$, a contradiction. Thus, $P(x) - P'(x)$ has the $n - 1$ distinct real zeros $0, \beta_1, \ldots, \beta_{n-2}$, so the nth zero, β_{n-1}, must be real and non-zero.

Suppose now that $\beta_{n-1} = \beta_i$ for some i, $1 \leq i \leq n - 2$. Then β_{n-1} will be a multiple zero of $P(x) - P'(x)$, so $P(\beta_{n-1}) = P'(\beta_{n-1}) = P''(\beta_{n-1})$. Also, $P(\beta_{n-1}) \neq 0$, since otherwise $P(\beta_{n-1}) = P'(\beta_{n-1}) = 0$, which says that β_{n-1} is a double zero of $P(x)$, a contradiction.

Also, from $P(x) = x^2 \prod_{j=1}^{n-2} (x - \alpha_j)$, we get $P'(x) = P(x)\left\{ \dfrac{2}{x} + \sum_{j=1}^{n-2} \dfrac{1}{x - \alpha_j} \right\}$

and

$$ P''(x) = \frac{2}{x^2}[x P'(x) - P(x)] + \sum_{j=1}^{n-2} \frac{P'(x)}{x - \alpha_j} - \sum_{j=1}^{n-2} \frac{P(x)}{(x - \alpha_j)^2}. $$

Substituting $x = \beta_{n-1}$ and canceling P and its derivatives evaluated at β_{n-1}, we obtain $\displaystyle\sum_{j=1}^{n-2} \frac{1}{\beta_{n-1} - \alpha_j} = 1 - \frac{2}{\beta_{n-1}}$ and

$$ 1 = \frac{2}{\beta_{n-1}^2}(\beta_{n-1} - 1) + \sum_{j=1}^{n-2} \frac{1}{\beta_{n-1} - \alpha_j} - \sum_{j=1}^{n-2} \frac{1}{(\beta_{n-1} - \alpha_j)^2}. $$

These results imply $\displaystyle\sum_{j=1}^{n-2} \frac{1}{(\beta_{n-1} - \alpha_j)^2} = -\frac{2}{\beta_{n-1}^2}$, a contradiction. Thus, $P(x) - P'(x)$ has distinct real zeros.

Next, set $Q(x) = P(x) - P'(x)$, so $Q(x) + Q'(x) = P(x) - P''(x)$. Then $0, \beta_1, \ldots, \beta_{n-1}$ are also the zeros of the function $e^x Q(x)$. By Rolle's theorem, the function $\dfrac{d}{dx}[e^x Q(x)]$ has at least $n - 1$ distinct, real non-zero zeros $\gamma_1, \gamma_2, \ldots, \gamma_{n-1}$, as does the nth degree polynomial $Q(x) + Q'(x) = e^{-x}\dfrac{d}{dx}[e^x Q(x)]$. The remaining zero, γ_n, will be real as well. The last part of the proof is the same as that in the final two paragraphs of the proof of Theorem 12.15. $\square$

Theorem 12.17. For $n \geq 0$, the polynomials $\lambda_{2n+1}(x)$ and $\dfrac{\lambda_{2n+2}(x)}{x}$ have real, distinct zeros.

Proof. The result is true for $\lambda_1(x) = x$. Assume that $\lambda_{2n+1}(x)$ is monic and has real, distinct zeros for some $n \geq 0$. Then by (12.4) and Theorem 12.15, it follows that $\dfrac{\lambda_{2n+2}(x)}{x} = \lambda_{2n+1}(x) - \lambda_{2n+1}''(x)$ is monic with real, distinct zeros as well. From (12.37) and the remark following this equation, we have that $\lambda_{2n+2}(x)$ has distinct zeros, except for the double zero at $x = 0$. But then Theorem 12.16 implies that $\lambda_{2n+2}(x) - \lambda''_{2n+2}(x)$ has non-zero real, distinct zeros. From (12.4), it then follows that $\lambda_{2n+3}(x)$ has real, distinct zeros. $\square$

Chapter 13

The Orthogonal Sequences $\{A_m(z)\}$ and $\{B_m(z)\}$

This chapter deals with the $G_m(z)$ and $H_m(z)$ polynomials (in Definition 1.19) that are derived from the function $f(t) = \tanh^{-1} t$. In this special case we will denote these polynomials by $A_m(z)$ and $B_m(z)$. Since these polynomials have not been studied before, we will develop some of their properties here. The latter will be used in establishing the orthogonality results in Chapter 14. Important among the results are the structure constants for $\{A_m(z)\}_{m=0}^{\infty}$ in Theorem 13.15 and for $\{B_m(z)\}_{m=0}^{\infty}$, which are found in the next chapter.

Theorem 13.1. For $m \geq 0$, we have that

$$(13.1) \quad A_{m+2}(z) =$$
$$\left(z + 2(2m+3)^2\right) A_{m+1}(z) - 4(m+1)^2(2m+1)(2m+3) A_m(z),$$

where $A_0(z) = 1$ and $A_1(z) = z + 2$,
and

$$(13.2) \quad B_{m+2}(z) =$$
$$\left(z + 8(m+2)^2\right) B_{m+1}(z) - 4(m+1)(m+2)(2m+3)^2 B_m(z),$$

where $B_0(z) = 1$ and $B_1(z) = z + 8$.

Proof. The recursions follow from (3.14) and (3.15), noting that $a_1 = 1$, $a_3 = \dfrac{1}{3}$, and $a_5 = \dfrac{1}{5}$, so $c_1 = 2$ and $c_2 = 1$. $\square$

Specializing (1.36)–(1.39), we also have for $m \geq 0$ that

$$(13.3) \qquad A_m(z) = \sum_{j=0}^{m} d_{2j+1}^{(2m+1)} z^j,$$

$$(13.4) \qquad B_m(z) = \sum_{j=0}^{m} d_{2j+2}^{(2m+2)} z^j,$$

$$(13.5) \qquad \delta_{2m+1}(x) = x A_m(x^2),$$

$$(13.6) \qquad \delta_{2m+2}(x) = x^2 B_m(x^2).$$

It is evident from (13.3) and (13.4) that $\deg(A_m(z)) = \deg(B_m(z)) = m$ and using (11.2), (11.8), and (11.9) that the polynomials are monic with integer coefficients. That they are also non-lacunary with positive integral coefficients follows from (11.9).

Table 13.1 $A_m(z)$, $0 \leq m \leq 9$

m	$A_m(z)$
0	1
1	$z + 2$
2	$z^2 + 20z + 24$
3	$z^3 + 70z^2 + 784z + 720$
4	$z^4 + 168z^3 + 6384z^2 + 52352z + 40320$
5	$z^5 + 330z^4 + 29568z^3 + 804320z^2 + 5360256z + 3628800$
6	$z^6 + 572z^5 + 99528z^4 + 6296576z^3 + 136804096z^2$ $+ 782525952z + 479001600$
7	$z^7 + 910z^6 + 272272z^5 + 33141680z^4 + 1656182528z^3$ $+ 30459752960z^2 + 154594381824z + 87178291200$
8	$z^8 + 1360z^7 + 643552z^6 + 133802240z^5 + 12765978368z^4$ $+ 535086755840z^3 + 8632830664704z^2$ $+ 39746508226560z + 20922789888000$
9	$z^9 + 1938z^8 + 1364352z^7 + 446370496z^6 + 72329756928z^5$ $+ 5750333382144z^4 + 209797380112384z^3$ $+ 3041109959196672z^2 + 129024832993689960z$ $+ 6402373705728000$

In studying these polynomials it will be useful to have the following relationships. Similar identities do not hold for general G's and H's.

Theorem 13.2. For $m \geq 0$,

$$(13.7) \qquad A_{m+1}(z) = B_{m+1}(z) - (2m+2)(2m+3)B_m(z),$$

$$(13.8) \qquad zB_m(z) = A_{m+1}(z) - 2(m+1)(2m+1)A_m(z).$$

Also, for $m \geq 1$,

$$(13.9) \qquad A_m(z) = (2m)! \left\{ z \sum_{k=0}^{m-1} \frac{B_k(z)}{(2k+2)!} + 1 \right\},$$

and for $m \geq 0$,

$$(13.10) \qquad B_m(z) = (2m+1)! \sum_{k=0}^{m} \frac{A_k(z)}{(2k+1)!}.$$

Proof. **(13.7)** Let $P_{m+1}(z) = B_{m+1}(z) - (2m+2)(2m+3)B_m(z)$, $m \geq 0$. For $m = 0$ and 1, $P_{m+1}(z) = A_{m+1}$. Also, $P_{m+1}(z)$ satisfies recursion (13.1) (with m replaced by $m+1$), using (13.2). Thus, $P_{m+1}(z) = A_{m+1}(z)$.

(13.8) Replace m by $m+1$ in (13.8) and substitute the resulting B terms away using (13.7). The result is true, since it is equation (13.2).

(13.9) Putting $a_{2r+1} = \dfrac{1}{2r+1}$ into (1.42) gives the result.

(13.10) The same proof as in (13.9), but using (1.43). $\square$

Note. Equations (13.7) and (13.8) can also be proved from (10.8) using (13.5) and (13.6).

Theorem 13.3. For $m \geq 1$, we have that

$$(13.11) \quad (2zD+1)A_m(z) =$$

$$(2m)! \left\{ 1 + (2m+1)z \sum_{k=0}^{m-1} \frac{B_k(z)}{(2m-2k-1)(2k+2)!} \right\}$$

and for $m \geq 0$ that

$$(13.12) \qquad (zD+1)B_m(z) = \frac{(2m+2)!}{2} \sum_{k=0}^{m} \frac{A_k(z)}{(2m-2k+1)(2k+1)!}.$$

Proof. Substitute $A_m(z)$ for $G_m(z)$ and $B_m(z)$ for $H_m(z)$ in Theorem 1.23 and use $a_{2n+1} = \dfrac{1}{2n+1}$ and (13.24). $\square$

Corollary 13.4. For $m \geq 1$, we have

(13.13) $(2zD + 1)A_m(z)$

$$= (2m)! \left\{ 1 + \frac{z}{2} \sum_{j=0}^{m-1} \frac{H(m) - H(j) + 2\xi_1(m - j - 1)}{(2j + 1)!} A_j(z) \right\},$$

where

$$H(r) = \begin{cases} \sum_{k=1}^{r} \frac{1}{k}, & r \geq 1, \\ 0, & r = 0, \end{cases}$$

and

(13.14) $(zD + 1)B_m(z) =$

$$(2m + 1)! \left\{ \xi_1(m) + \frac{z}{2} \sum_{j=0}^{m-1} \frac{\xi_1(m) - \xi_1(j) + \xi_1(m - j - 1)}{(2j + 2)!} B_j(z) \right\}.$$

Proof. (13.13) From (13.11), we have using (13.10) that

$$\frac{1}{z} \left\{ \frac{1}{(2m)!} (2zD + 1)A_m(z) - 1 \right\} = (2m + 1) \sum_{k=0}^{m-1} \frac{B_k(z)}{(2m - 2k - 1)(2k + 2)!}$$

$$= \frac{2m + 1}{2} \sum_{k=0}^{m-1} \frac{1}{(2m - 2k - 1)(k + 1)} \sum_{j=0}^{k} \frac{A_j(z)}{(2j + 1)!}$$

$$= \frac{2m + 1}{2} \sum_{j=0}^{m-1} \frac{A_j(z)}{(2j + 1)!} \sum_{k=j}^{m-1} \frac{1}{(2m - 2k - 1)(k + 1)}.$$

Now,

$$\sum_{k=j}^{m-1} \frac{1}{(k + 1)(2m - 2k - 1)} = \frac{1}{2m + 1} \sum_{k=j}^{m-1} \left\{ \frac{1}{k + 1} + \frac{2}{2m - 2k - 1} \right\}$$

$$= \frac{1}{2m + 1} \left\{ \sum_{r=j+1}^{m} \frac{1}{r} + 2 \sum_{r=0}^{m-j-1} \frac{1}{2r + 1} \right\}$$

$$= \frac{1}{2m + 1} \{ H(m) - H(j) + 2\xi_1(m - j - 1) \},$$

where we used (11.17). Combining the results proves the formula.

Table 13.2 $B_m(z)$, $0 \le m \le 9$

m	$B_m(z)$
0	1
1	$z + 8$
2	$z^2 + 40z + 184$
3	$z^3 + 112z^2 + 2464z + 8448 = (z + 24)(z^2 + 88z + 352)$
4	$z^4 + 240z^3 + 14448z^2 + 229760z + 648576$
5	$z^5 + 440z^4 + 55968z^3 + 2393600z^2 + 30633856z + 74972160$
6	$z^6 + 728z^5 + 168168z^4 + 15027584z^3 + 510205696z^2$ $\quad + 5561407488z + 12174658560$
7	$z^7 + 1120z^6 + 425152z^5 + 68456960z^4 + 4811975168z^3$ $\quad + 137602949120z^2 + 1322489954304z + 2643856588800$
8	$z^8 + 1632z^7 + 948192z^6 + 249443584z^5 + 3138671488z^4$ $\quad + 1843944001536z^3 + 46060832825344z^2$ $\quad + 399463775797248z + 740051782041600$
9	$z^9 + 2280z^8 + 1922496z^7 + 770652160z^6 + 157639462656z^5$ $\quad + 16484438231040z^4 + 840426228637696z^3$ $\quad + 18793914785464320z^2 + 1495190946220277776z$ $\quad + 259500083163955200$

(13.14) From (13.12), we obtain using (13.9) that

$$\frac{2}{(2m+2)!}(zD+1)B_m(z) = \sum_{k=0}^{m} \frac{A_k(z)}{(2m-2k+1)(2k+1)!}$$

$$= \frac{1}{2m+1} + \sum_{k=1}^{m} \frac{1}{(2k+1)(2m-2k+1)}\left\{1 + z\sum_{j=0}^{k-1} \frac{B_j(z)}{(2j+2)!}\right\}$$

$$= \sum_{k=0}^{m} \frac{1}{(2k+1)(2m-2k+1)}$$

$$\qquad + z\sum_{j=0}^{m-1} \frac{B_j(z)}{(2j+2)!} \sum_{k=j+1}^{m} \frac{1}{(2k+1)(2m-2k+1)}.$$

Now,

$$\sum_{k=0}^{m} \frac{1}{(2k+1)(2m-2k+1)}$$

$$= \frac{1}{2m+2}\sum_{k=0}^{m}\left\{\frac{1}{2k+1} + \frac{1}{2m-2k+1}\right\} = \frac{1}{m+1}\xi_1(m),$$

using (11.17). Also,

$$\sum_{k=j+1}^{m} \frac{1}{(2k+1)(2m-2k+1)} = \frac{1}{2m+2} \sum_{k=j+1}^{m} \left\{ \frac{1}{2k+1} + \frac{1}{2m-2k+1)} \right\}$$

$$= \frac{1}{2m+2} \left\{ \xi_1(m) - \xi_1(j) + \sum_{r=0}^{m-j-1} \frac{1}{2r+1} \right\}$$

$$= \frac{1}{2m+2} \left\{ \xi_1(m) - \xi_1(j) + \xi_1(m-j-1) \right\}.$$

Combining the results gives the theorem. □

The next theorem gives the generating functions for $A_m(z)$ and $B_m(z)$.

Theorem 13.5. If $z < 0$, then

$$(13.15) \qquad \frac{1}{\sqrt{|z|}} \sin\left(\sqrt{|z|} \, \tanh^{-1} t \right) = \sum_{m=0}^{\infty} A_m(z) \frac{t^{2m+1}}{(2m+1)!}$$

and

$$(13.16) \qquad \frac{1}{z} \left\{ \cos\left(\sqrt{|z|} \, \tanh^{-1} t \right) - 1 \right\} = \sum_{m=0}^{\infty} B_m(z) \frac{t^{2m+2}}{(2m+2)!}.$$

Proof. These formulas are specializations of Theorem 1.24 with $f(t) = \tanh^{-1} t$. □

It seems there are no simple formulas with a double-argument, although, as the next theorems show, there are quadruple-argument formulas.

Theorem 13.6. For $m \geq 0$,

$$(13.17) \qquad A_m(4z) = - \sum_{j=0}^{m} 2^{2j+1} \frac{(2m+1)!}{(m+j+1)!} L_{m+j+1,2j+1} \, A_j(z)$$

and

$$(13.18) \qquad B_m(4z) = \sum_{j=0}^{m} 2^{2j+2} \frac{(2m+2)!}{(m+j+2)!} L_{m+j+2,2j+2} \, B_j(z).$$

Proof. (13.17) This follows from (10.58) and (13.5).
(13.18) This follows from (10.59) and (13.6). □

Corollary 13.7. For $m \geq 0$,

$$(13.19) \qquad A_m\left(\frac{z}{4} \right) = \frac{1}{2^{2m}} \sum_{j=0}^{m} \frac{(2m)!}{(2j)!} \binom{2m+1}{m-j} A_j(z)$$

and

$$(13.20) \qquad B_m\left(\frac{z}{4}\right) = \frac{1}{2^{2m}} \sum_{j=0}^{m} \frac{(2m+1)!}{(2j+1)!} \binom{2m+2}{m-j} B_j(z).$$

Proof. These results follow from Theorem 10.24 using (13.5) and (13.6). ☐

Theorem 13.8. We have

$$(13.21) \qquad B_m(4z) = \frac{1}{2} \sum_{j=0}^{m} \binom{2m+2}{2j+1} A_j(z) A_{m-j}(z), \ m \geq 0,$$

$$(13.22) \quad A_m(4z) = A_m(z) + z \sum_{j=0}^{m-1} \binom{2m+1}{2j+1} A_j(z) B_{m-j-1}(z), \ m \geq 1,$$

and

$$(13.23) \quad B_m(4z) = B_m(z) + \frac{z}{2} \sum_{j=0}^{m-1} \binom{2m+2}{2j+2} B_j(z) B_{m-j-1}(z), \ m \geq 1.$$

Proof. These results are obtained by specializing Theorem 1.23. ☐

Note that (13.21) is an example of triples of orthogonal polynomials discussed in [AsI, p. 16].

We can also evaluate $A_m(z)$ and $B_m(z)$ at certain values of z.

Theorem 13.9. For $m \geq 0$ with $c_m = \left(\frac{(2m)!}{2^m m!}\right)^2$, we have

$$(13.24) \qquad A_m(0) = (2m)!, \ B_m(0) = (2m+1)! \, \xi_1(m),$$

$$(13.25) \qquad A_m(1) = (2m+1) c_m, \ B_m(1) = c_{m+1},$$

$$(13.26) \qquad A_m(4) = (2m+1)!, \ B_m(4) = \frac{1}{2}(2m+2)!,$$

$$(13.27) \quad A_m(9) = \frac{1}{3}(2m+1)(8m+3) c_m, \ B_m(9) = \frac{1}{9}(8m+9) c_{m+1},$$

and

$$(13.28) \quad A_m(16) = (2m+1)(2m+1)!, \ B_m(16) = (m+1)^2(2m+1)!.$$

Proof. (13.24) Put $x = 0$ into (13.3) and (13.4) and use (11.5) and (11.19), respectively.

(13.25) Put $x = 1$ into (13.5) and (13.6) and use (10.109).

(13.26)–(13.28) The proof is similar, putting $x = 2, 3$, and 4. $\square$

Theorem 10.17 can now be used to express $A_m(z)$ and $B_m(z)$ in terms of falling factorial polynomials. These representations are particularly interesting, since the coefficients are simple and explicit.

Corollary 13.10. For $m \geq 0$ and $z \in (-\infty, 0]$, we have that

(13.29)
$$A_m(z) = - \sum_{k=1}^{2m+1} L_{2m+1,k} \left(i\sqrt{|z|} - 2 \right)_{k-1,2}$$

and

(13.30)
$$B_m(z) = \frac{1}{i\sqrt{|z|}} \sum_{k=1}^{2m+2} L_{2m+2,k} \left(i\sqrt{|z|} - 2 \right)_{k-1,2}.$$

Proof. (13.29) Combining (13.5) with (10.39), we obtain

$$x A_m(x^2) = \delta_{2m+1}(x) = - \sum_{k=0}^{2m+1} L_{2m+1,k} (x)_{k,2} = - \sum_{k=1}^{2m+1} L_{2m+1,k} (x)_{k,2},$$

using (10.36). But

(13.31)
$$(x)_{k,2} = x(x-2)_{k-1,2}, \ k \geq 1,$$

so

$$A_m(x^2) = - \sum_{k=1}^{2m+1} L_{2m+1,k} (x-2)_{k-1,2}.$$

Setting $x = i\sqrt{|z|}$, we obtain the result.

(13.30) Combining (13.6) with (10.39), we get

$$x^2 B_m(x^2) = \delta_{2m+2}(x) = \sum_{k=0}^{2m+2} L_{2m+2,k} (x)_{k,2}.$$

But using (13.31), this becomes

$$x B_m(x^2) = \sum_{k=1}^{2m+2} L_{2m+2,k} \left(x-2 \right)_{k-1,2}.$$

Setting $x = i\sqrt{|z|}$, we obtain the result. $\square$

Theorem 13.11. Let $z \in (-\infty, 0]$. Then for $m \geq 0$ we have

$$(13.32) \quad A_m(z) = \frac{1}{2} \sum_{k=0}^{2m+1} \binom{2m+1}{k} \left(\frac{i}{2}\sqrt{|z|} + k - 1\right)_{2m}$$

$$= (-1)^m (2m+1)! \sum_{k=0}^{m} \frac{(-1)^k}{2k+1} \binom{\frac{i}{2}\sqrt{|z|}}{m-k} \binom{i\sqrt{|z|} + 2k}{2k}$$

and

$$(13.33) \quad B_m(z) = \frac{1}{2i\sqrt{|z|}} \sum_{k=0}^{2m+2} \binom{2m+2}{k} \left(\frac{i}{2}\sqrt{|z|} + k - 1\right)_{2m+1}$$

$$= \frac{(-1)^{m+1}(2m+2)!}{z} \sum_{k=0}^{m+1} (-1)^k \binom{\frac{i}{2}\sqrt{|z|}}{m+1-k} \binom{i\sqrt{|z|} + 2k - 1}{2k}.$$

Proof. (13.32) Replacing x by $\frac{x}{2}$ in (10.50), we have

$$\delta_{2m+1}(x) = \frac{x}{2} \sum_{k=0}^{2m+1} \binom{2m+1}{k} \left(\frac{x}{2} + k - 1\right)_{2m}.$$

Combining this with (13.5) gives

$$A_m(x^2) = \frac{1}{2} \sum_{k=0}^{2m+1} \binom{2m+1}{k} \left(\frac{x}{2} + k - 1\right)_{2m}.$$

Setting $x = i\sqrt{|z|}$, so $x^2 = -|z| = z$, we obtain the result.

Next, combining (10.56) (with $n = 2m+1$) and (13.5) gives

$$xA_m(x^2) = \delta_{2m+1}(x) = (-1)^m (2m+1)! \sum_{k=0}^{m} (-1)^k \binom{\frac{x}{2}}{m-k}\binom{x+2k}{2k+1}$$

$$= (-1)^m (2m+1)! \, x \sum_{k=0}^{m} \frac{(-1)^k}{2k+1} \binom{\frac{x}{2}}{m-k}\binom{x+2k}{2k},$$

so

$$A(x^2) = (-1)^m (2m+1)! \sum_{k=0}^{m} \frac{(-1)^k}{2k+1} \binom{\frac{x}{2}}{m-k}\binom{x+2k}{2k}.$$

Setting $x = i\sqrt{|z|}$ gives the result.

(13.33) From (10.50), we have

$$\delta_{2m+2}(x) = \frac{x}{2} \sum_{k=0}^{2m+1} \binom{2m+2}{k} \left(\frac{x}{2} + k - 1\right)_{2m+1}.$$

Combining this with (13.6) gives

$$B_m(x^2) = \frac{1}{2x} \sum_{k=0}^{2m+1} \binom{2m+2}{k} \left(\frac{x}{2} + k - 1\right)_{2m+1}.$$

The result follows by setting $x = i\sqrt{|z|}$.

Finally, combining (10.56) (with $n = 2m + 2$) and (13.6) gives

$$x^2 B_m(x^2) = (-1)^{m+1}(2m+2)! \sum_{k=0}^{m+1} (-1)^k \binom{\frac{x}{2}}{m+1-k} \binom{x+2k-1}{2k}.$$

The result follows by setting $x = i\sqrt{|z|}$. $\square$

We can now derive determinant formulas for both $A_m(z)$ and $B_m(z)$. We will also find the values of two interesting determinants whose entries are tangent numbers. These determinants and others were previously evaluated in a different way using properties of certain continued fractions (see [CaA]).

Theorem 13.12. For $m \geq 1$,

$$(13.34) \qquad A_m(z) = \frac{1}{r_m} \begin{vmatrix} C_1 & C_3 & \cdots & C_{2m+1} \\ C_3 & C_5 & \cdots & C_{2m+3} \\ \vdots & \vdots & \ddots & \vdots \\ C_{2m-1} & C_{2m+1} & \cdots & C_{4m-1} \\ 1 & z & \cdots & z^m \end{vmatrix},$$

where

$$(13.35) \qquad r_m = \begin{vmatrix} C_1 & C_3 & \vdots & C_{2m-1} \\ C_3 & C_5 & \cdots & C_{2m+1} \\ \vdots & \vdots & \ddots & \vdots \\ C_{2m-1} & C_{2m+1} & \cdots & C_{4m-3} \end{vmatrix} = (-1)^m \prod_{j=1}^{2m-1} j!,$$

and

$$(13.36) \qquad B_m(z) = \frac{1}{s_m} \begin{vmatrix} C_3 & C_5 & \cdots & C_{2m+3} \\ C_5 & C_7 & \cdots & C_{2m+5} \\ \vdots & \vdots & \ddots & \vdots \\ C_{2m+1} & C_{2m+3} & \cdots & C_{4m+1} \\ 1 & z & \cdots & z^m \end{vmatrix},$$

where

$$(13.37) \qquad s_m = \begin{vmatrix} C_3 & C_5 & \cdots & C_{2m+1} \\ C_5 & C_7 & \cdots & C_{2m+3} \\ \vdots & \vdots & \ddots & \vdots \\ C_{2m+1} & C_{2m+3} & \cdots & C_{4m-1} \end{vmatrix} = \prod_{j=1}^{2m} j!.$$

Proof. Substituting $k = 1$ into Theorem 8.12 gives (13.34) and (13.35), using (9.3), and (13.36) and (13.37), using (9.4). $\square$

The next formula gives the inversions of (13.3) and (13.4).

Theorem 13.13. For $n \geq 0$,

$$(13.38) \qquad z^n = \sum_{j=0}^{n} \lambda_{2j+1}^{(2n+1)} A_j(z) = \sum_{j=0}^{n} \lambda_{2j+2}^{(2n+2)} B_j(z).$$

Proof. This theorem is a specialization of (2.9). $\square$

We conclude this chapter by finding the structure constants for $\{A_m(z)\}_{m=0}^{\infty}$ sequence.

Let $r(u, v) \overset{\text{def}}{=} 2\min\{u, v\}$, $u, v \geq 0$.

Definition 13.14. For $m, n \geq 0$ and $0 \leq j \leq r(m, n)$, let

$$(13.39) \qquad C_j^{(m,n)} = (-1)^j (j\,!)^2 \binom{2m}{j} \binom{2n}{j} \binom{2m + 2n - j + 1}{j}.$$

The next theorem gives the structure constants the A_m sequence and a linearization in terms of the B_m sequence (cf. Theorem 14.13).

Theorem 13.15. For $m, n \geq 0$, we have that

$$(13.40) \qquad A_m(z) A_n(z) = \sum_{j=0}^{r(m,n)} C_j^{(m,n)} A_{m+n-j}(z).$$

Also, for $0 \leq m \leq n$ and $B_{-1}(z) = 0$, we have that

(13.41)

$$A_m(z) A_n(z) = (2m + 1)(2n + 1) \sum_{j=0}^{2m} \frac{C_j^{(m,n)}}{(2m - j + 1)(2n - j + 1)} B_{m+n-j}(z)$$
$$- 2(n - m)(2n - 2m + 1) C_{2m}^{(m,n)} B_{n-m-1}(z).$$

Proof. (13.40) From (10.116), we obtain

$$(13.42) \quad \delta_{2m+1}(x) \delta_{2n+1}(x) = \sum_{j=0}^{\rho(2m+1,2n+1)} \Delta_j^{(2m+1,2n+1)} \delta_{2m+2n+2-2j}(x),$$

where $\rho(2m+1, 2n+1) = 2\min\{m, n\} + 1 - \delta_{m,n} = r(m, n) + 1 - \delta_{m,n}$ and

$$\Delta_j^{(2m+1,2n+1)} = (-1)^j (j\,!)^2 \binom{2m+1}{j} \binom{2n+1}{j} \binom{2m + 2n + 1 - j}{j}.$$

Also, replacing z by x^2 in (13.40) and using (13.5), we obtain

$$(13.43) \qquad \delta_{2m+1}(x)\,\delta_{2n+1}(x) = \sum_{j=0}^{r(m,n)} C_j^{(m,n)} x\, \delta_{2m+2n+1-2j}(x).$$

But using (10.8) on the right side of this equation gives

$$(13.44) \qquad \sum_{j=0}^{r(m,n)} C_j^{(m,n)} \Big(\delta_{2m+2n+2-2j}(x)$$

$$- (2m + 2n - 2j)(2m + 2n + 1 - 2j)\,\delta_{2m+2n-2j}(x)\Big)$$

$$= \sum_{j=0}^{r(m,n)} C_j^{(m,n)} \delta_{2m+2n+2-2j}(x)$$

$$- \sum_{j=1}^{r(m,n)+1} C_{j-1}^{(m,n)} (2m + 2n + 2 - 2j)(2m + 2n + 3 - 2j)\,\delta_{2m+2n+2-2j}(x)$$

$$= \delta_{2m+2n+2}(x) + \sum_{j=1}^{r(m,n)} \Big(C_j^{(m,n)} - (2m + 2n + 2 - 2j)$$

$$\times (2m + 2n + 3 - 2j)\, C_{j-1}^{(m,n)}\Big)\delta_{2m+2n+2-2j}(x) - (2m + 2n - 2r(m,n))$$

$$\times (2m + 2n + 1 - 2r(m,n))\, C_{r(m,n)}^{(m,n)} \delta_{2m+2n-2r(m,n)}(x).$$

Now, the $j = 0$ term on the right side of (13.42) is $\delta_{2m+2n+2}(x)$, which is the first term on the right of (13.44). For $1 \le j \le r(m,n)$, it is routine to show that

$$C_j^{(m,n)} - (2m + 2n + 2 - 2j)(2m + 2n + 3 - 2j)\, C_{j-1}^{(m,n)} = \Delta_j^{(2m+1,2n+1)},$$

so again the two sides agree. Finally, when $m = n$, we find that $j = \rho(2m + 1, 2n + 1) = r(m,n) = 2m$, so there is no further term in (13.42), which is the case on the right side of (13.44) where the final term is 0. Also, since (13.44) is symmetric in m and n, we need only consider the case $m < n$. In this case, the term for $j = 2m + 1$ on the left is

$$\Delta_{2m+1}^{(2m+1,2n+1)} \delta_{2n-2m}(x) = -\frac{(2n)!\,(2n+1)!}{(2n - 2m - 1)!\,(2n - 2m)!}\,\delta_{2n-2m}(x),$$

which equals $-(2n - 2m)(2n + 1 - 2m)\, C_{2m}^{(m,n)} \delta_{2n-2m}(x)$, the final term on the right when $r(m,n) = 2m$.

(13.41) Substituting (13.7) into (13.40), we find

$$A_m A_n =$$

$$\sum_{j=0}^{2m} C_j^{(m,n)}\left[B_{m+n-j} - 2(m+n-j)(2m+2n-2j+1)B_{m+n-j-1}\right]$$

$$= \sum_{j=0}^{2m} C_j^{(m,n)} B_{m+n-j}$$

$$- 2 \sum_{j=1}^{2m+1} (m+n-j+1)(2m+2n-2j+3)\, C_{j-1}^{(m,n)} B_{m+n-j}$$

$$= \sum_{j=1}^{2m} \left[C_j^{(m,n)} - 2(m+n-j+1)(2m+2n-2j+3)C_{j-1}^{(m,n)}\right] B_{m+n-j}$$

$$+ B_{m+n} - 2(n-m)(2n-2m+1)\, C_{2m}^{(m,n)} B_{n-m-1}.$$

But

$$C_{j-1}^{(m,n)} =$$

$$-\frac{j\,(2m+2n-j+2)}{2(2m-j+1)(2n-j+1)(m+n-j+1)(2m+2n-2j+3)}\, C_j^{(m,n)},$$

so we obtain

$$A_m A_n = \sum_{j=1}^{2m} \left[1 + \frac{j\,(2m+2n-j+2)}{(2m-j+1)(2n-j+1)}\right] C_j^{(m,n)} B_{m+n-j}$$

$$+ B_{m+n} - 2(n-m)(2n-2m+1)\, C_{2m}^{(m,n)} B_{n-m-1},$$

from which the result follows. $\square$

Corollary 13.16. For $0 \le m \le n$, we have

$$(13.45) \qquad \sum_{j=0}^{2m}(-1)^j \binom{2m+2n-j+1}{j}\binom{2m+2n-2j}{2m-j} = 1$$

and

$$(13.46) \quad \sum_{j=0}^{2m}(-4)^j (2m+2n-2j+1)\binom{2m+2n-j+1}{j}$$

$$\times \binom{2m+2n-2j}{2m-j}\binom{2m+2n-2j}{m+n-j}$$

$$= (2m+1)(2n+1)\binom{2m}{m}\binom{2n}{n}.$$

Proof. The results follow from substituting $x = 0$ and 1, respectively, into (13.40) and then using (13.24) and (13.25). $\square$

Finally, we give a more standard proof of a generalization of the interesting identity (13.45).

Theorem 13.17. For $n \geq 0$,

$$(13.47) \qquad \sum_{j=0}^{n}(-1)^j\binom{x-j+1}{j}\binom{x-2j}{n-j} = \frac{1}{2}[1+(-1)^n].$$

Proof. Let

$$S(x,n) = \sum_{j=0}^{n}(-1)^j\binom{x-j+1}{j}\binom{x-2j}{n-j}, \quad n \geq 0.$$

Then, using [Go, p. iv],

$$(13.48) \qquad \binom{-x}{n} = (-1)^n\binom{x+n-1}{n},$$

we have

$$\binom{x-j+1}{j} = (-1)^j\binom{-x-2+2j}{j},$$

so $S(x,n) = \displaystyle\sum_{j=0}^{n}\binom{-x-2+2j}{j}\binom{x-2y}{n-j}$. Next, using [Go, p. 41, 3.144],

we get

$$\sum_{k=0}^{n}\binom{x+kz}{k}\binom{y-kz}{n-k} = \sum_{k=0}^{n}\binom{x+y-k}{n-k}z^k,$$

so $S(x,n) = \displaystyle\sum_{j=0}^{n}\binom{-2-j}{n-j}2^j$ (note there is no x). Again using (13.48), we

find that

$$\binom{-2-j}{n-j} = (-1)^{n-j}\frac{n+1}{j+1}\binom{n}{j},$$

so $S(x,n) = (-1)^n(n+1)\displaystyle\sum_{j=0}^{n}\binom{n}{j}\frac{(-2)^j}{j+1}$. Finally, from [Go, p. 5, 1.37], we

obtain the result

$$\sum_{k=0}^{m}\binom{n}{k}\frac{x^k}{k+1} = \frac{(x+1)^{n+1}-1}{(n+1)x}. \quad \square$$

Chapter 14

The Weight Functions for $\{A_m(z)\}$ and $\{B_m(z)\}$

In this chapter we will find the weight functions $w_A(z)$ and $w_B(z)$ for the orthogonal sequences $\{A_m(z)\}_{m=0}^{\infty}$ and $\{B_m(z)\}_{m=0}^{\infty}$ and verify that the evaluations $(A_m, A_n)_A = (m + n)!\,(m + n + 1)!\,\delta_{m,n}$ and $(B_m, B_n)_B = \frac{1}{2}(m + n + 1)!\,(m + n + 2)!\,\delta_{m,n}$ which were obtained by specializing (7.12) and (7.13), are those that are obtained when these inner products are computed relative to their respective weight functions. Also included in this chapter (Theorem 14.9) is the formula for the Laplace transform of $w_A(z)$. Weight functions are also constructed for the secondary sequences derived from any function in Class III. The final result in the chapter is the formula giving the structure constants for $\{B_m(z)\}_{m=0}^{\infty}$.

The weight functions $w_A(z)$ and $w_B(z)$, associated with $\{A_m(z)\}_{m=0}^{\infty}$ and $\{B_m(z)\}_{m=0}^{\infty}$ on the interval $(-\infty, 0)$, respectively, are the following (defined on all of $\mathbb{R}$).

Definition 14.1 For $z \in \mathbb{R}$, let

$$(14.1) \qquad w_A(z) = \frac{1}{2}\operatorname{csch}\left(\frac{\pi}{2}\sqrt{|z|}\right)$$

and

$$(14.2) \qquad w_B(z) = \frac{1}{4}|z|\operatorname{csch}\left(\frac{\pi}{2}\sqrt{|z|}\right).$$

The fact that $(-\infty, 0]$ is the orthogonality interval and the relationships in (13.5) and (13.6) imply that the zeros of $\delta_n(x)$ are pure imaginary.

We begin by representing the A and B moments as integrals with respect to these weights and by giving their values. **Theorem 14.2.** For $n \geq 0$,

$$(14.3) \qquad \mu_A(n) = \int_{-\infty}^{0} z^n w_A(z)dz = -C_{2n+1}$$

and

$$(14.4) \qquad \mu_B(n) = \int_{-\infty}^{0} z^n w_B(z)dz = \frac{1}{2} C_{2n+3}.$$

Proof. (14.3) From (12.27) and (7.24), we have $\mu_A(n) = -C_{2n+1}$. But, from Nörlund [No, p. 75] (cf. [Ob, p. 61, (7.2)]), we have that

$$(14.5) \qquad \int_{0}^{\infty} x^{2n+1} \operatorname{csch}\left(\frac{\pi x}{2}\right)dx = (-1)^{n+1}C_{2n+1}.$$

Setting $x = \sqrt{-z}$ in (14.5) and using (14.1), we obtain

$$\int_{-\infty}^{0} z^n w_A(z)dz = -C_{2n+1},$$

which gives (14.3). Since $w_B(z) = -\frac{1}{2}z\, w_A(z)$ and using (14.3), (12.28), and (7.24), we have

$$\int_{-\infty}^{0} z^n w_B(z)dz = -\frac{1}{2} \int_{-\infty}^{0} z^{n+1} w_A(z)dz = \frac{1}{2}C_{2n+3} = \mu_B(n). \quad \square$$

Note for $n = 0$ that (14.3) and (14.4) become

$$(14.6) \qquad \mu_A(0) = \mu_B(0) = 1,$$

which is as it should be.

We can also find formulas generalizing the evaluations in (14.3) and (14.4). From [Ob, p. 61, (6.2)], we have

$$\int_{0}^{\infty} x^{z-1} \operatorname{csch}(ax)dx = 2a^{-z}(1 - 2^{-z})\Gamma(z)\,\zeta(z),$$

where $Re(z) > 1$. Then for $p > -\frac{1}{2}$,

$$\int_{-\infty}^{0} |z|^p w_A(z)dz = \frac{2(2^{2p+2} - 1)}{\pi^{2p+2}}\,\Gamma(2p+2)\,\zeta(2p+2),$$

and for $p > -\dfrac{3}{2}$,

$$(14.7) \qquad \int_{-\infty}^{0} |z|^p w_B(z)\,dz = \frac{2^{2p+4}-1}{\pi^{2p+4}}\Gamma(2p+4)\,\zeta(2p+4),$$

using the transformation $x = \sqrt{-z}$ in both cases.

Lemma 14.3. For $n \geq 1$,

$$(14.8) \qquad \sum_{j=1}^{n} \lambda_{2j+1}^{(2n+1)} \int_{-\infty}^{0} A_j(z)\,w_A(z)\,dz = 0$$

and

$$(14.9) \qquad \sum_{j=1}^{n} \lambda_{2j+2}^{(2n+2)} \int_{-\infty}^{0} B_j(z)\,w_B(z)\,dz = 0.$$

Proof. (14.8) From (13.38), Table 13.1, (14.3), and (14.6) we have

$$\sum_{j=1}^{n} \lambda_{2j+1}^{(2n+1)} A_j(z)\,w_A(z)\,dz$$

$$= \int_{-\infty}^{0} \left\{ \sum_{j=0}^{n} \lambda_{2j+1}^{(2n+1)} A_j(z) - \lambda_{1}^{(2n+1)} A_0(z) \right\} w_A(z)\,dz$$

$$= \int_{-\infty}^{0} \left\{ z^n - \lambda_{1}^{(2n+1)} \right\} w_A(z)\,dz = \mu_A(n) - \lambda_{1}^{(2n+1)}\mu_A(0)$$

$$= \lambda_{1}^{(2n+1)} - \lambda_{1}^{(2n+1)} = 0.$$

(14.9) The proof is similar to that of (14.8). $\square$

We next prove that each $A_m(z)$ is orthogonal to $A_0(z) = 1$ relative to $w_A(z)$ and the comparable result for $B_m(z)$.

Lemma 14.4. For $n \geq 0$,

$$(14.10) \qquad \int_{-\infty}^{0} A_n(z)\,w_A(z)\,dz = \delta_{n,0}$$

and

$$(14.11) \qquad \int_{-\infty}^{0} B_n(z)\,w_B(z)\,dz = \delta_{n,0}.$$

Proof. (14.10) True for $n = 0$ by (14.6) and also for $n = 1$, since, if we put $n = 1$ in (14.8), we get

$$\lambda_{3}^{(3)} \int_{-\infty}^{0} A_1(z)\,w_A(z)\,dz = 0,$$

where $\lambda_3^{(3)} = 1 \neq 0$ by (12.19).

If we now assume that equation (14.10) is true for each m, $1 \leq m \leq n$, then (14.8) implies for $n + 1$ that

$$\lambda_{2n+3}^{(2n+3)} \int_{-\infty}^0 A_{n+1}(z)\, w_A(z)\, dz = -\sum_{j=1}^n \lambda_{2j+1}^{(2n+3)} \int_{-\infty}^0 A_j(z)\, w_A(z)\, dz = 0,$$

which implies the result by (12.19) and induction. Equation (14.11) is proved similarly. $\square$

Lemma 14.5. For $0 \leq m < n$,

$$(14.12) \qquad \int_{-\infty}^0 z^m A_n(z)\, w_A(z)\, dz = 0$$

and

$$(14.13) \qquad \int_{-\infty}^0 z^m B_n(z)\, w_B(z)\, dz = 0.$$

Proof. (14.12) We prove this equation by a double induction in which $m \geq 0$ is specified first and then $n > m$ is assumed.

The equation is true for $m = 0$ and $n > 0$ by (14.10). Assume it is true for some $m \geq 0$ and all $n > m + 1$. Then using (13.1), we have for $m + 1$ and $n > m + 1$ that

$$\int_{-\infty}^0 z^{m+1} A_n(z)\, w_A(z)\, dz = \int_{-\infty}^0 z^m [A_{n+1}(z) - 2(2n + 1)^2 A_n(z)$$
$$+ 4n^2(2n - 1)(2n + 1)A_{n-1}(z)]w_A(z)\, dz = 0,$$

by the induction assumption. Equation (14.13) is proved in a similar way. $\square$

The next theorem verifies that the two integrals in Theorem 14.6 yield the inner products given in the introduction to this chapter.

Theorem 14.6. For $m, n \geq 0$,

$$(14.14) \qquad \int_{-\infty}^0 A_m(z)A_n(z)w_A(z)dz = (m + n)!\,(m + n + 1)!\,\delta_{m,n}$$

and

$$(14.15) \qquad \int_{-\infty}^0 B_m(z)B_n(z)w_B(z)dz = \frac{1}{2}(m + n + 1)!\,(m + n + 2)!\,\delta_{m,n}.$$

Proof. (14.14) Without loss of generality we may assume that $m \leq n$. For $m < n$, we have by (13.3) and (14.12) that

$$\int_{-\infty}^0 A_m(z)A_n(z)w_A(z)dz = \sum_{j=0}^m d_{2j+1}^{(2m+1)} \int_{-\infty}^0 z^j A_n(z)w_A(z)dz = 0.$$

For $m \geq 0$ and $m = n$, let

$$a_m = \int_{-\infty}^{0} [A_m(z)]^2 \, w_A(z) dz.$$

We will then show that $a_{m+1} = 4(m + 1)^2(2m + 1)(2m + 3) a_m$ for $m \geq 0$, which will imply the result in (14.14) (cf. [Ch, p. 75]). (Note the symmetric form of (14.14) readily reduces to the expression in m by setting $m = n$ allowed by $\delta_{m,n}$.) Certainly $a_0 = 1$ by (14.6). Also, using (14.12) we have

$$a_{m+1} = \int_{-\infty}^{0} [A_{m+1}(z)]^2 \, w_A(z) dz = \int_{-\infty}^{0} z^{m+1} A_{m+1}(z) w_A(z) dz.$$

By adding some terms to the integrand that are zero by (13.16), we obtain the equation

$$a_{m+1} = \int_{-\infty}^{0} z^m \Big([z + 2(2m + 3)^2] A_{m+1}(z) - A_{m+2}(z) \Big) w_A(z) dz$$

$$= 4(m + 1)^2 (2m + 1)(2m + 3) \int_{-\infty}^{0} z^m A_m(z) w_A(z) dz$$

$$= 4(m + 1)^2 (2m + 1)(2m + 3) a_m,$$

using (13.1). The result in (14.15) is proved similarly. $\square$

We have thus established the following result.

Corollary 14.7. For $P(z), Q(z) \in \mathbb{R}[z]$, we have

(14.16) $$(P, Q)_A = \int_{-\infty}^{0} P(z)Q(z) w_A(z) dz$$

and

(14.17) $$(P, Q)_B = \int_{-\infty}^{0} P(z)Q(z) w_B(z) dz.$$

Proof. Since the A and B sequences are simple, each is a basis for the vector space $\mathbb{R}[z]$. Also, since $(*, *)_A$ and $(*, *)_B$ and the two integrals on the right are proper bilinear inner products on $\mathbb{R}[z] \times \mathbb{R}[z]$ that agree on the two bases, respectively, they are equal by extending linearly. $\square$

In the next theorem we compute some inner products.

Theorem 14.8. We have that

(14.18) $$(z^n, z^k)_A = -C_{2n+2k+1}, \quad 0 \leq k \leq n,$$

(14.19) $$(z^n, z^k)_B = \frac{1}{2} C_{2n+2k+3}, \quad 0 \leq k \leq n,$$

$$(14.20) \qquad (z^n, A_k)_A = \begin{cases} 0, \ 0 \le n < k, \\ (2n)! \, (2n+1)!, \ n = k, \ n \ge 0, \\ (2k)! \, (2k+1)! \, \lambda_{2k+1}^{(2n+1)}, \ 0 \le k \le n, \end{cases}$$

$$(14.21) \qquad (z^n, B_k)_B = \begin{cases} 0, \ 0 \le n < k, \\ (n+1)[(2n+1)!]^2, \ n = k, \ n \ge 0, \\ (k+1)[(2k+1)!]^2 \, \lambda_{2k+2}^{(2n+2)}, \ 0 \le k \le n, \end{cases}$$

$(14.22) \quad (z^n, A_k)_B =$

$$\begin{cases} -6, \ n = 0, \ k = 1, \\ 0, \ k \ge 2, \ 0 \le n < k-1, \\ -\dfrac{2k+1}{2}[(2k)!]^2, \ n = k-1, \ k \ge 2, \\ \dfrac{2k+1}{2}[(2k)!]^2 \left[2(k+1)(2k+1)\lambda_{2k+2}^{(2n+2)} - \lambda_{2k}^{(2n+2)} \right], \ 0 \le k \le n, \end{cases}$$

$$(14.23) \qquad (z^n, B_k)_A = \begin{cases} (2k+1)!, \ n = 0, \ k \ge 0, \\ 0, \ 1 \le n \le k, \\ -2(k+1)[(2k+1)!]^2 \, \lambda_{2k+2}^{(2n)}, \ 0 \le k \le n-1, \end{cases}$$

$(14.24) \quad (A_m, B_n)_B =$
$$\frac{1}{2}(2m+1)[(2m)!]^2 [2(m+1)(2m+1)\delta_{m,n} - \delta_{m,n+1}], \ m,n \ge 0,$$

$$(14.25) \qquad (A_m, B_n)_A = \begin{cases} (2m)! \, (2n+1)!, \ 0 \le m \le n, \\ 0, \ 0 \le n < m, \end{cases}$$

$$(14.26) \qquad (B_m, B_n)_A = (2m+1)! \, (2n+1)! \, \xi_1(m), \ 0 \le m \le n,$$

and

$(14.27) \quad (A_m, A_n)_B = (m+n+1)(2m+1)! \, (2n+1)! \, \delta_{m,n}$
$$-\frac{1}{2}(2m+1)! \, (2n+2)! \, \delta_{m,n+1} - \frac{1}{2}(2n+1)! \, (2m+2)! \, \delta_{n,m+1}, \ m,n \ge 0.$$

Proof. (14.18) Using (14.16) and (14.3), we have

$$(z^n, z^k)_A = \int_{-\infty}^{0} z^{n+k} w_A(z) dz = -C_{2n+2k+1}.$$

Equation (14.19) is proved similarly from (14.17) and (14.4).

(14.20) From (13.38), (14.16), and (14.14), we have

$$(z^n, A_k)_A = \left(\sum_{j=0}^{n} \lambda_{2j+1}^{(2n+1)} A_j(z), A_k(z) \right)_A$$

$$= \sum_{j=0}^{n} \lambda_{2j+1}^{(2n+1)} \int_{-\infty}^{0} A_j(z) A_k(z) w_A(z) dz$$

$$= \sum_{j=0}^{n} \lambda_{2j+1}^{(2n+1)} (j+k)!\, (j+k+1)!\, \delta_{j,k}$$

$$= (2k)!\, (2k+1)!\, \lambda_{2k+1}^{(2n+1)}.$$

Equation (14.21) is proved similarly.

(14.22)

(1) Putting $m = 0$ into (13.7) gives $A_1(z) = B_1(z) - 6B_0(z)$, so $(1, A_1)_B = (1, B_1)_B - 6(1, B_0)_B = -6$.

(2) By (13.7), we have

$$(14.28) \qquad\qquad A_k = B_k - 2k(2k+1)B_{k-1},$$

so

$$(z^n, A_k)_B = (z^n, B_k)_B - 2k(2k+1)(z^n, B_{k-1})_B.$$

But, by (14.13), we have that $(z^m, B_n)_B = 0, 0 \le m < n$, which gives the result.

(3) Using (14.28), (14.21), and (12.19), we have

$$(z^{k-1}, A_k)_B = (z^{k-1}, B_k)_B - 2k(2k+1)(z^{k-1}, B_{k-1})_B$$

$$= -2k^2(2k+1)[(2k-1)!]^2 \lambda_{2k}^{(2k)} = -\frac{2k+1}{2}[(2k)!]^2.$$

(4) For $k = 0$, we find using (14.21) that $(z^n, A_0)_B = (z^n, 1)_B = (z^n, B_0)_B = \lambda_2^{(2n+2)}$, which agrees with the RHS using (12.18). Otherwise, using (14.28) we have that

$$(z^n, A_k)_B = (z^n, B_k)_B - 2k(2k+1)(z^n, B_{k-1})_B$$

$$= (k+1)[(2k+1)!]^2 \lambda_{2k+2}^{(2n+2)} - 2k^2(2k+1)[(2k-1)!]^2 \lambda_{2k}^{(2n+2)}$$

$$= \frac{2k+1}{2}[(2k)!]^2 \left(2(k+1)(2k+1)\lambda_{2k+2}^{(2n+2)} - \lambda_{2k}^{(2n+2)} \right).$$

(14.23)

(1) For $n = 0$, we have from (13.10) that

$$B_k = (2k+1)! \sum_{j=0}^{k} \frac{1}{(2j+1)!} A_j,$$

so

$$(1, B_k)_A = (2k+1)! \sum_{j=0}^{k} \frac{1}{(2j+1)!}(1, A_j)_A = (2k+1)!(1,1)_A = (2k+1)!.$$

(2) For $1 \le n \le k$, since $z\, w_A(z) = -2w_B(z)$ from (14.1) and (14.2), we have that

$$(z^n, B_k)_A = \int_{-\infty}^{0} z^n B_k(z)w_A(z)dz = -\int_{-\infty}^{0} z^{n-1}B_k(z)w_B(z)w_B(z)dz$$
$$= -2(z^{n-1}, B_k)_B = 0,$$

using (14.13).

(3) By the argument in (2) and (14.21), we have

$$(z^n, B_k)_A = -2(z^{n-1}, B_k)_k = -2(k+1)[(2k+1)!]^2\lambda_{2k+2}^{(2n)}.$$

(14.24) Equation is true for $m = 0$. For $m \ge 1$, we have from (13.7) that

$$A_m = B_m - 2m(2m+1)B_{m-1}.$$

Using (14.15), we find that

$$(A_m, B_n)_B = (B_m, B_n)_B - 2m(2m+1)(B_{m-1}, B_n)_B$$
$$= (m+1)[(2m+1)!]^2\delta_{m,n} - 2m^2(2m+1)[(2m-1)!]^2\delta_{m-1,n}$$
$$= \frac{1}{2}(2m+1)[(2m)!]^2[2(m+1)(2m+1)\delta_{m,n} - \delta_{m,n+1}]. \quad \square$$

(14.25) From (13.10), we have that

(14.29) $$B_n(z) = (2n+1)! \sum_{k=0}^{n} \frac{1}{(2k+1)!}A_k(z),$$

and using (14.14), we find that

$$(A_m, B_n)_A = (2n+1)! \sum_{k=0}^{n} \frac{1}{(2k+1)!}(A_m, A_k)_A$$
$$= (2n+1)! \sum_{k=0}^{n} \frac{1}{(2k+1)!}(2m)!(2m+1)!\delta_{m,k}$$
$$= (2m)!(2m+1)!(2n+1)! \sum_{k=0}^{n} \frac{\delta_{m,k}}{(2k+1)!}.$$

Now, if $m > n$, then for $0 \le k \le n$, we have $(A_m, B_n)_A = 0$. On the other hand, if $0 \le m \le n$, then $(A_m, B_n)_A = (2m)!(2n+1)!.$

(14.26) Substituting (13.10) into (14.25) gives

$$(B_m, B_n)_A = (2m+1)! \sum_{k=0}^{m} \frac{1}{(2k+1)!}(A_k, B_n)_A$$

$$= (2m+1)! \sum_{k=0}^{m} \frac{1}{(2k+1)!} (2k)!(2n+1)!$$

$$= (2m+1)!(2n+1)!\xi_1(m),$$

where we have used (11.17).

(14.27) For $m = 0$, we have from (13.27) that

$$(A_m, A_0)_B = (1, A_m)_B = \begin{cases} 1, & m = 0, \\ -6, & m = 1, \\ 0, & m \geq 2. \end{cases}$$

For $m \geq 1$, we have

$$(A_m, A_n)_B = (A_m, B_n)_B - 2n(2n+1)(A_m, B_{n-1})_B$$

$$= \frac{1}{2}(2m+1)[(2m)!]^2 [2(m+1)(2m+1)\delta_{m,n} - \delta_{m,n+1}]$$

$$- 2n(2n+1)\left\{ \frac{1}{2}(2m+1)[(2m)!]^2 [2(m+1)(2m+1)\delta_{m,n-1} - \delta_{m,n}] \right\}$$

$$= -\frac{1}{2}(2m+1)[(2m)!]^2 [4(m+1)^2(2m+1)(2m+3)\delta_{m+1,n}$$

$$- 2(2m+1)^2\delta_{m,n} + \delta_{m,n+1}].$$

This becomes the final result when the RHS is transformed into an expression containing only m by replacing n by m, $m-1$, and $m+1$, respectively, a substitution permitted by the δ's. □

We next derive the formula for the Laplace transform of $w_A(z)$. We first prove the following result.

Theorem 14.9. For $s > 0$, we have

$$(14.30) \quad \int_0^\infty e^{-st} csch(\sqrt{t})\, dt = \sqrt{\frac{\pi}{s}} + \pi^{\frac{3}{2}} \sum_{n=1}^{\infty} (-1)^n n\, e^{\pi^2 n^2 s}\Gamma\left(-\frac{1}{2}, \pi^2 n^2 s\right),$$

where $\Gamma(*, *)$ is an incomplete gamma function [ObB, p. 420].

Proof. From the expansion [Mark, p. 36, Ex. 3], [Han, p. 104, 6.1.33]

$$\operatorname{csch} x = \frac{1}{x} + 2x \sum_{n=1}^{\infty} \frac{(-1)^n}{x^2 + \pi^2 n^2}, \quad x \neq 0,$$

we obtain

$$(14.31) \quad \operatorname{csch}(\sqrt{t}) = \frac{1}{\sqrt{t}} + 2\sqrt{t} \sum_{n=1}^{\infty} \frac{(-1)^n}{t + \pi^2 n^2}, \quad t > 0.$$

Substituting this result into the LHS of (14.30) gives

$$(14.32) \quad \int_0^\infty e^{-st} csch(\sqrt{t}) \, dt$$

$$= \int_0^\infty \frac{e^{-st}}{\sqrt{t}} \, dt + 2 \int_0^\infty \sqrt{t} e^{-st} \sum_{n=1}^\infty \frac{(-1)^n}{t + \pi^2 n^2} \, dt.$$

From [ObB, p. 21, 3.3], we find that the first integral on the RHS of (14.32) is equal to $\sqrt{\frac{\pi}{s}}$. Also, we can interchange the order of integration and summation in the second term on the RHS to obtain

$$(14.33) \quad \sum_{n=1}^\infty (-1)^n 2 \int_0^\infty \frac{\sqrt{t} e^{-st}}{t + \pi^2 n^2} \, dt.$$

To justify this interchange, we use Lebesgue's dominated convergence theorem:

Let $\{f_N(t)\}$ be defined by

$$f_N(t) = \sqrt{t} e^{-st} \sum_{n=1}^N \frac{(-1)^n}{t + \pi^2 n^2}, \quad s > 0, \ N \geq 1,$$

and let $(t) = g(t, s) = g \frac{\sqrt{t}}{6} e^{-st}$, $s > 0$ and $t \geq 0$. (Note for $s > 0$ that $g \in L_1([0, \infty), B([0, \infty)), \lambda)$.) We then have that $|f_N(t)| \leq g(t)$, since

$$|f_N(t)| \leq \frac{\sqrt{t} e^{-st}}{\pi^2} \sum_{n=1}^\infty \frac{1}{n^2 + t\pi^{-2}} < \frac{\sqrt{t} e^{-st}}{\pi^2} \sum_{n=1}^\infty \frac{1}{n^2} = \frac{\sqrt{t} e^{-st}}{6} = g(t).$$

Thus, the function

$$(14.34) \quad f(t) = \sqrt{t} e^{-st} \sum_{n=1}^\infty \frac{(-1)^n}{t + \pi^2 n^2} = \lim_{N \to \infty} f_N(t)$$

is integrable and the interchange is justified.

To evaluate the integral in (14.33), we use the formula [ObB, p. 22, 3.8]

$$2 \int_0^\infty \frac{\sqrt{t} e^{-st}}{t + a} \, dt = \sqrt{\pi a} \, e^{as} \Gamma \left(-\frac{1}{2}, as \right), \quad a, s > 0,$$

which, by setting $a = \pi^2 n^2$, gives from (14.34) that

$$\int_0^\infty f(t) \, dt = \pi^{\frac{3}{2}} \sum_{n=1}^\infty (-1)^n n \, e^{\pi^2 n^2 s} \Gamma \left(-\frac{1}{2}, \pi^2 n^2 s \right),$$

which completes the proof. $\square$

Corollary 14.10. For $s > 0$,

$$(14.35) \quad \int_{-\infty}^{0} e^{sz} w_A(z)dz = \frac{2}{\sqrt{\pi}}\left\{\frac{1}{\sqrt{s}}+2\sum_{n=1}^{\infty}(-1)^n n\, e^{4n^2 s}\Gamma\left(-\frac{1}{2}, 4n^2 s\right)\right\}.$$

Proof. This follows from (14.30) by setting $t = -\frac{\pi^2 z}{4}$, replacing s by $\frac{4s}{\pi^2}$, and using (14.1). $\square$

There is a similar, but more complicated, formula for the Laplace transform of $w_B(z)$.

We next derive the weight functions and intervals of orthogonality for the secondary sequences $\{\tilde{A}_m(z)\}_{m=0}^{\infty}$ and $\{\tilde{B}_m(z)\}_{m=0}^{\infty}$ derived from any function f in Class III.

Recall at the end of Chapter 7 that we obtained the general form of the functions in the three classes in $\mathcal{F}_2$, parameterized in (7.2) by $a_1, c_1, c_2 \in \mathbb{R}$, where $a_1 \neq 0$ and $c_2 > 0$. There are two forms in Class III, where $c_1 \neq 0$, $\Delta = 0$, and $c_2 = \frac{1}{4}c_1^2$ by (3.4), which are given by (7.43) and (7.44), viz.,

$$(14.36) \quad f(t) = \begin{cases} a_1\rho\tanh^{-1}\left(\frac{t}{\rho}\right), & c_1 > 0, \\ a_1\rho\tan^{-1}\left(\frac{t}{\rho}\right), & c_1 < 0, \end{cases}$$

where $\rho = \sqrt{\dfrac{2}{|c_1|}}$.

For each function f in (14.36), we will denote G_m in (1.36) by $\tilde{A}_m$ and H_m in (1.37) by $\tilde{B}_m$.

Theorem 14.11. For each f in (14.36), the terms of the secondary sequences $\tilde{A}_m(z)\}_{m=0}^{\infty}$ and $\{\tilde{B}_m(z)\}_{m=0}^{\infty}$ are given by the formulas

$$(14.37) \quad \tilde{A}_m(z) = a_1\left(\frac{c_1}{2}\right)^m A_m\left(\frac{2a_1^2}{c_1}z\right)$$

and

$$(14.38) \quad \tilde{B}_m(z) = a_1^2\left(\frac{c_1}{2}\right)^m B_m\left(\frac{2a_1^2}{c_1}z\right).$$

If

$$I = \begin{cases} (-\infty, 0], & c_1 > 0, \\ [0, \infty), & c_1 < 0, \end{cases}$$

then the functions

$$(14.39) \quad w_{\tilde{A}}(z) = \frac{2}{|c_1|}w_A\left(\frac{2a_1^2}{c_1}z\right)$$

and

$$(14.40) \qquad w_{\tilde{B}}(z) = \frac{2}{a_1^2 |c_1|} w_B\left(\frac{2a_1^2}{c_1} z\right)$$

are weight functions for these sequences, respectively. Also,

$$(14.41) \quad \int_I \tilde{A}_m(z)\, \tilde{A}_n(z)\, w_{\tilde{A}}(z)\, dz$$

$$= \left(\frac{c_1}{2}\right)^{m+n} (m+n)!\,(m+n+1)!\,\delta_{m,n}$$

and

$$(14.42) \quad \int_I \tilde{B}_m(z)\, \tilde{B}_n(z)\, w_{\tilde{B}}(z)\, dz$$

$$= \frac{1}{2}\left(\frac{c_1}{2}\right)^{m+n} (m+n+1)!\,(m+n+2)!\,\delta_{m,n}.$$

Proof.
Case $c_1 > 0$. From (14.36) we have that

$$\sum_{n=0}^{\infty} \tilde{\delta}_n(x)\frac{t^n}{n!} = exp(xf(t)) = exp\left(a_1\rho x \tanh^{-1}\left(\frac{t}{\rho}\right)\right)$$

$$= \sum_{n=0}^{\infty} \rho^{-n}\delta_n(a_1\rho x)\frac{t^n}{n!},$$

so by (11.1),

$$\tilde{\delta}_n(x) = \rho^{-n}\delta_n(a_1\rho x) = \sum_{j=0}^{n} \rho^{-n}(a_1\rho)^j\, d_j^{(n)} x^j.$$

Then from the above and (13.3) we obtain

$$\tilde{A}_m(z) = \sum_{j=0}^{m} \rho^{-(2m+1)}(a_1\rho)^{2j+1} d_{2j+1}^{(2m+1)} z^j = \frac{a_1}{\rho^{2m}} \sum_{j=0}^{m} d_{2j+1}^{(2m+1)}(a_1^2\rho^2 z)^j$$

$$= \frac{a_1}{\rho^{2m}} A_m(a_1^2\rho^2 z) = a_1 \left(\frac{c_1}{2}\right)^m A_m\left(\frac{2a_1^2}{c_1} z\right),$$

which is (14.37).

Using this result, (14.39), and (14.14), and setting $u = \frac{2a_1^2}{c_1} z$, we also obtain

$$\int_{-\infty}^{0} \tilde{A}_m(z)\, \tilde{A}_n(z)\, w_{\tilde{A}}(z)\, dz = \left(\frac{c_1}{2}\right)^{m+n} \int_{-\infty}^{0} A_m(u)\, A_n(u)\, w_A(u)\, du$$

$$= \left(\frac{c_1}{2}\right)^{m+n} (m+n)!\,(m+n+1)!\,\delta_{m,n},$$

which is (14.41).

Equations (14.38) and (14.42) can be established in a similar way using (13.4) and (14.15).

Case $c_1 < 0$. If we use the formula $tan^{-1}(t) = -i\,tanh^{-1}(i\,t)$, the demonstrations of (14.37)–(14.42) are similar. □

Remarks. We observe that the special cases $f^*(t,0) = tanh^{-1}t$ and $f^*(t,1) = tan^{-1}t$ of the Class II function $f^*(t,k)$ in (8.49) are both Class III functions covered by the above discussion.

Also, note that the zeros of $\delta_n(x)$ are pure imaginary and are distinct, except for $x = 0$, which is a double zero of $\delta_{2n}(x)$, $n \geq 1$. From (10.108), we have that $\delta_n(0) = 0$, $n \geq 1$. The multiplicity of $x = 0$ follows inductively from the recursion (10.8). Formulas (13.5) and (13.6) show that a non-zero zero of $\delta_n(x)$ is the square root of a zero of $A_m(z)$ or $B_m(z)$, whose respective zeros are distinct and lie in the interval $(-\infty, 0)$. It is interesting to observe that on the first side the zeros of $\delta_n(x)$ lie on the imaginary axis and on the second side the zeros of $\lambda_n(x)$ lie on the real axis (cf. Theorem 12.4).

We conclude this chapter by deriving the structure constants for $\{B_m(z)\}_{m=0}^{\infty}$ (cf. Theorem 13.15).

Definition 14.12. For $m, n \geq 0$ and $0 \leq j \leq m+n$, write that $\rho(m,n) = min\{m,n\} - \delta_{m,n}$ and $\hat{s}(j,k) = min\{j, m+n-k+1\}$, where $0 \leq k \leq \rho(2m+2, 2n+2)$. Then define

$$(14.43) \quad D_j^{(m,n)} = \frac{1}{(2j+2)!} \times$$

$$\sum_{k=0}^{\rho(2m+2,2n+2)} (-1)^{k+1} k!\,(2m+2n-k+3)!\binom{2m+2}{k}\binom{2n+2}{k}\xi_1[\hat{s}(j,k)],$$

where ξ_1 is defined in (11.17).

Theorem 14.13. For $m, n \geq 0$, we have that

$$(14.44) \quad B_m(z)B_n(z) = \sum_{j=0}^{m+n} D_j^{(m,n)} B_j(z).$$

Proof. From (10.116) we have

$$\delta_{2m+2}(x)\,\delta_{2n+2}(x) = \sum_{k=0}^{\rho(2m+2,2n+2)} \Delta_k^{(2m+2,2n+2)}\delta_{2m+2n-2k+4}(x).$$

Using (13.6), we find that

$$(14.45) \quad z\,B_m(z)B_n(z) = \sum_{k=0}^{\rho(2m+2,2n+2)} \Delta_k^{(2m+2,2n+2)} B_{m+n-k+1}(z).$$

Since $\{B_m(z)\}_{m=0}^{\infty}$ is a simple sequence and $deg(B_m B_n) = m + n$, there exists the unique real sequence of structure constants $\{D_k^{(m,n)}\}_{k=0}^{m+n}$, where

$$B_m(z)B_n(z) = \sum_{k=0}^{m+n} D_k^{(m,n)} B_k(z).$$

Substituting this sum into (14.45), we find that

$$(14.46) \quad \sum_{k=0}^{m+n} D_k^{(m,n)} z B_k(z) = \sum_{k=0}^{\rho(2m+2,2n+2)} \Delta_k^{(2m+2,2n+2)} B_{m+n-k+1}(z).$$

If for $0 \leq j \leq m + n$ we multiply both sides of (14.46) by the product $B_j(z)w_A(z)$, integrate from $-\infty$ to 0, and use the relationship $zw_A(z) = -2w_B(z)$ obtained from (14.1) and (14.2), we obtain

$$\int_{-\infty}^{0} \sum_{k=0}^{\rho(2m+2,2n+2)} \Delta_k^{(2m+2,2n+2)} B_j(z)B_{m+n-k+1}(z)w_A(z)dz$$

$$= \int_{-\infty}^{0} \sum_{k=0}^{m+n} D_k^{(m,n)} B_k(z)B_j(z)z\, w_A(z)dz$$

$$= -2\int_{-\infty}^{0} \sum_{k=0}^{m+n} D_k^{(m,n)} B_k(z)B_j(z)w_B(z)dz.$$

Thus, we have from Corollary 14.7 that

$$\sum_{k=0}^{\rho(2m+2,2n+2)} \Delta_k^{(2m+2,2n+2)} (B_j, B_{m+n-k+1})_A = -2 \sum_{k=0}^{m+n} D_k^{(m,n)} (B_k, B_j)_B.$$

Using (14.26) and (14.15), we obtain

$$- \sum_{k=0}^{m+n} D_k^{(m,n)} (j+k+1)!(j+k+2)!\delta_{j,k}$$

$$= \sum_{k=0}^{\rho(2m+2,2n+2)} \Delta_k^{(2m+2,2n+2)} (2j+1)!(2m+2n-2k+3)!\xi_1[\hat{s}(j,k)].$$

Thus, since j is fixed,

$$-D_j^{(m,n)}(2j+2)! = \sum_{k=0}^{\rho(2m+2,2n+2)} \Delta_k^{(2m+2,2n+2)} (2m+2n-2k+3)!\xi_1[\hat{s}(j,k)],$$

which gives (14.43), using (10.117). □

Chapter 15

Miscellaneous Results

The following formulas can be used to calculate the C_{2n+1}'s recursively using (11.10) to calculate the d's (cf. (8.30); see also [KnB]).

Theorem 15.1. For $n \geq 0$,

(15.1)
$$\sum_{j=0}^{n} d_{2j+1}^{(2n+1)} C_{2j+1} = -\delta_{n,0}$$

and

(15.2)
$$\sum_{j=0}^{n} d_{2j+2}^{(2n+2)} C_{2j+3} = 2\delta_{n,0}.$$

Proof. (15.1) For $n \geq 0$, by (14.18), we have

$$\sum_{j=0}^{n} C_{2j+1}\, d_{2j+1}^{(2n+1)} = -\sum_{j=0}^{n} (y^j, 1)_A d_{2j+1}^{(2n+1)} = -\left(\sum_{j=0}^{n} d_{2j+1}^{(2n+1)} y^j, 1\right)_A$$
$$= -(A_n, 1)_A = -\delta_{n,0}.$$

(15.2) For $n \geq 0$, by (14.19) we have

$$\sum_{j=0}^{n} C_{2j+3}\, d_{2j+2}^{(2n+2)} = 2\sum_{j=0}^{n} (y^j, 1)_B\, d_{2j+2}^{(2n+2)} = 2\left(\sum_{j=0}^{n} d_{2j+2}^{(2n+2)} y^j, 1\right)_B$$
$$= 2(B_m, 1)_B = 2\delta_{n,0}. \quad \square$$

The next theorem is a generalization of Theorem 15.1.

Theorem 15.2. For $m, n \geq 0$,

$$(15.3) \quad \sum_{j=0}^{m} \sum_{k=0}^{n} d_{2j+1}^{(2m+1)} d_{2k+1}^{(2n+1)} C_{2j+2k+1} = -(m+n+1)(2m)!\,(2n)!\,\delta_{m,n}$$

and

$$(15.4) \quad \sum_{j=0}^{m} \sum_{k=0}^{n} d_{2j+2}^{(2m+2)} d_{2k+2}^{(2n+2)} C_{2j+2k+3}$$

$$= (m+n+2)(2m+1)!\,(2n+1)!\,\delta_{m,n}.$$

Proof. (15.3) By (14.14), (13.3), and (14.18) we have

$$(m+n+1)(2m)!\,(2n)!\,\delta_{m,n} = (A_m, A_n)_A$$

$$= \sum_{j=0}^{m} \sum_{k=0}^{n} d_{2j+1}^{(2m+1)} d_{2k+1}^{(2n+1)} (z^j, z^k)_A = -\sum_{j=0}^{m} \sum_{k=0}^{n} d_{2j+1}^{(2m+1)} d_{2k+1}^{(2n+1)} C_{2j+2k+1}.$$

Equation (15.4) is proved similarly. □

Theorem 15.3. For $0 \leq k \leq n$,

$$(15.5) \quad \lambda_{2k+1}^{(2n+1)} = \frac{-1}{(2k)!\,(2k+1)!} \sum_{j=0}^{k} d_{2j+1}^{(2k+1)} C_{2n+2j+1}$$

and

$$(15.6) \quad \lambda_{2k+2}^{(2n+2)} = \frac{1}{(2k+1)!\,(2k+2)!} \sum_{j=0}^{k} d_{2j+2}^{(2k+2)} C_{2n+2j+3}.$$

Proof. (15.5) By (14.20), (13.3), and (14.18) we have

$$(2k)!\,(2k+1)!\,\lambda_{2k+1}^{(2n+1)} = (A_k, z^n)_A$$

$$= \sum_{j=0}^{k} d_{2j+1}^{(2k+1)} (z^j, z^n)_A = -\sum_{j=0}^{k} d_{2j+1}^{(2k+1)} C_{2n+2j+1},$$

which proves the result. Equation (15.6) is proved similarly. □

Note that substituting $k = 0$ into Theorem 15.3 gives Theorem 12.11.

There is another pair of formulas that connects the C's and the λ's.

Theorem 15.4. For $0 \leq k \leq n$,

$$(15.7) \quad C_{2n+2k+1} = -\sum_{j=0}^{k} (2j)!\,(2j+1)!\,\lambda_{2j+1}^{(2k+1)} \lambda_{2j+1}^{(2n+1)}$$

and

(15.8) $$C_{2n+2k+3} = \sum_{j=0}^{k}(2j+1)!(2j+2)!\lambda_{2j+2}^{(2k+2)}\lambda_{2j+2}^{(2n+2)}.$$

Proof. (15.7) By (15.5), we have for $0 \le j \le n$ that

$$-(2j)!(2j+1)!\lambda_{2j+1}^{(2n+1)} = \sum_{s=0}^{j} d_{2s+1}^{(2j+1)} C_{2n+2s+1}.$$

If we multiply each side by $\lambda_{2j+1}^{(2k+1)}$ and sum over j for $0 \le j \le k$, then

$$- \sum_{j=0}^{k}(2j)!(2j+1)!\lambda_{2j+1}^{(2k+1)}\lambda_{2j+1}^{(2n+1)} = \sum_{j=0}^{k}\sum_{s=0}^{j} d_{2s+1}^{(2j+1)} C_{2n+2s+1}\lambda_{2j+1}^{(2k+1)}$$

$$= \sum_{s=0}^{k} C_{2n+2s+1}\sum_{j=s}^{k} d_{2s+1}^{(2j+1)}\lambda_{2j+1}^{(2k+1)} = \sum_{s=0}^{k} C_{2n+2s+1}\delta_{k,s} = C_{2n+2k+1},$$

using (12.13). Equation (15.8) is proved similarly. $\square$

The next four results are derived from Jacobi polynomial identities (cf. (10.91)).

Theorem 15.5. For $n \ge 1$,

(15.9) $$\delta_n(x) = (-1)^{n-1}(n-1)!\,2^{n-1}x\sum_{k=0}^{n-1}\frac{1}{2^k}\binom{n}{k}\binom{-1-\frac{x}{2}}{n-k-1}.$$

Proof. From [GrR, p. 1035, 8.960, 2nd eq.], we have

$$P_n^{(\alpha,\beta)}(x) = \frac{1}{2^n}\sum_{k=0}^{n}\binom{n+\alpha}{k}\binom{n+\beta}{n-k}(x-1)^{n-k}(x+1)^k.$$

Then, from (10.91), we have that

$$\delta_k(x) = (n-1)!\,x\,P_{n-1}^{(1,-n-\frac{x}{2})}(-3)$$

$$= \frac{(n-1)!\,x}{2^{n-1}}\sum_{k=0}^{n-1}\binom{n}{k}\binom{-1-\frac{x}{2}}{n-k-1}(-4)^{n-k-1}(-2)^k$$

$$= (-1)^{n-1}(n-1)!\,2^{n-1}x\sum_{k=0}^{n-1}\frac{1}{2^k}\binom{n}{k}\binom{-1-\frac{x}{2}}{n-k-1}. \quad \square$$

Corollary 15.6. For $n \ge 1$,

(15.10) $$\delta_n(x) = \sum_{k=1}^{n}(-1)^k L_{n,k}\,(x+2k-2)_{k,2}.$$

Proof. From (10.31), we have for $r \geq 0$ that

$$\binom{-1-\frac{x}{2}}{r} = \frac{1}{r!}\left(-1-\frac{x}{2}\right)_r = \frac{(-1)^r}{2^r r!}(x+2r)_{r,2}.$$

Substituting this result into (15.9) and using (10.36) gives

$$\delta_n(x) = (n-1)!\, x \sum_{k=0}^{n-1} \frac{(-1)^k}{(n-k-1)!}\binom{n}{k}(x+2(n-k-1))_{n-k-1,2}$$

$$= (-1)^n(n-1)! \sum_{k=1}^{n} \frac{(-1)^k}{(k-1)!}\binom{n}{k}(x+2k-2)_{k,2}$$

$$= \sum_{k=1}^{n}(-1)^k L_{n,k}(x+2k-2)_{k,2}. \quad \square$$

Theorem 15.7. For $n \geq 0$,

(15.11) $\quad \delta_{n+1}(x) =$

$$\frac{(n+1)!}{2^n} \sum_{k=0}^{n}\binom{n}{k}\frac{\delta_{k+1}(2k+2)}{(n+k+2)!}(x)_{k+1,2}(x+2n+2)_{n-k,2}.$$

Proof. From [Han, p. 294, 45.1.14], we have

$$(-1)^n \frac{n!}{[c]_n}(n+a+1)P_n^{(a,c)}(x)$$

$$= \sum_{k=0}^{n}(-1)^k \frac{[-n]_k[a+c+n+1]_k}{[n+a+2]_k[1-c-n]_k}(2k+a+1)P_k^{(a,0)}(x),$$

so putting $a=1$ and $x=-3$, we obtain

$$(-1)^n \frac{n!}{[c]_n}(n+2)P_n^{(1,c)}(-3)$$

$$= 2\sum_{k=0}^{n}(-1)^k \frac{[-n]_k[c+n+2]_k}{[n+3]_k[1-c-n]_k}(k+1)P_k^{(1,0)}(-3).$$

Using (10.32), we find that

$$\frac{[-n]_k}{[n+3]_k} = (-1)^k \frac{n!(n+2)!}{(n-k)!(n+k+2)!}.$$

Thus,

$$\frac{(-1)^n}{[c]_n}P_n^{(1,c)}(-3)$$

$$= 2(n+1)! \sum_{k=0}^{n} \frac{k+1}{(n-k)!(n+k+2)!}\frac{[c+n+2]_k}{[1-c-n]_k}P_k^{(1,0)}(-3).$$

Setting $c = -n - 1 - \frac{x}{2}$ gives

$$\frac{(-1)^n}{[-n-1-\frac{x}{2}]_n} P_n^{(1,-n-1-\frac{x}{2})}(-3)$$

$$= 2(n+1)! \sum_{k=0}^{n} \frac{k+1}{(n-k)!(n+k+2)!} \frac{[1-\frac{x}{2}]_k}{[2+\frac{x}{2}]_k} P_k^{(1,0)}(-3).$$

Now, for $k \geq 0$ we have that

$$\frac{[-n-1-\frac{x}{2}]_n}{[2+\frac{x}{2}]_k}$$

$$= (-1)^n 2^{k-n} \frac{(x+2n+2)_{n,2}}{(x+2k+2)_{k,2}} = (-1)^n 2^{k-n} (x+2n+2)_{n-k,2}.$$

Thus,

$$P_n^{(1,-n-1-\frac{x}{2})}(-3) = 2^{-n+1}(n+1)! \times$$

$$\sum_{k=0}^{n} (-1)^k \frac{k+1}{(n-k)!(n+k+2)!} (x-2)_{k,2} (x+2n+2)_{n-k,2} P_k^{(1,0)}(-3).$$

Using (10.91), this becomes

$$\delta_{n+1}(x) = \frac{(n+1)!}{2^n} x \sum_{k=0}^{n} \binom{n}{k} \frac{\delta_{k+1}(2k+2)}{(n+k+2)!} (x-2)_{k,2} (x+2n+2)_{n-k,2},$$

from which the result follows using $x(x-2)_{k,2} = (x)_{k+1,2}$. □

Theorem 15.6. For $n \geq 1$,

$$(15.12) \quad \delta_n(x) = \frac{(n-1)!\, n!\, 2^{n-1}}{(x+2n-2)_{n-1,2}} \times$$

$$\sum_{k=0}^{n-1} \frac{1}{2^k k!(k+1)!} \frac{x+2n-4k-2}{x+2n-2k}(x+2n-2)_{k,2}\, \delta_{k+1}(x+2n-2k).$$

Proof. From [Han, p. 294, 45.1.12] we have

$$(-1)^{n-1} a \frac{[a+b+1]_n}{[a]_n} P_{n-1}^{(a,b+1)}(x)$$

$$= \sum_{k=0}^{n-1} (-1)^k (2k+a+b+1) \frac{[a+b+1]_k}{[a+1]_k} P_k^{(a,b)}(x),$$

so putting $a = 1$ and $x = -3$, we obtain

$$(-1)^{n-1} \frac{[b+2]_n}{[1]_n} P_{n-1}^{(1,b+1)}(-3) = \sum_{k=0}^{n-1} (-1)^k (b+2k+2) \frac{[b+2]_k}{[2]_k} P_k^{(1,b)}(-3).$$

By (10.32), $[1]_n = n!$ and $[2]_k = (k+1)!$, so setting $b = -n - 1 - \dfrac{x}{2}$, we obtain

$$\frac{(-1)^{n-1}}{n!}\left[-n-1-\frac{x}{2}\right]_n P_{n-1}^{(1,-n-\frac{x}{2})}(-3)$$

$$= \sum_{k=0}^{n-1} \frac{(-1)^k}{(k+1)!}\left(-n-\frac{x}{2}+2k+1\right)\left[-n+1-\frac{x}{2}\right]_k P_k^{(1,-n-1-\frac{x}{2})}(-3)$$

$$= \frac{1}{2}\sum_{k=0}^{n-1} \frac{(-1)^{k-1}}{(k+1)!}(x+2n-4k-2)\left[-n+1-\frac{x}{2}\right]_k P_k^{(1,-n-1-\frac{x}{2})}(-3).$$

Now, in general, $\left[-n+1-\dfrac{x}{2}\right]_r = \dfrac{(-1)^r}{2^r}(x+2n-2)_{r,2}$, $r \geq 0$, so the equation becomes

$$\frac{1}{n!2^{n-1}}(x+2n-2)_{n,2}P_{n-1}^{(1,-n-\frac{x}{2})}(-3)$$

$$= \sum_{k=0}^{n-1} \frac{1}{2^k(k+1)!}(x+2n-4k-2)(x+2n-2)_{k,2}P_k^{(1,-n-1-\frac{x}{2})}(-3).$$

The result follows from (10.91) and the formula $(x+2n-2)_{n,2} = x(x+2n-2)_{n-1,2}$. □

Corollary 15.9. For $n \geq 2$,

$$(15.13) \quad (x-2n+2)\delta_n(x+2) - (x+2)\delta_n(x)$$

$$= -\frac{(n-1)!\,n!\,2^{n-1}}{(x+2n-2)_{n-2,2}}\sum_{k=0}^{n-2}\frac{1}{2^k\,k!\,(k+1)!}\frac{x+2n-4k-2}{x+2n-2k}$$

$$\times\,(x+2n-2)_{k,2}\,\delta_{k+1}(x+2n-2k).$$

Proof. Observe that (15.12) can be written as

$$\delta_n(x) = \frac{x-2n+2}{x+2}\delta_n(x+2) + \frac{(n-1)!\,n!\,2^{n-1}}{(x+2n-2)_{n-1,2}}\,\times$$

$$\sum_{k=0}^{n-2}\frac{1}{2^k k!\,(k+1)!}\frac{x+2n-4k-2}{x+2n-2k}(x+2n-2)_{k,2}\,\delta_{k+1}(x+2n-2k).$$

Clearing fractions and using the simple relation $(x+2n-2)_{n-1,2} = (x+2)(x+2n-2)_{n-2,2}$ gives the result. □

Theorem 15.10. [Ca4, p. 123] Let p be an odd prime. Then

$$(15.14) \quad A_{\frac{p-1}{2}}(z) \equiv B_{\frac{p-1}{2}}(z) \equiv z^{\frac{p-1}{2}} - 1 \equiv \prod_{k=1}^{\frac{p-1}{2}}(z-r_k) \pmod{p},$$

where the r_k's are the quadratic non-residues (mod p), and

(15.15) $$A_{\frac{p+1}{2}}(z) \equiv z A_{\frac{p-1}{2}}(z) \ (\text{mod } p).$$

Proof. Replacing m by $\dfrac{p-3}{2}$ in (13.7) gives

$$A_{\frac{p-1}{2}}(z) = B_{\frac{p-1}{2}}(z) - p(p-1)B_{\frac{p-3}{2}}(z) \equiv B_{\frac{p-1}{2}}(z) \ (\text{mod } p).$$

By (13.3), we have

(15.16) $$A_{\frac{p-1}{2}}(z) = \sum_{j=0}^{\frac{p-1}{2}} d_{2j+1}^{(p)} z^j.$$

But (11.4) gives $\mathrm{R}\ d_p^{(p)} = 1$. Also, substituting $n = p, j = 1$, and $k = 2j$ in (12.23), we have for $1 \leq j \leq \dfrac{p-3}{2}$ that

$$d_{2j+1}^{(p)} = \frac{1}{2j+1} \sum_{s=0}^{p-1-2j} \binom{p}{s+1} d_1^{(s+1)} d_{2j}^{(p-1-s)} \equiv 0 \ (\text{mod } p),$$

since $p \left| \dbinom{p}{s+1} \right.$ for $0 \leq s \leq p-1-2j$. Further, by (12.5), we have that $d_1^{(p)} = (p-1)! \equiv -1 \ (\text{mod } p)$ by Wilson's theorem. With these results we find that (15.16) becomes

$$A_{\frac{p-1}{2}}(z) \equiv z^{\frac{p-1}{2}} - 1 \ (\text{mod } p).$$

But, since the quadratic non-residues of p are the zeros of $z^{\frac{p-1}{2}} - 1$, the final product in (15.14) follows. Replacing m by $\dfrac{p-3}{2}$ in (13.1) leads to the congruence (15.15). $\square$

Chapter 16

Uniqueness and Completion Results

In this short chapter we give some results on the uniqueness of measures and the density of $\mathcal{P}$ (cf. Definition 6.1) in the corresponding L_2 spaces. This chapter is related to Chapter 6.

Lemma 16.1. Consider the orthogonal sequences $\{G_m(z)\}_{m=0}^{\infty}$ and $\{H_m(z)\}_{m=0}^{\infty}$ in Definition 1.19 and assume that $c_2 > 0$, $2\sqrt{c_2} \leq |c_1|$, and $(1,1)_G = (1,1)_H = 1$. Also, for $m \geq 0$, let

$$\hat{G}_m = \frac{G_m}{\sqrt{(G_m, G_m)_G}} \quad \text{and} \quad \hat{H}_m = \frac{H_m}{\sqrt{(H_m, H_m)_H}}.$$

Then

(16.1)
$$\sum_{m=0}^{\infty} [\hat{G}_m(0)]^2 = \sum_{m=0}^{\infty} [\hat{H}_m(0)]^2 = \infty.$$

Proof. By (1.40) and (7.12) with $(1,1)_G = 1$, we have for $m \geq 0$ that

$$\left[\hat{G}_m(0)\right]^2 = \frac{(2m+1)\, a_{2m+1}^2}{a_1^2 c_2^m}.$$

Using (3.34), with $S_m(x,y) = y^m P_m\left(\dfrac{x}{y}\right)$, we obtain

$$\left[\hat{G}_m(0)\right]^2 = \frac{1}{2m+1}\left[P_m\left(\frac{c_1}{2\sqrt{c_2}}\right)\right]^2 = \frac{1}{2m+1}\left[P_m\left(\frac{|c_1|}{2\sqrt{c_2}}\right)\right]^2,$$

since $P_m(-x) = (-1)^m P_m(x)$.

By hypothesis, $r \overset{\text{def}}{=} \dfrac{|c_1|}{2\sqrt{c_2}} \geq 1$, so from [MaO, p. 52] (modified to include the case $r = 1$), we have that

(16.2) $$1 = P_0(r) \leq P_1(r) \leq P_2(r) \leq \cdots.$$

Then

$$\sum_{m=0}^{\infty} [\hat{G}_m(0)]^2 = \sum_{m=0}^{\infty} \frac{1}{2m+1} [P_m(r)]^2 \geq \sum_{m=0}^{\infty} \frac{1}{2m+1} = \infty.$$

Next, we have from (1.41) and (3.34) that

$$H_m(0) = \frac{1}{2}(2m+2)! \sum_{k=0}^{m} a_{2k+1}\, a_{2(m-k)+1}$$

$$= \frac{(-1)^m}{2}(2m+2)!\, a_1^2\, c_2^{\frac{m}{2}} \sum_{k=0}^{m} \frac{1}{(2k+1)\,[2(m-k)+1]}\, P_k(r)\, P_{m-k}(r),$$

again using $P_m(-x) = (-1)^m P_m(x)$.

Next, we find from (7.13) with $(1,1)_H = 1$ that

$$[\hat{H}_m(0)]^2 = \frac{[H_m(0)]^2}{(H_m, H_m)_H}$$

$$= (m+1)\Big(\sum_{k=0}^{m} \frac{1}{(2k+1)\,[2(m-k)+1]}\, P_k(r)\, P_{m-k}(r) \Big)^2.$$

Finally, we have from (16.2) that

$$\sum_{m=0}^{\infty} (\hat{H}_m(0))^2 = \sum_{m=0}^{\infty} (m+1) \left[\sum_{k=0}^{m} \frac{1}{(2k+1)\,(2(m-k)+1)}\, P_k(r)\, P_{m-k}(r) \right]^2$$

$$\geq \sum_{m=0}^{\infty} (m+1) \left[\sum_{k=0}^{m} \frac{1}{(2k+1)(2m-2k+1)} \right]^2$$

$$= \sum_{m=0}^{\infty} (m+1) \left[\sum_{k=0}^{m} \frac{1}{2m+2} \left(\frac{1}{2k+1} + \frac{1}{2m-2k+1} \right) \right]^2$$

$$= \sum_{m=0}^{\infty} (m+1) \left[\frac{1}{m+1} \sum_{k=0}^{m} \frac{1}{2k+1} \right]^2 \geq \sum_{m=0}^{\infty} \frac{1}{m+1} = \infty. \quad \Box$$

Theorem 16.2. Let $f \in \mathcal{F}_2$ be such that $2\sqrt{c_2} \leq |c_1|$. Then there are respective unique Borel probability measures $\mu_G, \mu_H : \underline{\mathcal{B}}(\mathbb{R}) \longrightarrow [0, 1]$ such that $\displaystyle\int_{\mathbb{R}} G_m\, G_n\, d\mu_G = 0$ and $\displaystyle\int_{\mathbb{R}} H_m\, H_n\, d\mu_H = 0$, $m, n \geq 0$, $m \neq n$.

Proof. By Favard's theorem (Theorem 6.9), there are inner products $(*, *)_G$ and $(*, *)_H$ mapping $\mathcal{P} \times \mathcal{P} \longrightarrow \mathbb{R}$ such that $(G_m, G_n)_G = (H_m, H_n)_H = 0$, $m, n \geq 0$, $m \neq n$, and $(1, 1)_G = (1, 1)_H = 1$. By Theorem 6.4, there are Borel probability measures $\mu_G, \mu_H : \underline{B}(\mathbb{R}) \longrightarrow [0, 1]$ such that

$$(f, g)_G = \int_{\mathbb{R}} fg \, d\mu_G \text{ and } (f, g)_H = \int_{\mathbb{R}} fg \, d\mu_H.$$

By Definition 6.8, the respective sequences $\{\mu_n\}_0^\infty$ of moments are given by

$$\mu_n = (p_n, 1)_G = \int_{-\infty}^{\infty} x^n d\mu_G(x) \text{ and } \mu_n = (p_n, 1)_H = \int_{-\infty}^{\infty} x^n d\mu_H(x).$$

Now observe that each of the sequences $\{\hat{G}_m(z)\}$ and $\{\hat{H}_m(z)\}$, defined in Lemma 16.1, is an example of the sequence $\{\hat{S}_m(y)\}$ in Theorem 6.11. This latter sequence, on the other hand, can be identified with the sequence $\{P_n(x)\}$ in [Ak1, p. 3], if the subscript of the D_n in [Ak1, p. 1] is increased by one. From [Ak1, p. 84, Prob. 10], we then find that the criteria for each of the measures μ_G and μ_H to be respectively unique are those in (16.2), which proves the theorem. $\square$

Comments. Some specializations of Theorem 16.2 are:
1. The two respective measures given by (8.44), (8.45) and (8.47), (8.48) for the sequences $\{A_m(z, k)\}$ and $\{B_m(z, k)\}$ are unique (up to normalization). (Here $c_1 = k^2 + 1$ and $c_2 = k^2$.) We cannot conclude the same uniqueness result for the sequences $\{A_m^*(z, k)\}$ and $\{B_m^*(z, k)\}$ in 8 (b), because the condition $2\sqrt{c_2} \leq |c_1|$ in Theorem 16.2 does not hold. (There $c_1 = 2(1 - 2k^2)$ and $c_2 = 1$, $0 < k < 1$.)

2. The weight functions w_A and w_B, defined respectively in Definition 14.1 for the sequences $\{A_m(z)\}_{m=0}^\infty$ and $\{B_m(z)\}_{m=0}^\infty$, are "essentially unique" (in that other weight functions must be equal to these λ-a.e., λ being Lebesgue measure).

Corollary 16.3. Let $f \in \mathcal{F}_2$ be such that $|c_1| \geq 2\sqrt{c_2}$. Also, for $m, n \geq 0$, $m \neq n$, let $\mu_G, \mu_H : \underline{B}(\mathbb{R}) \longrightarrow [0, 1]$ be the unique Borel probability measures for which $\int_{\mathbb{R}} G_m G_n \, d\mu_G = 0$ and $\int_{\mathbb{R}} H_m H_n \, d\mu_G = 0$, respectively. Then $\mathcal{P}$ is norm-dense in each of the two real Hilbert spaces $L_2^{\mathbb{R}}(\mathbb{R}, \underline{B}(\mathbb{R}), \mu_G)$ and $L_2^{\mathbb{R}}(\mathbb{R}, \underline{B}(\mathbb{R}), \mu_H)$.

Proof. By Theorem 16.2, the measures μ_G and μ_H are unique. The density of $\mathcal{P}$ then follows from [Ak1, p. 45, Cor. 2.3.3]. $\square$

Comment. As a result of Corollary 16.3, each of the four sequences $\{A_m(z, k)\}_{m=0}^\infty$, $\{B_m(z, k)\}_{m=0}^\infty$ (cf. Tables 8.3 and 8.4), and $\{A_m(z)\}_{m=0}^\infty$, $\{B_m(z)\}_{m=0}^\infty$ in (13.3) and (13.4) is an orthogonal basis of the corresponding L_2 space.

Chapter 17

Polynomial Inequalities

In Theorem 10.14, we proved a useful inequality for $\delta_n(x)$. In this chapter we obtain more inequalities for $\delta_n(x)$, $A_m(z)$, and $B_m(z)$, including one similar to Turán's inequality.

We begin with some basic inequalities.

Theorem 17.1. We have that

$$(17.1) \qquad A_m(x) > 0 \text{ and } B_m(x) > 0, \ x \geq 0, \ m \geq 0,$$

$$(17.2) \qquad \delta_{2n}(x) \geq 0 \text{ and } \operatorname{sgn}(x)\,\delta_{2n+1}(x) \geq 0, \ x \in \mathbb{R}, \ n \geq 0,$$

with equality only when $x = 0$;

$$(17.3) \qquad \delta_m(x) \leq \delta_n(x), \ x \geq 1, \ 0 \leq m \leq n,$$

with equality only when $m = n$ or $\delta_0(1) = \delta_1(1) = \delta_2(1)$;

$$(17.4) \qquad \delta_m(x) \leq \delta_m(y), \ 0 \leq x \leq y, \ m \geq 0,$$

with equality only when $x = y$ or $m = 0$;

$$(17.5) \qquad A_m(x) < A_{m+1}(x) \text{ and } B_m(x) < B_{m+1}(x), \ x \geq 0, \ m \geq 0,$$

$$(17.6) \qquad A_m(x) \leq A_m(y), \ 0 \leq x \leq y, \ m \geq 0;$$

with equality only when $x = y$ or $m = 0$;

(17.7) $B_m(x) \le B_m(y)$, $0 \le x \le y$, $m \ge 0$,

with equality only when $x = y$ or $m = 0$;

(17.8) $A_m(x) \le B_m(x)$, $x \ge 0$, $m \ge 0$,

with equality only when $m = 0$, and

(17.9) $B_m(x) < A_{m+1}(x)$, $x \ge 0$, $m \ge 0$.

Proof. **(17.1)** The coefficients of $A_m(x)$ and $B_m(x)$ in (13.3) and (13.4) are positive from (11.9).

(17.2) These inequalities follow from (13.6), (17.1) and (13.5), (17.1), respectively.

(17.3) Using Table 10.1, we have that $\delta_0(x) = 1 \le x = \delta_1(x)$. Also, since by (10.8), $\delta_n(x) = x\delta_{n-1}(x) + (n-2)(n-1)\delta_{n-2}(x)$ for $n \ge 2$ and $x \ge 1$, then (17.3) holds using this reduction repeatedly. For $m \ge 3$ and $x \ge 1$, then $\delta_n(x) > x\delta_{n-1}(x) \ge \delta_{n-1}(x)$. Thus, $\delta_m(x) < \delta_n(x)$ for $x \ge 1$ and $2 \le m \le n$.

It remains to consider $m = 0$ and $m = 1$. For $x > 1$, we have $\delta_0(x) = 1 < x = \delta_1(x) < x^2 = \delta_2(x)$, $n \ge 3$. For $x = 1$, we have $\delta_0(1) = \delta_1(1) = \delta_2(1) < \delta_3(1) < \delta_4(1) < \cdots$ from the above.

(17.4) This follows from (11.1), (11.8), and (11.9).

(17.5) We have that $A_0(x) = B_0(x) = 1$. Then, for $m \ge 0$, from (13.8) and (17.1), $A_{m+1}(x) = (2m+1)(2m+2)A_m(x) + xB_m(x) > A_m(x)$. By (13.7) and (17.1) we have $B_{m+1}(x) = A_{m+1}(x) + (2m+2)(2m+3)B_m(x) > B_m(x)$.

(17.6), **(17.7)** These follow from (13.3), (13.4), and (11.9), respectively.

(17.8) We have equality for $m = 0$. For $m \ge 1$ and $x \ge 0$, we have from (13.7) and (17.1) that $A_m(x) = B_m(x) - 2m(2m+1)B_{m-1}(x) < B_m(x)$, since the second term on the right is > 0.

(17.9) Note first by (11.9) that for $m \ge 0$, the coefficients of $A_m(x)$ in (13.3) and $B_m(x)$ in (13.4) are positive. Thus, for $x \ge 0$, we have $A'_{m+1}(x) \ge 0$ and $B'_m(x) \ge 0$. Hence, by (13.24) and (13.25),

$$\min_{x \in [0,1]} A_{m+1}(x) = A_{m+1}(0) = (2m+2)!$$

and

$$\max_{x \in [0,1]} B_m(x) = B_m(1) = \frac{(2m+2)!}{2^{2m+2}} \binom{2m+2}{m+1}.$$

Suppose next that $m \geq 0$. Then for $x \geq 1$, we obtain from (13.8) that $A_{m+1}(x) = xB_m(x) + 2(m+1)(2m+1)A_m(x) > B_m(x)$. On the other hand, for $0 \leq x \leq 1$,

$$\frac{A_{m+1}(x)}{B_m(x)} \geq \frac{\min\limits_{x \in [0,1]} A_{m+1}(x)}{\max\limits_{x \in [0,1]} B_m(x)} = \frac{2^{2m+2}}{\binom{2m+2}{m+1}}.$$

But it is readily proved by induction that $\binom{2n}{n} \leq \dfrac{2^{2n}}{\sqrt{3n+1}}$, $n \geq 0$, so

$$\frac{A_{m+1}(x)}{B_m(x)} \geq \sqrt{3m+4} > 1. \quad \square$$

The next result is similar to Turán's inequality [Tu].

Theorem 17.2. For $x \in \mathbb{R}$ and $n \geq 2$, we have that

$$(17.10) \qquad W_n(x) \overset{\text{def}}{=} \begin{vmatrix} \delta_{n-2}(x) & \delta_n(x) \\ \delta_n(x) & \delta_{n+2}(x) \end{vmatrix} \geq 0,$$

with equality only when $x = 0$.

Proof. That the formula $W_n(0) = 0$ is true for $n \geq 2$ follows from (10.108). If $x \neq 0$, then the truth of $W_n(x) > 0$ for $n \geq 2$ will be shown in the two cases that follow.

Case 1. n even. Replacing n by $2n$ in (10.10), we obtain using Table 10.1 that

$$(17.11) \quad \delta_{2n+2}(x) =$$
$$(\delta_2(x) + 8n^2)\delta_{2n}(x) - 4n(n-1)(2n-1)^2 \delta_{2n-2}(x), \quad n \geq 1.$$

Also, putting $m = 2$ and replacing n by $2n-2$ in (10.116), we obtain for $n \geq 3$ that

$$(17.12) \quad \delta_2(x)\delta_{2n-2}(x) =$$
$$\delta_{2n}(x) - 8(n-1)^2 \delta_{2n-2}(x) + 4(n-1)(n-2)(2n-3)^2\delta_{2n-4}(x).$$

If we now multiply (17.11) by $\delta_{2n-2}(x)$, we get by using (17.12) and dropping the x that

$$\delta_{2n-2}\delta_{2n+2} = \delta_{2n-2}\Big((\delta_2 + 8n^2)\delta_{2n} - 4n(n-1)(2n-1)^2\delta_{2n-2}\Big)$$
$$= \delta_{2n}(\delta_2\,\delta_{2n-2}) + 8n^2\,\delta_{2n-2}\,\delta_{2n} - 4n(n-1)(2n-1)^2\delta_{2n-2}^2$$
$$= \delta_{2n}\Big(\delta_{2n} - 8(n-1)^2\,\delta_{2n-2} + 4(n-1)(n-2)(2n-3)^2\,\delta_{2n-4}\Big)$$
$$\quad + 8n^2\delta_{2n-2}\,\delta_{2n} - 4n(n-1)(2n-1)^2\,\delta_{2n-2}^2$$
$$= \delta_{2n}^2 + 8(2n-1)\delta_{2n-2}\,\delta_{2n}$$
$$\quad + 4(n-1)\Big((n-2)(2n-3)^2\delta_{2n-4}\,\delta_{2n} - n(2n-1)^2\,\delta_{2n-2}^2\Big),$$

so

$$W_{2n} = 8(2n-1)\delta_{2n-2}\delta_{2n} + 4(n-1)(n-2)(2n-3)^2\left(\delta_{2n-4}\delta_{2n} - \delta_{2n-2}^2\right)$$
$$- 8(n-1)(8n^2 - 16n + 9)\delta_{2n-2}^2$$
$$= 4(n-1)(n-2)(2n-3)^2 W_{2n-2}$$
$$+ 8\delta_{2n-2}\left((2n-1)\delta_{2n} - (n-1)(8n^2 - 16n + 9)\delta_{2n-2}\right).$$

But, using (10.8) with n replaced by $2n-2$, we find that

$$(2n-1)\delta_{2n}(x) - (n-1)(8n^2 - 16n + 9)\delta_{2n-2}(x)$$
$$= (2n-1)\left[x\delta_{2n-1}(x) + 2(n-1)(2n-1)\delta_{2n-2}(x)\right]$$
$$- (n-1)(8n^2 - 16n + 9)\delta_{2n-2}(x)$$
$$= (2n-1)x\delta_{2n-1}(x) + (n-1)(8n-7)\delta_{2n-2}(x).$$

Thus, we obtain a recursion formula for $W_{2n}(x)$, $n \geq 1$:

$$(17.13)\quad W_{2n}(x) = 4(n-1)(n-2)(2n-3)^2 W_{2n-2}(x)$$
$$+ 8(2n-1)x\delta_{2n-2}(x)\delta_{2n-1}(x) + 8(n-1)(8n-7)\delta_{2n-2}^2(x).$$

To complete this case by induction on n, first note for $n=1$ from (17.10) and Table 10.1 that $W_2(x) = 8x^2 \geq 0$, where $W_2(x) > 0$ exactly when $x \neq 0$. If we assume that $W_{2n-2}(x) \geq 0$ for some $n \geq 2$, with equality only at $x = 0$, then it follows from (17.13) that $W_{2n}(x) \geq 0$, with equality only at $x = 0$, because for $x \in \mathbb{R}$, $\delta_{2n-2}(x) \geq 0$ by (17.2) and $x\delta_{2n-1} \geq 0$ by (17.2). Thus, $W_{2n}(x) > 0$ when $x \neq 0$, since all the terms on the right are ≥ 0 and certainly the third term exceeds 0.

Case 2. n odd. Putting $2n+1$ for n in (10.10) gives

$$(17.14)\quad \delta_{2n+3}(x) =$$
$$\left(\delta_2(x) + 2(2n+1)^2\right)\delta_{2n+1}(x) - 4n^2(4n^2-1)\delta_{2n-1}(x), \quad n \geq 1.$$

Also, putting $m=2$ and replacing n by $2n-1$ in (10.116), we obtain for $n \geq 2$ that

$$(17.15)\quad \delta_2(x)\delta_{2n-1}(x) = \delta_{2n+1}(x)$$
$$- 2(2n-1)^2\delta_{2n-1}(x) + 4(n-1)^2[4(n-1)^2 - 1]\delta_{2n-3}(x), \quad n \geq 2.$$

If we multiply (17.14) by $\delta_{2n-1}(x)$, we get using (17.15) and dropping the x that

$$\delta_{2n-1}\delta_{2n+3} = \delta_{2n+1}\left(\delta_2\delta_{2n-1}\right)$$
$$+ 2(2n+1)^2\delta_{2n-1}\delta_{2n+1} - 4n^2(4n^2-1)\delta_{2n-1}^2$$
$$= \delta_{2n+1}\left(\delta_{2n+1} - 2(2n-1)^2\delta_{2n-1} + 4(n-1)^2[4(n-1)^2-1]\delta_{2n-3}\right)$$
$$+ 2(2n+1)^2\delta_{2n-1}\delta_{2n+1} - 4n^2(4n^2-1)\delta_{2n-1}^2$$
$$= \delta_{2n+1}^2 + 16n\delta_{2n-1}\delta_{2n+1} + 4(n-1)^2(2n-1)(2n-3)\delta_{2n-3}\delta_{2n+1}$$
$$- 4n^2(4n^2-1)\delta_{2n-1}^2,$$

so

$$W_{2n+1} = 16n\, \delta_{2n-1}\, \delta_{2n+1}$$
$$+ 4(n-1)^2(2n-1)(2n-3)\left(\delta_{2n-3}\,\delta_{2n+1} - \delta_{2n-1}^2\right)$$
$$- 4(2n-1)(8n^2 - 8n + 3)\,\delta_{2n-1}^2$$
$$= 4(n-1)^2(2n-1)(2n-3)W_{2n-1}$$
$$+ 4\,\delta_{2n-1}\left(4n\,\delta_{2n+1} - (2n-1)(8n^2 - 8n + 3)\,\delta_{2n-1}\right).$$

But, using (10.8) with n replaced by $2n-1$, we find that

$$4n\,\delta_{2n+1} - (2n-1)(8n^2 - 8n + 3)\,\delta_{2n-1} = 4n\left(x\,\delta_{2n} + 2n(2n-1)\,\delta_{2n-1}\right)$$
$$- (2n-1)(8n^2 - 8n + 3)\,\delta_{2n-1} = 4nx\,\delta_{2n} + (2n-1)(8n-3)\,\delta_{2n-1}.$$

Thus, for $n \geq 1$, a recursion formula for $W_{2n+1}(x)$ is

$$(17.16) \quad W_{2n+1}(x) = 4(n-1)^2(2n-1)(2n-3)W_{2n-1}(x)$$
$$+ 16n\, x\, \delta_{2n-1}(x)\, \delta_{2n}(x) + 4(2n-1)(8n-3)\,\delta_{2n-1}^2(x).$$

From Table 10.1, we find that $W_3(x) = \delta_1(x)\,\delta_5(x) - \delta_3^2(x) = 4x^2(4x^2+5) \geq 0$ with equality only at $x = 0$. If we assume $W_{2n-1}(x) \geq 0$ for some $n \geq 2$ with equality only at $x = 0$, then it follows from (17.16) that $W_{2n+1}(x) \geq 0$, because $x\,\delta_{2n-1}(x)\,\delta_{2n}(x) \geq 0$ by (17.2) so the two terms on the right are non-negative. If $x \neq 0$, then the third term is positive by (17.2), so $W_{2n+1}(x) > 0$. $\square$

Many of the techniques used in the above proof are to be found in [As1, 131–138].

Corollary 17.3. For $x \geq 0$ and $n \geq 1$, we have

$$(17.17) \qquad \begin{vmatrix} A_{n-1}(x) & A_n(x) \\ A_n(x) & A_{n+1}(x) \end{vmatrix} > 0$$

and

$$(17.18) \qquad \begin{vmatrix} B_{n-1}(x) & B_n(x) \\ B_n(x) & B_{n+1}(x) \end{vmatrix} > 0.$$

Proof. (17.17) For $x = 0$, we have from (13.24) that

$$A_{n-1}(0)\, A_{n+1}(0) - A_n^2(0) = (2n-2)!\,(2n+2)! - [(2n)!]^2$$
$$= 2\,(4n+1)(2n-2)!\,(2n)! > 0.$$

For $x > 0$, from (13.5) and (17.10) we have that

$$A_{n-1}(x)\, A_{n+1}(x) - A_n^2(x) = \frac{1}{x}\left\{\delta_{2n-1}(\sqrt{x})\,\delta_{2n+3}(\sqrt{x}) - \delta_{2n+1}^2(\sqrt{x})\right\}$$
$$= \frac{1}{x}\, W_{2n+1}(\sqrt{x}) > 0.$$

(17.18) For $x = 0$, we have from (13.24) for $n \geq 1$ that

$$B_{n-1}(0)\, B_{n+1}(0) - B_n^2(0) = (2n-1)!\, \xi_1(n-1)(2n+3)!\, \xi_1(n+1)$$
$$- \left[(2n+1)!\, \xi_1(n)\right]^2,$$

Now, from (11.17), we have that $\xi_1(n) = \sum\limits_{k=0}^{n} \dfrac{1}{2k+1}$, $k \geq 0$, from which it follows that

$$\xi_1(n-1)\, \xi_1(n+1) = \xi_1^2(n) - \frac{1}{(2n+1)(2n+3)}\left(2\,\xi_1(n) + 1\right).$$

Thus, using the simple estimates $1 \leq \xi_1(n) \leq n+1$, $n \geq 0$, we have

$$B_{n-1}(0)\, B_{n+1}(0) - B_n^2(0) = 2(2n-1)!\,(2n+1)!\,(4n+3)\,\xi_1^2(n)$$
$$- 2(n+1)(2n-1)!\,(2n)!\,(2\,\xi_1(n) + 1)$$
$$\geq 2(2n-1)!\,(2n+1)!\,(4n+3) - 2(n+1)(2n-1)!\,(2n)!\,(2(n+1)+1)$$
$$= 2n\,(6n+5)(2n-1)!\,(2n)! > 0.$$

For $x > 0$, we have from (13.6) and (17.10) that

$$B_{n-1}(x)\, B_{n+1}(x) - B_n^2(x) = \frac{1}{x^2}\left\{ \delta_{2n}(\sqrt{x})\, \delta_{2n+4}(\sqrt{x}) - \delta_{2n+2}^2(\sqrt{x}) \right\}$$
$$= \frac{1}{x^2} W_{2n+2}(\sqrt{x}) > 0. \quad \square$$

Theorem 17.4. For $x \in \mathbb{R}$ and $n \geq 1$, we have

(17.19)
$$\begin{vmatrix} A_n(x) & A_n'(x) \\ A_n'(x) & A_n''(x) \end{vmatrix} \leq -\frac{1}{n}\left[A_n'(x)\right]^2$$

and

(17.20)
$$\begin{vmatrix} B_n(x) & B_n'(x) \\ B_n'(x) & B_n''(x) \end{vmatrix} \leq -\frac{1}{n}\left[B_n'(x)\right]^2.$$

Proof. From [Lo], [Mit, 3.3.29], we have the result of Love: If the zeros of a real polynomial $P(x)$ of degree n are real, then $(n-1)[P'(x)]^2 - nP(x)P''(x) \geq 0$. Using this inequality, we have

$$A_n(x)A_n''(x) - [A_n'(x)]^2 \leq \frac{n-1}{n}[A_n'(x)]^2 - [A_n'(x)]^2 = -\frac{1}{n}[A_n'(x)]^2.$$

The proof of (17.20) is similar. $\quad \square$

Corollary 17.5. For $n \geq 0$, the functions $-\log[A_n(x)]$ and $-\log[B_n(x)]$ are convex on $(0, \infty)$.

Proof. If $n = 0$, then $A_0(x) = 1$, so $-\log[A_0(x)] = 0$. For $n \geq 1$, we have

$$\frac{d^2}{dx^2}\left\{-\log[A_n(x)]\right\} = -\frac{d}{dx}\left\{\frac{A_n'(x)}{A_n(x)}\right\} = -\frac{1}{[A_n(x)]^2}\left\{A_n(x)A_n''(x)-\right.$$

$$\left.[A_n'(x)]^2\right\} = -\frac{1}{[A_n(x)]^2}\begin{vmatrix} A_n(x) & A_n'(x) \\ A_n'(x) & A_n''(x) \end{vmatrix} \geq 0,$$

using (17.19). The proof for the second function is similar using (17.20). $\square$

Corollary 17.6. Let $n \geq 0$ and $0 \leq x \leq y$. Then

(17.21)
$$\begin{vmatrix} A_n(x) & A_n(y) \\ A_n'(x) & A_n'(y) \end{vmatrix} \leq 0$$

and

(17.22)
$$\begin{vmatrix} B_n(x) & B_n(y) \\ B_n'(x) & B_n'(y) \end{vmatrix} \leq 0.$$

Proof. For $x > 0$, using the convexity of the function $-\log[A_n(x)]$ in Corollary 17.5, we have that $\frac{d}{dx}\left\{-\log[A_n(x)]\right\} \leq \frac{d}{dy}\left\{-\log[A_n(y)]\right\}$, i.e., $\frac{A_n'(x)}{A_n(x)} \geq \frac{A_n'(y)}{A_n(y)}$, which is (17.21). The result for $x = 0$ is true by continuity. The proof of (17.22) is similar. $\square$.

Corollary 17.7. Let $n \geq 0$ and $x, y \geq 0$. Then we have

(17.23)
$$A_n(x)A_n(y) \leq \left[A_n\left(\frac{x+y}{2}\right)\right]^2$$

and

(17.24)
$$B_n(x)B_n(y) \leq \left[B_n\left(\frac{x+y}{2}\right)\right]^2.$$

Proof. For $x, y > 0$, the function $-\log[A_n(t)]$ is convex on $(0, \infty)$. Thus, it is mid-convex, i.e.,

$$-\log\left[A_n\left(\frac{x+y}{2}\right)\right] \leq \frac{1}{2}\{-\log[A_n(x)] - \log[A_n(y)]\}$$

so

$$\log[A_n(x)A_n(y)] \leq \log\left[A_n\left(\frac{x+y}{2}\right)\right]^2,$$

which implies the result. If $xy = 0$, then the inequality holds by continuity. The proof of (17.24) is similar. $\square$

Theorem 17.8. For $n \geq 2$, let $f(x) = \sum_{k=0}^{n} a_k x^k$, $a_n > 0$, be a real polynomial whose zeros are real and negative. Then

(17.25)
$$\frac{a_0}{a_1} < \frac{a_1}{a_2} < \cdots < \frac{a_{n-1}}{a_n}.$$

Proof. First note that a real polynomial with a positive leading coefficient and negative zeros has positive coefficients. For, if $g(x) = c_n x^n + \cdots + c_0$, $n \geq 1$, $c_n > 0$, has the zeros $\{-\gamma_1, \cdots, -\gamma_n\}$, $\gamma_i > 0$, then the expansion of the product $g(x) = c_n \prod_{i=1}^{n} (x + \gamma_i)$ is a polynomial with positive coefficients. Now let $n = 2$, and take $f(x) = a_2 x^2 + a_1 x + a_0$, $a_2 > 0$, with zeros $-\alpha_1$ and $-\alpha_2$. Then a_1 and a_0 are positive by the above result and $\alpha_1 + \alpha_2 = \dfrac{a_1}{a_2}$ and $\alpha_1 \alpha_2 = \dfrac{a_0}{a_2}$. Thus,

$$\frac{a_0}{a_1} = \frac{\alpha_1 \alpha_2}{\alpha_1 + \alpha_2} < \alpha_1 + \alpha_2 = \frac{a_1}{a_2}.$$

Next, assume that the theorem is true for some $n \geq 2$. Then consider the polynomial $f(x) = a_{n+1} x^{n+1} + \cdots + a_0$, $a_{n+1} > 0$, with the zeros $\{-\alpha_1, \cdots, -\alpha_{n+1}\}$, so all the coefficients of $f(x)$ are positive. Now, let $-\alpha$ be one of these zeros, so we can write

(17.26) $f(x) = (x + \alpha)(b_n x^n + \cdots + b_0).$

Equating corresponding coefficients on the two sides of (17.26) gives

(17.27) $a_k = \alpha b_k + b_{k-1}, \quad 0 \leq k \leq n+1,$

where we take $b_{-1} = b_{n+1} = 0$.

Since $b_n = a_{n+1} > 0$ and the zeros of the second polynomial factor in (17.26) are all negative, then $b_0, \cdots, b_{n-1}$ are positive and we have by assumption that $\dfrac{b_0}{b_1} < \cdots < \dfrac{b_{n-1}}{b_n}$, $n \geq 2$. From three consecutive ratios, we obtain for $0 \leq k \leq n - 1$ that

(17.28) $b_k b_{k+2} < b_{k+1}^2, \quad b_{k-1} b_{k+2} < b_k b_{k+1}, \quad \text{and} \quad b_{k-1} b_{k+1} < b_k^2.$

Finally, for $0 \leq k \leq n - 1$, $n \geq 1$, and using (17.27) and (17.28), we have that

$$\begin{aligned}
a_k a_{k+2} &= \left(\alpha b_k + b_{k-1}\right)\left(\alpha b_{k+2} + b_{k+1}\right) \\
&= \alpha^2 b_k b_{k+2} + \alpha b_k b_{k+1} + \alpha b_{k-1} b_{k+2} + b_{k-1} b_{k+1} \\
&< \alpha^2 b_{k+1}^2 + 2\alpha b_k b_{k+1} + b_k^2 = (\alpha b_{k+1} + b_k)^2 = a_{k+1}^2,
\end{aligned}$$

which proves the theorem. $\square$

Remark. It is worth mentioning that when the zeros of $f(x) = \sum_{k=0}^{n} a_k x^k$ lie in the complex left half-plane, then A. C. Hindmarsh [Hi] has shown that $\dfrac{a_0}{a_1} < \dfrac{a_2}{a_3} < \cdots < \dfrac{a_{n-1}}{a_n}$. The results in (17.25), where the zeros are restricted to the negative real axis, are stronger and imply the inequalities of Hindmarsh.

Corollary 17.9. For $n \geq 2$,

$$(17.29) \qquad \frac{d_1^{(2n+1)}}{d_3^{(2n+1)}} < \frac{d_3^{(2n+1)}}{d_5^{(2n+1)}} < \cdots < \frac{d_{2n-1}^{(2n+1)}}{d_{2n+1}^{(2n+1)}}$$

and

$$(17.30) \qquad \frac{d_2^{(2n+2)}}{d_4^{(2n+2)}} < \frac{d_4^{(2n+2)}}{d_6^{(2n+2)}} < \cdots < \frac{d_{2n}^{(2n+2)}}{d_{2n+2}^{(2n+2)}}.$$

Proof. All the zeros of $A_n(x)$ and $B_n(x)$, $n \geq 1$, are negative, since these polynomials are orthogonal on the interval $(-\infty, 0]$ (cf. Chapter 15). Theorem 17.8 implies (17.29) and (17.30), since these d's are the positive coefficients of $A_n(x)$ and $B_n(x)$ as in (13.3) and (13.4), respectively. □

Corollary 17.10. For $n \geq 2$,

$$(17.31) \qquad \frac{\xi_0(n)}{\xi_2(n)} < \frac{\xi_2(n)}{\xi_4(n)} < \cdots < \frac{\xi_{2n-2}(n)}{\xi_{2n}(n)}$$

and

$$(17.32) \qquad \frac{\xi_1(n)}{\xi_3(n)} < \frac{\xi_3(n)}{\xi_5(n)} < \cdots < \frac{\xi_{2n-1}(n)}{\xi_{2n+1}(n)}.$$

Proof. Apply equations (11.18) and (11.19) to the inequalities in Corollary 17.9. □

Corollary 17.11. For $n \geq 2$ and $1 \leq j < k \leq n$, we have

$$(17.33) \qquad \begin{vmatrix} d_{2j-1}^{(2n+1)} & d_{2j+1}^{(2n+1)} \\ d_{2k-1}^{(2n+1)} & d_{2k+1}^{(2n+1)} \end{vmatrix} < 0$$

and

$$(17.34) \qquad \begin{vmatrix} d_{2j}^{(2n+2)} & d_{2j+2}^{(2n+2)} \\ d_{2k}^{(2n+2)} & d_{2k+2}^{(2n+2)} \end{vmatrix} < 0.$$

Proof. These results follow from Corollary 17.9. □

Theorem 17.12. For $0 \leq x \leq y$ and $0 \leq m \leq n$, we have

$$(17.35) \qquad \begin{vmatrix} A_m(x) & A_m(y) \\ A_n(x) & A_n(y) \end{vmatrix} \geq 0$$

and

$$(17.36) \qquad \begin{vmatrix} B_m(x) & B_m(y) \\ B_n(x) & B_n(y) \end{vmatrix} \geq 0,$$

with equality in each only when $x = y$ or $m = n$.

Proof. Assume $0 \leq x < y$ and $0 \leq m < n$. Let

$$D_{m,n}(x,y) = \begin{vmatrix} A_m(x) & A_m(y) \\ A_n(x) & A_n(y) \end{vmatrix},$$

and set $\lambda_1 = 1$ and $\lambda_k = 4(k-1)^2(2k-3)(2k-1)$, $k \geq 2$. Then by the Christoffel-Darboux identity [Ch, p. 23, 4.9] and (17.1), we have that

$$(17.37) \qquad D_{m,m+1}(x,y) = \lambda_1 \cdots \lambda_{m+1}(y-x) \sum_{j=0}^{m} \frac{A_j(x)A_j(y)}{\lambda_1 \cdots \lambda_{j+1}} > 0.$$

Also by (17.1), we have that $f_{m,n}(x) \overset{\text{def}}{=} \dfrac{A_m(x)}{A_n(x)} > 0$. Then

$$(17.38) \qquad D_{m,n}(x,y) = A_n(x)A_n(y)\Big[f_{m,n}(x) - f_{m,n}(y)\Big],$$

which implies using (17.37) that $f_{m,m+1}(x) - f_{m,m+1}(y) > 0$, i.e., $f_{m,m+1}$ is a positive, strictly decreasing function. Now $f_{m,n}(x) = \prod_{j=m}^{n-1} f_{j,j+1}(x)$, so $f_{m,n}$ is a positive, strictly decreasing function as well. Thus, equation (17.38) implies that $D_{m,n}(x,y) > 0$. Finally, $D_{m,n}(x,y) = 0$ only when $x = y$ or $m = n$.

The proof of (17.36) is the same as for (17.35), except that $\lambda_1 = 1$ and $\lambda_k = 4(k-1)k(2k-1)^2$, $k \geq 2$. $\square$

Corollary 17.13. For $0 \leq m \leq n$ and $x^2 \leq y^2$, we have that

$$(17.39) \qquad \begin{vmatrix} \delta_{2m}(x) & \delta_{2m}(y) \\ \delta_{2n}(x) & \delta_{2n}(y) \end{vmatrix} \geq 0$$

and if $xy \geq 0$, then

$$(17.40) \qquad \begin{vmatrix} \delta_{2m+1}(x) & \delta_{2m+1}(y) \\ \delta_{2n+1}(x) & \delta_{2n+1}(y) \end{vmatrix} \geq 0.$$

Proof. For $m = 0$ we have by (17.4), since $\delta_{2n}(x)$ is an even function, that

$$\begin{vmatrix} \delta_0(x) & \delta_0(y) \\ \delta_{2n}(x) & \delta_{2n}(y) \end{vmatrix} = \delta_{2n}(y) - \delta_{2n}(x) \geq 0, \ x^2 \leq y^2.$$

Next, for $|x| \leq |y|$ and $m \geq 1$, we have from (13.6) and (17.36) that

$$\begin{vmatrix} \delta_{2m}(x) & \delta_{2m}(y) \\ \delta_{2n}(x) & \delta_{2n}(y) \end{vmatrix} = \begin{vmatrix} x^2 B_{m-1}(x^2) & y^2 B_{m-1}(y^2) \\ x^2 B_{n-1}(x^2) & y^2 B_{n-1}(y^2) \end{vmatrix}$$

$$= x^2 y^2 \begin{vmatrix} B_{m-1}(x^2) & B_{m-1}(y^2) \\ B_{n-1}(x^2) & B_{n-1}(y^2) \end{vmatrix} \geq 0.$$

Also, for $m \geq 0$ and $xy \geq 0$, we have from (13.5) and (17.35) that

$$\begin{vmatrix} \delta_{2m+1}(x) & \delta_{2m+1}(y) \\ \delta_{2n+1}(x) & \delta_{2n+1}(y) \end{vmatrix} = \begin{vmatrix} xA_m(x^2) & yA_m(y^2) \\ xA_n(x^2) & yA_n(y^2) \end{vmatrix}$$

$$= xy \begin{vmatrix} A_m(x^2) & A_m(y^2) \\ A_n(x^2) & A_n(y^2) \end{vmatrix} \geq 0. \quad \square$$

Corollary 17.14. For $x \geq 0$ and $0 \leq m \leq n$, we have

(17.41)
$$\begin{vmatrix} A_m(x) & A'_m(x) \\ A_n(x) & A'_n(x) \end{vmatrix} \geq 0$$

and

(17.42)
$$\begin{vmatrix} B_m(x) & B'_m(x) \\ B_n(x) & B'_n(x) \end{vmatrix} \geq 0.$$

Proof. Equality holds in each case for $m = n$. Assume $0 \leq m < n$. To prove (17.41), let $f_{m,n}(x) \overset{\text{def}}{=} \dfrac{A_m(x)}{A_n(x)}$, so we have

$$f'_{m,n}(x) = \frac{1}{[A_n(x)]^2} \left(A_n(x) A'_m(x) - A_m(x) A'_n(x) \right).$$

Since in the proof of (17.35) we proved that $f_{m,n}$ is a positive, strictly decreasing function, we have that $f'_{m,n}(x) \leq 0$, so

$$\begin{vmatrix} A_m(x) & A'_m(x) \\ A_n(x) & A'_n(x) \end{vmatrix} = -[A_n(x)]^2 f'_{m,n}(x) \geq 0.$$

A similar proof establishes (17.42). $\square$

Corollary 17.15. For $0 \leq x \leq y$ and $0 \leq m \leq n$, we have that

(17.43)
$$\begin{vmatrix} A_n(x) & A_n(y) \\ B_m(x) & B_m(y) \end{vmatrix} \leq 0.$$

Proof. Let $0 \leq x \leq y$ and $0 \leq m \leq n$. From (13.10) and (17.35) we have that

$$\begin{vmatrix} A_n(x) & A_n(y) \\ B_m(x) & B_m(y) \end{vmatrix} = (2m+1)! \sum_{k=0}^{m} \frac{1}{(2k+1)!} \begin{vmatrix} A_n(x) & A_n(y) \\ A_k(x) & A_k(y) \end{vmatrix} \leq 0. \quad \square$$

Theorem 17.16. If $n \geq 1$ and $x \geq 0$, then

(17.44)
$$\delta_{2n+1}(x) \geq x \left(x^2 + [(2n)!]^{\frac{1}{n}} \right)^n$$

and

(17.45) $$\delta_{2n+2}(x) \geq x^2 \left(x^2 + [(2n+1)!\,\xi_1(n)]^{\frac{1}{n}} \right)^n,$$

where $\xi_1(n)$ is defined in (11.17).

Proof. The results are trivial for $x = 0$, so assume $x > 0$.

(17.44) From the end of Chapter 15, we know that the non-zero zeros of $\delta_{2n+1}(x)$ are pure imaginary, say $\pm\lambda_1 i, \ldots, \pm\lambda_n i$, with $\lambda_j > 0$. By (11.1) and (11.5), the constant term of $\dfrac{\delta_{2n+1}(x)}{x}$ is $\prod\limits_{j=1}^{n} \lambda_j^2 = \prod\limits_{j=1}^{n} (\lambda_j i)(-\lambda_j i) =$ $d_1^{(2n+1)} = (2n)!$. Also, $\dfrac{\delta_{2n+1}(x)}{x} = \prod\limits_{j=1}^{n}(x^2 + \lambda_j^2)$, so using [Mit, p. 208, 3.2.34], we have that

$$\frac{\delta_{2n+1}(x)}{x^{2n+1}} = \prod_{j=1}^{n}\left[1 + \left(\frac{\lambda_j}{x}\right)^2\right] \geq \left(1 + \left[\prod_{j=1}^{n}\left(\frac{\lambda_j}{x}\right)^2\right]^{\frac{1}{n}}\right)^n = \left(1 + \frac{[(2n)!]^{\frac{1}{n}}}{x^2}\right)^n,$$

so $\delta_{2n+1}(x) \geq x\left(x^2 + [(2n)!]^{\frac{1}{n}}\right)^n$.

(17.45) From the end of Chapter 15, we know that the non-zero zeros of $\delta_{2n+2}(x)$ are pure imaginary, say $\pm\lambda_1 i, \ldots, \pm\lambda_n i$, with $\lambda_j > 0$, so from (11.1) and (11.6), the constant term of $\dfrac{\delta_{2n+2}(x)}{x^2}$ is $\prod\limits_{j=1}^{n} \lambda_j^2 =$ $\prod\limits_{j=1}^{n}(\lambda_j i)(-\lambda_j i) = d_2^{(2n+2)} = (2n+1)!\,\xi_1(n)$. Also, $\dfrac{\delta_{2n+2}(x)}{x^2} = \prod\limits_{j=1}^{n}(x^2 + \lambda_j^2)$, so, as before, we have that

$$\frac{\delta_{2n+2}(x)}{x^{2n+2}} = \prod_{j=1}^{n}\left[1 + \left(\frac{\lambda_j}{x}\right)^2\right] \geq \left(1 + \left[\prod_{j=1}^{n}\left(\frac{\lambda_j}{x}\right)^2\right]^{\frac{1}{n}}\right)^n$$

$$= \left(1 + \frac{[(2n+1)!\,\xi_1(n)]^{\frac{1}{n}}}{x^2}\right)^n,$$

so $\delta_{2n+2}(x) \geq x^2\left(x^2 + [(2n)!\,\xi_1(n)]^{\frac{1}{n}}\right)^n$. $\square$

Theorem 17.17. For $m, n \geq 0$ and $x \geq 0$, we have that

(17.46) $$\delta_{m+n}(x) \geq \delta_m(x)\,\delta_n(x)$$

and

(17.47) $$\delta'_{m+n}(x) \geq \begin{vmatrix} \delta_m(x) & \delta_n(x) \\ \delta'_m(x) & \delta'_n(x) \end{vmatrix}_+,$$

where the subscript "+" indicates the permanent function.

Proof. (17.46) The result holds for $m = n = 0$. Without loss of generality we may assume that $0 \leq m \leq n$, $m + n \neq 0$. Then using (10.128) and (17.2), we have

$$(17.48) \quad \delta_{m+n}(x) = \delta_m(x)\,\delta_n(x) + \frac{1}{m+n}\sum_{j=1}^{m}(j!)^2(m+n-2j)$$

$$\times \binom{m}{j}\binom{n}{j}\binom{m+n}{j}\delta_{m-j}(x)\,\delta_{n-j}(x) \geq \delta_m(x)\,\delta_n(x).$$

(17.47) Differentiating (17.48) gives

$$\delta'_{m+n}(x) = \begin{vmatrix} \delta_m(x) & \delta_n(x) \\ \delta'_m(x) & \delta'_n(x) \end{vmatrix}_+ + D\Big(\frac{1}{m+n}\sum_{j=1}^{m}\cdots\Big) \geq \begin{vmatrix} \delta_m(x) & \delta_n(x) \\ \delta'_m(x) & \delta'_n(x) \end{vmatrix}_+ ,$$

since the differentiated sum is non-negative because its terms are derivatives of increasing functions by (11.9). $\square$

Corollary 17.18. For $n \geq 0$ and $x \in \mathbb{R}$, we have

$$(17.49) \qquad\qquad [\delta_n(x)]^2 \leq \delta_{2n}(x).$$

Proof. Put $m = n$ in (17.46). The result is true for any $x \in \mathbb{R}$ by (17.2). $\square$

Theorem 17.19. For $m \geq 0$ and $0 \leq x \leq y$, we have that

$$(17.50) \qquad\qquad \begin{vmatrix} \delta_m(x) & \delta_m(y) \\ \delta'_m(x) & \delta'_m(y) \end{vmatrix} \leq 0.$$

Proof. Let $S = \{(x, y) : 0 \leq x \leq y\}$ and $S^0 = \{(x, y) : 0 < x < y\}$.
(a) $m = 2n + 1$, $n \geq 0$. For $(x, y) \in S$, let

$$(17.51) \qquad u_n(x, y) = \begin{vmatrix} \delta_{2n+1}(x) & \delta_{2n+1}(y) \\ \delta'_{2n+1}(x) & \delta'_{2n+1}(y) \end{vmatrix}$$

and

$$(17.52) \qquad f_n(x, y) = \begin{vmatrix} \delta_{2n+1}(x) & \delta_{2n+1}(y) \\ \delta''_{2n+1}(x) & \delta''_{2n+1}(y) \end{vmatrix}.$$

We begin by proving a useful result:

For $n \geq 0$, there exists at most one continuous function $v : S \longrightarrow \mathbb{R}$ such that

$$(17.53) \quad \begin{cases} (1) & v \mid S^0 \text{ is a } C^\infty \text{ function} \\ (2) & \dfrac{\partial v(x, y)}{\partial x} + \dfrac{\partial v(x, y)}{\partial y} = f_n(x, y), \ (x, y) \in S^0 \\ (3) & v(0, y) = -(2n)!\,\delta_{2n+1}(y), \ y \geq 0. \end{cases}$$

To see this, suppose there exist two such functions v_1 and v_2 and let $w = v_1 - v_2$. Then $w : S \longrightarrow \mathbb{R}$ is continuous and

$$\begin{cases} (1') & w \mid S^0 \text{ is a } C^\infty \text{ function} \\ (2') & \dfrac{\partial w(x,y)}{\partial x} + \dfrac{\partial w(x,y)}{\partial y} = 0, \ (x,y) \in S^0 \\ (3') & w(0,y) = 0, \ y \geq 0. \end{cases}$$

Now put $x = \frac{1}{2}(\xi-\eta)$ and $y = \frac{1}{2}(\xi+\eta)$, so $w(x,y) = w\big(\frac{1}{2}(\xi-\eta), \frac{1}{2}(\xi+\eta)\big) = \widehat{w}(\xi,\eta)$. Then, $\dfrac{\partial \widehat{w}}{\partial \xi} = \dfrac{\partial \widehat{w}}{\partial x}\dfrac{\partial x}{\partial \xi} + \dfrac{\partial \widehat{w}}{\partial y}\dfrac{\partial y}{\partial \xi} = \dfrac{1}{2}\big(\dfrac{\partial w}{\partial x} + \dfrac{\partial w}{\partial y}\big) = 0$. Thus, there exists a function $\widetilde{w}$ for which $\widehat{w}(\xi,\eta) = \widetilde{w}(\eta)$, so $w(x,y) = \widetilde{w}(y-x) = w(0, y-x) = 0$ on S^0 by $(3')$. By continuity, $w = 0$ on all of S.

Next, $u_n : S \longrightarrow \mathbb{R}$ and, being a polynomial, u_n is continuous on S and is a C^∞ function on S^0. Also, from (17.51) and (17.52), we find that

$$\Big(\frac{\partial}{\partial x} + \frac{\partial}{\partial y}\Big)u_n(x,y) = \Big(\frac{\partial}{\partial y} + \frac{\partial}{\partial x}\Big)\Big(\frac{\partial}{\partial y} - \frac{\partial}{\partial x}\Big)\delta_{2n+1}(x)\,\delta_{2n+1}(y)$$

$$. = \Big(\frac{\partial^2}{\partial y^2} - \frac{\partial^2}{\partial x^2}\Big)\delta_{2n+1}(x)\,\delta_{2n+1}(y) = f_n(x,y), \ (x,y) \in S^0,$$

and using (11.1), (11.4), and (11.5), that $u_n(0,y) = -(2n)!\,\delta_n(y)$, $y \geq 0$. Thus, $u_n(x,y)$ is the unique function satisfying (17.53).

Now, let v_n be the function defined for $n \geq 0$ by

$$(17.54) \quad v_n(x,y) = \frac{1}{2}\int_{y-x}^{y+x} f_n\Big(\tfrac{1}{2}(t+x-y), \tfrac{1}{2}(t-x+y)\Big)dt$$
$$- (2n)!\,\delta_{2n+1}(y-x)$$

Then, $v_n : S \longrightarrow \mathbb{R}$. To prove this, we can easily show for any $(x,y) \in S$ that $\big(\frac{1}{2}(t+x-y), \frac{1}{2}(t-x+y)\big) \in S$, since $y-x \leq t \leq y+x$. Also, v_n is continuous on S and $v_n \mid S^0$ is a C^∞ function.

Next, using the Leibniz formula [Buc, p. 122, eq. [3-4], p. 295] [St, p. 380, ex. 9b],

$$\frac{d}{dx}\int_{a(x)}^{b(x)} f(t,x)\,dt = b'(x)f(b(x),x) - a'(x)f(a(x),x) + \int_{a(x)}^{b(x)} f_x(t,x)\,dt,$$

we find that

$$2\frac{\partial v_n(x,y)}{\partial x} = f_n(x,y) + f_n(0, y-x)$$
$$+ \frac{1}{2}\int_{y-x}^{y+x}\Big(\frac{\partial f_n}{\partial x} - \frac{\partial f_n}{\partial y}\Big)dt + 2(2n)!\,\delta'_{2n+1}(y-x)$$

and

$$2\frac{\partial v_n(x,y)}{\partial y} = f_n(x,y) - f_n(0,y-x)$$

$$+ \frac{1}{2}\int_{y-x}^{y+x}\left(-\frac{\partial f_n}{\partial x} + \frac{\partial f_n}{\partial y}\right)dt - 2(2n)!\,\delta'_{2n+1}(y-x),$$

so

$$\frac{\partial v_n(x,y)}{\partial x} + \frac{\partial v_n(x,y)}{\partial y} = f_n(x,y), \quad (x,y) \in S^0.$$

Finally, from (17.54) we see that $v_n(0,y) = -(2n)!\,\delta_{2n+1}(y)$, $y \geq 0$. Thus, $v_n(x,y)$ also satisfies (17.53), so we obtain the integral representation

$$(17.55)\quad u_n(x,y) = \frac{1}{2}\int_{y-x}^{y+x} f_n\left(\tfrac{1}{2}(t+x-y), \tfrac{1}{2}(t-x+y)\right)dt$$

$$- (2n)!\,\delta_{2n+1}(y-x)$$

But, for $n \geq 0$ and $(x,y) \in S$,

$$(17.56)\qquad\qquad f_n(x,y) \leq 0, \ n \geq 0, \ (x,y) \in S.$$

To prove this, note that $f_0(x,y) = 0$, and for $n \geq 1$, we have by (10.75) and (17.40) that

$$f_n(x,y) = (2n+1)!\sum_{k=1}^{n}\frac{\xi_1(n-k)}{(n-k+1)(2k-1)!}\begin{vmatrix}\delta_{2n+1}(x) & \delta_{2n+1}(y) \\ \delta_{2k-1}(x) & \delta_{2k-1}(y)\end{vmatrix} \leq 0.$$

Thus, by (17.56), the value of the integral in (17.55) is non-positive, as is the value of the final term by (17.2). Hence, $u_n(x,y) \leq 0$, $n \geq 0$ and $(x,y) \in S$.

(b) Inequality (17.50) is true for $m = 0$. Assume $m = 2n+2$, $n \geq 0$. For $(x,y) \in S$ and using notation comparable to that in part (a), let

$$(17.57)\qquad\qquad u_n(x,y) = \begin{vmatrix}\delta_{2n+2}(x) & \delta_{2n+2}(y) \\ \delta'_{2n+2}(x) & \delta'_{2n+2}(y)\end{vmatrix}$$

and

$$(17.58)\qquad\qquad f_n(x,y) = \begin{vmatrix}\delta_{2n+2}(x) & \delta_{2n+2}(y) \\ \delta''_{2n+2}(x) & \delta''_{2n+2}(y)\end{vmatrix}.$$

As in part (a), we prove a basic result. For $n \geq 0$, there exists at most one continuous function $v : S \longrightarrow \mathbb{R}$ such that

$$(17.59)\qquad\begin{cases}(1) & v\mid S^0 \text{ is a } C^\infty \text{ function} \\ (2) & \dfrac{\partial v(x,y)}{\partial x} + \dfrac{\partial v(x,y)}{\partial y} = f_n(x,y), \ (x,y) \in S^0 \\ (3) & v(0,y) = 0, \ y \geq 0.\end{cases}$$

The proof of this is essentially the same as in part (a), as is the verification that $u_n(x, y)$ satisfies (17.59). As before, we define the (simpler) function

$$v_n(x, y) = \frac{1}{2} \int_{y-x}^{y+x} f_n\left(\tfrac{1}{2}(t + x - y), \tfrac{1}{2}(t - x + y)\right) dt, \ n \geq 0,$$

and show in the same way that $v_n : S \longrightarrow \mathbb{R}$ is a continuous map that satisfies (17.59). Thus, for $n \geq 0$ and $(x, y) \in S$,

$$u_n(x, y) = \frac{1}{2} \int_{y-x}^{y+x} f_n\left(\tfrac{1}{2}(t + x - y), \tfrac{1}{2}(t - x + y)\right) dt,$$

from which it follows from (17.56) that $u_n(x, y) \leq 0$. $\square$

Corollary 17.20. The function $f(x) = -\log[\delta_n(x)]$ is convex for $n \geq 0$ and $x > 0$.

Proof. Since $\delta_n(x) > 0$ by (17.2), the function $f(x)$ is defined for $x > 0$. It suffices to show that $f'(x) = -\dfrac{\delta_n'(x)}{\delta_n(x)}$ is non-decreasing for $x > 0$, i.e., that $-\dfrac{\delta_n'(x)}{\delta_n(x)} \leq -\dfrac{\delta_n'(y)}{\delta_n(y)}$, when $0 < x \leq y$. But this inequality holds by (17.50). $\square$

Corollary 17.21. For $n \geq 0$, we have that

(17.60)
$$\begin{vmatrix} \delta_n(x) & \delta_n'(x) \\ \delta_n'(x) & \delta_n''(x) \end{vmatrix} \leq 0.$$

Proof. By Corollary 17.20, the function $-\log[\delta_n(x)]$ is convex for $x > 0$, so for $x > 0$, $\dfrac{d^2}{dx^2}\left\{-\log[\delta_n(x)]\right\} \geq 0$, which implies that

$$\frac{1}{[\delta_n(x)]^2} \begin{vmatrix} \delta_n(x) & \delta_n'(x) \\ \delta_n'(x) & \delta_n''(x) \end{vmatrix} \leq 0.$$

For $x < 0$, then using (10.6) we have

$$\begin{vmatrix} \delta_n(x) & \delta_n'(x) \\ \delta_n'(x) & \delta_n''(x) \end{vmatrix} = \begin{vmatrix} (-1)^n \delta_n(-x) & (-1)^{n+1} \delta_n'(-x) \\ (-1)^{n+1} \delta_n'(-x) & (-1)^n \delta_n''(x) \end{vmatrix}$$

$$= \begin{vmatrix} \delta_n(-x) & \delta_n'(-x) \\ \delta_n'(-x) & \delta_n''(-x) \end{vmatrix} \leq 0,$$

by the preceding case.

In the case $x = 0$, the result is true by the two preceding cases and continuity. $\square$

Note. Inequality (17.60) implies that $\dfrac{d^2}{dx^2}[\delta_n(x)]^2 \leq 4[\delta_n'(x)]^2$.

Corollary 17.22. For $n \geq 0$ and $x, y \geq 0$, we have

(17.61)
$$\delta_n(x)\,\delta_n(y) \leq \left[\delta_n\left(\frac{x+y}{2}\right)\right]^2.$$

Proof. True for $n = 0$. Also, $\delta_n(x)\,\delta_n(y) = 0$ if $n \geq 1$ and $x = 0$ or $y = 0$. For $x, y > 0$, we have $f(x) = -\log[\delta_n(x)]$ is convex by Corollary 17.20. Thus, $f(x)$ is mid-convex, i.e., $-\log\left[\delta_n\left(\frac{x+y}{2}\right)\right] \leq \frac{1}{2}\left\{-\log[\delta_n(x)] - \log[\delta_n(y)]\right\}$, which implies the result. $\square$

The next result is a generalization of Soble's inequality [So, Th. 1, p. 640], which applies to $\delta_n(x)$, $A_m(z)$, and $B_m(z)$.

Theorem 17.23. Let $f_n(x)$ be a polynomial of degree $n \geq 0$ with non-negative coefficients. Then, if $0 \leq r \leq n$ and $x > 0$, it follows that

(17.62)
$$f_n^{(r)}(x) \leq \frac{n!}{(n-r)!}\,\frac{f_n(x)}{x^r}.$$

Proof. Clear for $r = 0$. For $r = 1$, equation (17.62) is Soble's inequality. Assume for some r, where $2 \leq r \leq n$, that $f_n^{(r-1)} \leq \dfrac{n!}{(n-r+1)!}\,\dfrac{f_n(x)}{x^{r-1}}$. (Here $f_n^{(r-1)}(x)$ also has non-negative coefficients.) Then putting $f_n^{(r-1)}(x)$ into Soble's inequality gives

$$f_n^{(r)}(x) \leq \frac{(n-r+1)}{x}\,f_n^{(r-1)}(x) \leq \frac{(n-r+1)}{x}\,\frac{n!}{(n-r+1)!}\,\frac{f_n(x)}{x^{r-1}}$$
$$= \frac{n!}{(n-r)!}\,\frac{f_n(x)}{x^r}. \quad \square$$

Chapter 18

Some Concluding Questions

We conclude with some open questions about various aspects of this work.

1. The structure constants for the sequences $\{\delta_n(x)\}_{n=0}^{\infty}$, $\{A_m(z)\}_{m=0}^{\infty}$, and $\{B_m(z)\}_{m=0}^{\infty}$ are given in (10.116), (13.40), and (14.44), respectively. What are the structure constants for the other sequences of elliptic polynomials?

2. It is determined in Theorem 4.10 exactly when x and y divide $P_n(x, y)$ and $Q_n(x, y)$. When these factors are divided out, are the quotient polynomials irreducible? Also, when $P_n(x, y)$ and $Q_n(x, y)$ are specialized to produce the coefficients in the Maclaurin expansion of $f^{-1}(t)$, are the resulting polynomials more factorable than $P_n(x, y)$ and $Q_n(x, y)$? For example, in Theorem 8.11, are there other non-trivial factors of $C_{2n+1}(k)$ and $D_{2n+2}(k)$ besides $k^2 + 1$ and $k^4 + 14k^2 + 1$ and are any factors of these ever multiple factors?

3. Are there other identities connecting $A_m(z)$ and $B_m(z)$ besides those in Chapter 13?

4. It was observed in the remarks following Theorem 7.3 that the sequences $\{\bar{G}_m(z)\}_{m=0}^{\infty}$ and $\{\bar{H}_m(z)\}_{m=0}^{\infty}$ are not orthogonal. When do these polynomials have real zeros? Are there cases in which their zeros separate each other? Do the zeros of the barred polynomials separate the zeros of the unbarred polynomials?

5. Where do the zeros of the secondary sequences obtained from functions $f \in \mathcal{F}_1$ with $c_2 < 0$ lie? What properties of orthogonal polynomials do these generalized orthogonal polynomials have (cf. Theorem 7.3)?

6. $(n-2)(n-1)\,\delta_{n-2}(x)\,\delta_{n+2}(x) \leq n(n+1)[\delta_n(x)]^2$, $n \geq 2$, x real (cf. Theorem 17.2).

References

[AbS] Abramowitz, M. and Stegun, I., *Handbook of Mathematical Functions*, Applied Math Series 55, National Bureau of Standards, Washington, D.C., 1972.

[AgJ] Agrawal, B. M. and Jain R., *Lie theory and generating functions of some classical polynomials (Hindi)*, Vijnana Parishad Anusandhan Patrika, **26, No. 3** (1983), 235–242.

[Ai] Aigner, M., *Combinatorial Theory*, Springer-Verlag, New York, 1979.

[Ak1] Akhiezer, N. I., *The Classical Moment Problem*, Hafner Publ. Co., New York, 1965.

[Ak2] Akhiezer, N. I., *Lectures on Integral Transforms*, Translations of Mathematical Monographs, vol. 70, AMS, Providence, RI, 1988.

[Al] Al-Salam, W. A., *Characterization of certain classes of orthogonal polynomials related to elliptic functions*, Annali di Matematica pura ed applicata, **67** (1965), 75–94.

[As1] Askey, R., *Linearization of the product of orthogonal polynomials*, Problems in Analysis, Ed. by Gunning, R. C., Princeton, NJ, 1970.

[As2] Askey, R., *Orthogonal Polynomials and Special Functions*, SIAM, Philadelphia, PA, 1975.

[AsI] Askey, R. and Ismail, M., *Recurrence relations, continued fractions, and orthogonal polynomials*, Memoirs of the AMS, No. 300, AMS, Providence, RI, 1984.

[Ay] Ayoub, R., *The lemniscate and Fagnano's contributions to elliptic integrals*, Arch. for History of Exact Sciences, **29** (1984), 131–149.

[Ba] Bacry, H., *A unitary representation of SL(2,R)*, J. Math. Phys., **31** (1990), 2061–2077.

[BaB] Bacry, H. and Boon, M., *Liens entre certains polynomes et quelques fonctions transcendentes*, C. R. Acad. Sci. Paris, **301** (1985), 273–276.

[Bat] Bateman, H., *The polynomial of Mittag-Leffler*, Proc. Nat. Acad. Sci., **26** (1940), 491–496.

[Be] Belorizky, D., *Représentation de certaines fonctions par des séries particulières de polynomes*, C. R. Acad. Sci. Paris, **195** (1932), 1222–1224.

[BeB] Berndt, B. and Bhargava, S., *Ramanujan's inversion formulas for the lemniscate and allied functions*, J. Math. Anal. and Appl., **160** (1991), 504–524.

[BeMP] Bender, C. M., Mead, R., and Pinsky, S. S., *Continuous Hahn polynomials and the Heisenberg algebra*, J. Math. Phys., **28** (1987), 509–513.

[BoB] Borwein, J. and Borwein, P., *Pi and the AGM*, Wiley & Sons, New York, 1987.

[BoBu] Boas, R. P. and Buck, R. C., *Polynomial Expansions of Analytic Functions*, Academic Press, New York, 1964.

[Buc] Buck, R. C., *Advanced Calculus*, 2nd Ed., McGraw-Hill, New York, 1965.

[Bur] Burchnall, J. L., *A note on the polynomials of Hermite*, Quart. J. Math., **12** (1941), 9–11.

[ByF] Byrd, P. F. and Friedman, M. D., *Handbook of Elliptic Integrals for Engineers and Scientists*, 2nd Ed., vol. 67, Die Grundlehren der mathematischen Wissenschaft, Springer-Verlag, New York, 1971.

[Ca] Carlson, B. C., *Special Functions of Applied Mathematics*, Academic Press, New York, 1977.

[Ca1] Carlitz, L., *Some arithmetic properties of a special sequence of polynomials*, Duke. Math. J., **26** (1959), 583–590.

[Ca2] Carlitz, L., *Some orthogonal polynomials related to elliptic functions*, Duke Math. J., **27** (1960), 443–460.

[Ca3] Carlitz, L., *Some arithmetic properties of the lemniscate coefficients*, Math. Nachrichten, **22** (1960), 237–249.

[Ca4] Carlitz, L., *Some orthogonal polynomials related to elliptic functions II. Arithmetic properties*, Duke Math. J., **28** (1961), 107–124.

[Ca5] Carlitz, L., *The coefficients of the lemniscate function*, Math. Comp., **80** (1962), 475–478.

[CaA] Carlitz, L. and Al-Salam, A., *Some determinants of Bernoulli, Euler, and related numbers*, Portugal. Math., **18** (1959), 91–99.

[Ch] Chihara, T. S., *An Introduction to Orthogonal Polynomials*, Gordon and Breach, New York, 1978.

[Co] Comtet, L., *Advanced Combinatorics*, Reidel Publishing Co, Boston, 1974.

[Cro] Cronin, J., *Differential Equations*, Marcel Dekker, New York, 1980.

[Cry] Cryer, C. W., *Rodrigues' formula and the classical orthogonal polynomials*, Bull. Un. Mat. Ital., **25** (1970), 1–11.

[Du] Dumont, D., *A combinatorial interpretation for the Schett recurrence of the Jacobian elliptic function*, Math. Comp., 33 (1979), 1293–1297.

[Du1] Dumont, D., *Une approche combinatoire des fonctions elliptiques de Jacobi*, Advances in Math., 41 (1981), 1–39.

[Er] Erdélyi, A., *Bateman Manuscript Project. Higher Transcendental Functions*, McGraw-Hill, New York, 1955.

[Fa] Favard, J., *Sur les polynomes de Tchebicheff*, C. R. Acad. Sci. Paris, 200 (1935), 2052–2053.

[FaS] Faddeev, D. K. and Sominskii, I. S., *Problems in Higher Algebra*, Freeman & Co., San Francisco, 1965.

[Fe] Feldheim, E., *Relations entre les polynomes de Jacobi, Laguerre, et Hermite*, Acta Math., 75 (1943), 117–138.

[Go] Gould, H. W., *Combinatorial Identities*, revised edition, WV University, Morgantown, WV, 1972.

[Gr] Greub, W. H., *Linear Algebra*, Springer-Verlag, Berlin, 1963.

[GrH] Gröbner, W. and Hofreiter, N., *Integraltafel*, Part 2, Springer-Verlag, New York, 1965.

[GrR] Gradshteyn, I. S. and Ryzhik, I. M., *Table of Integrals, Series, and Products*, 4th Ed., Academic Press, New York, 1965.

[Ha] Hancock, H., *Lectures on the Theory of Elliptic Functions*, vol. 1, Wiley & Sons, New York, 1910.

[Han] Hansen, E. R., *A Table of Series and Products*, Prentice-Hall, Englewood Cliffs, NJ, 1975.

[Har] Hardy, G. H., *Notes on special systems of orthogonal functions (III)*, Proc. Camb. Phil. Soc., 36 (1940), 1–8.

[HeS] Hewitt, E. and Stromberg, K., *Real and Abstract Analysis*, Springer-Verlag, New York, 1965.

[Hi] Hindmarsh, A. C., *Solution to problem 5939*, Amer. Math. Monthly, 82 (1975), 533–536.

[Hu] Hurwitz, A., *Über die Entwicklungskoeffizienten der lemniskatischen Funktionen*, Math. Annalen, 51 (1899), 196–226.

[Iv] Ivanoff, V. F., *Derivatives of a composite function*, Amer. Math. Monthly, 65 (1958), 212; 68 (1961), 69–70.

[Jo] Jordan, C., *Calculus of Finite Differences*, 2nd Ed., Chelsea, New York, 1960.

[Ka] Kamber, F., *Formules exprimant les valeurs des coefficients des séries de puissances inverses*, Acta Math., 78 (1946), 193–204.

[Ke] Kelisky, R. P., *Inverse elliptic functions and Legendre polynomials*, Amer. Math. Monthly, 66 (1959), 480–483.

[Ki] Kiper, A., *Fourier series coefficients for powers of the Jacobian elliptic functions*, Math. Comp., 43 (1984), 247–259.

[KnB]   Knuth, D. E. and Buckholz, T. J., *Computation of tangent, Euler, and Bernoulli numbers*, Math. Comp., 21 (1967), 663–688.

[KrP]   Krantz, S. and Park, H., *A Primer of Real Analytic Functions*, Birkhäuser, Basel, 1992.

[Le]   Lebedev, N. N., *Special Functions and Their Applications*, Dover, New York, 1972.

[Lo]   Love, J. B., *Problem E1534*, Amer. Math. Monthly, 70 (1963), 443.

[MaO]   Magnus, W. and Oberhettinger, F., *Formulas and Theorems for the Special Functions of Mathematical Physics*, Chelsea, New York, 1949.

[Mard]   Marden, M., *Geometry of Polynomials*, AMS, Providence, RI, 1966.

[Mark]   Markushevich, A. I., *Theory of Functions of a Complex Variable*, vol. 2, Prentice-Hall, Englewood Cliffs, NJ, 1965.

[Me]   Meixner, J., *Orthonormale Polynomsysteme mit einer besonderen Gestalt der erzeugenden Funktion*, J. London Math. Soc., 9 (1934), 6–13.

[Mil]   Milne-Thomson, L. M., *The Calculus of Finite Differences*, Macmillan and Co., New York, 1960.

[Mit]   Mitrinović, D. S., *Analytic Inequalities*, Springer-Verlag, New York, 1970.

[MJ]   Mitrinović, D. S. and Janić, R. R., *Uvod u specijalne funkcije (Introduction to Special Functions)*, 2nd Ed., Izdavičko Preduzeće Gradevinska Knjiga, Belgrade. MR 55 #10735, 1975.

[ML]   Mittag-Leffler, G., *Sur la représentation analytique d'une branche uniforme d'une fonction monogene*, Acta Math., 24 (1901), 205–244.

[Na]   Natanson I. P., *Constructive Function Theory*, vol. II, Ungar, New York, 1965.

[Ne]   Nevai, P. G., *Orthogonal Polynomials*, Memoirs of the AMS, No. 213, AMS, Providence, RI, 1979.

[No]   Nörlund, N. E., *Vorlesungen über Differenzenrechnung*, Chelsea, New York, 1954.

[Ob]   Oberhettinger, F., *Tables of Mellin Transforms*, Springer-Verlag, New York, 1974.

[ObB]   Oberhettinger, F. and Badii, L., *Tables of Laplace Transforms*, Springer-Verlag, New York, 1973.

[Ol]   Olver, F. W. J., *Asymptotics and Special Functions*, Academic Press, New York, 1974.

[Pi]   Pidduck, F. B., *On the propagation of a disturbance in a fluid under gravity*, Proc. Roy. Soc. (London), A83 (1910), 347–356.

[Po]   Pourahmadi, M., *Taylor expansion of* $exp\left(\sum_{k=0}^{\infty} a_k x^k\right)$ *and some applications*, Amer. Math. Monthly, **91** (1984), 303–307.

[Pr]   Proskuryakov, I. V., *Problems in Linear Algebra*, Mir Publishers, Moscow, 1985.

[PWZ] Petkovšek, M., Wilf, H. S., and Zeilberger, D., *A=B*, A. K. Peters, Ltd., Wellesley, MA, 1996.

[Ra1] Rainville, E. D., *Special Functions*, Macmillan, New York, 1960.

[Ra2] Rainville, E. D., *Infinite Series*, Macmillan, New York, 1967.

[Ri1] Riordan, J., *An Introduction to Combinatorial Analysis*, John Wiley & Sons, New York, 1964.

[Ri2] Riordan, J., *Combinatorial Identities*, John Wiley & Sons, New York, 1968.

[Ro]   Rota, G.-C., *Finite Operator Calculus*, Academic Press, New York, 1975.

[Rom] Roman, S., *The Umbral Calculus*, Academic Press, New York, 1984.

[Sc1] Schett, A., *Properties of the Taylor series expansion coefficients of the Jacobian elliptic functions*, Math. Comp., **30** (1976), 143–147.

[Sc2] Schett, A., *Recurrence formula of the Taylor series expansion coefficients of the Jacobian elliptic functions*, Math. Comp., **31** (1977), 1003–1005.

[ShT] Shohat, J. A. and Tamarkin, J. D., *The Problem of Moments*, Math. Surveys, No. 1, AMS, Providence, RI, 1943.

[So]   Soble, A. B., *Majorants of polynomial derivatives*, Amer. Math. Monthly, **64** (1957), 639–643.

[St]   Stromberg, K. R., *Introduction to Classical Real Analysis*, Wadsworth, Belmont, CA, 1981.

[StC] Stanton, R. G. and Cowan, D. D., *Note on a "square" functional equation*, SIAM Review, **12** (1970), 277–279.

[Su]   Subbotin, Y. N., *Encyclopaedia of Mathematics*, vol. 3, Kluwer Academic, Boston, 1989.

[SW]   Stroeker, R. J. and de Weger, B. M. M., *On integral zeroes of binary Krawtchouk polynomials*, Nieuw Archief voor Wiskunde, **17** (1999), 175–186.

[Sz]   Szegö, G., *Collected Papers*, Ed. R. Askey, vol. 3, Birkhäuser, Boston, 1982.

[TaM] Tannery, J. and Molk J., *Eléments de la Théorie des Fonctions Elliptiques*, vols. 3 and 4, Chelsea, New York, 1972.

[Tu]   Turán, P., *On the zeros of the polynomials of Legendre*, Phil. Časopis Pěst. Mat., **75** (1950), 113–122.

[Va]   Van Orstrand, C. E., *Reversion of power series*, Phil. Mag., 19 (1910), 366–376.

[Vi]   Viennot, G., *Une Theorie Combinatoire des Polynomes Orthogonaux Generaux*, Notes de Conférences Données au Département de Mathématiques et d'Informatique, Université du Québec á Montréal, 1983.

[Wa]   Ward, M., *The reversion of a power series*, Rendiconti del Circolo Matematico di Palermo, 54 (1930), 42–46.

[WhW] Whittaker, E. T. and Watson, G. N., *A Course in Modern Analysis*, 4th Ed., Cambridge University Press, London, 1962.

[Wr]   Wrigge, S., *Calculation of the Taylor series expansion coefficients of the Jacobian elliptic function* sn *(x,k)*, Math. Comp., 36 (1981), 555–564.

Name Index

Subject Index

Symbol Index

.

Milton Keynes UK
Ingram Content Group UK Ltd.
UKHW021622071024
449327UK00020BA/1145